Mathematical Analysis and Applications

Mathematical Analysis and Applications

Special Issue Editor

Hari Mohan Srivastava

MDPI • Basel • Beijing • Wuhan • Barcelona • Belgrade

Special Issue Editor
Hari Mohan Srivastava
University of Victoria
Canada

Editorial Office
MDPI
St. Alban-Anlage 66
4052 Basel, Switzerland

This is a reprint of articles from the Special Issue published online in the open access journal *Axioms* (ISSN 2075-1680) in 2018 (available at: https://www.mdpi.com/journal/axioms/special_issues/mathematical_analysis)

For citation purposes, cite each article independently as indicated on the article page online and as indicated below:

LastName, A.A.; LastName, B.B.; LastName, C.C. Article Title. *Journal Name* **Year**, *Article Number*, Page Range.

ISBN 978-3-03897-400-0 (Pbk)
ISBN 978-3-03897-401-7 (PDF)

Contents

About the Special Issue Editor

Hari Mohan Srivastava has held the position of Professor Emeritus in the Department of Mathematics and Statistics at the University of Victoria in Canada since 2006, having joined the faculty there in 1969, first as an Associate Professor (1969–1974) and then as a Full Professor (1974–2006). He began his university-level teaching career right after having received his M.Sc. degree in 1959 at the age of 19 years from the University of Allahabad in India. He earned his Ph.D. degree in 1965 while he was a full-time member of the teaching faculty at the Jai Narain Vyas University of Jodhpur in India. He has held numerous visiting research and honorary chair positions at many universities and research institutes in different parts of the world. Having received several D.Sc. (honoris causa) degrees as well as honorary memberships and honorary fellowships of many scientific academies and learned societies around the world, he is also actively associated editorially with numerous international scientific research journals. His current research interests include several areas of Pure and Applied Mathematical Sciences, such as Real and Complex Analysis, Fractional Calculus and Its Applications, Integral Equations and Transforms, Higher Transcendental Functions and Their Applications, q-Series and q-Polynomials, Analytic Number Theory, Analytic and Geometric Inequalities, Probability and Statistics, and Inventory Modelling and Optimization. He has published 27 books, monographs, and edited volumes, 30 book (and encyclopedia) chapters, 45 papers in international conference proceedings, and more than 1100 scientific research articles in peer-reviewed international journals, as well as forewords and prefaces to many books and journals, and so on. He is a Clarivate Analytics [Thomson-Reuters] (Web of Science) Highly Cited Researcher. For further details about his other professional achievements and scholarly accomplishments, as well as honors, awards, and distinctions, including the lists of his most recent publications such as journal articles, books, monographs and edited volumes, book chapters, encyclopedia chapters, papers in conference proceedings, forewords to books and journals, et cetera), the interested reader should look into the following regularly updated website.

axioms

Editorial

Mathematical Analysis and Applications

Hari M. Srivastava [1,2]

[1] Department of Mathematics and Statistics, University of Victoria, Victoria, BC V8W 3R4, Canada; harimsri@math.uvic.ca

[2] Department of Medical Research, China Medical University Hospital, China Medical University, Taichung 40402, Taiwan

Received: 7 November 2018; Accepted: 7 November 2018; Published: 12 November 2018

Website: http://www.math.uvic.ca/faculty/harimsri/

This volume contains the invited, accepted and published submissions (see [1–15]) to a Special Issue of the MDPI journal ***Axioms*** on the subject-area of "Mathematical Analysis and Applications".

In recent years, investigations involving the theory and applications of mathematical analytic tools and techniques have become remarkably widespread in many diverse areas of the mathematical, physical, chemical, engineering and statistical sciences. In this Special Issue, we invite and welcome review, expository and original research articles dealing with the recent advances in mathematical analysis and its multidisciplinary applications.

The suggested topics of interest for the call for papers of this Special Issue included, but were not limited to, the following keywords:

- Mathematical (or Higher Transcendental) Functions and Their Applications
- Fractional Calculus and Its Applications
- q-Series and q-Polynomials
- Analytic Number Theory
- Special Functions of Mathematical Physics and Applied Mathematics
- Geometric Function Theory of Complex Analysis

Here, in this Editorial, we briefly describe the status of the Special Issue as follows:

1. Publications: 15;
2. Rejections: 24;
3. Article Average Processing Time: 52 days;
4. Article Type: Research Article (15); Review (0); Correction (0)

Authors' geographical distribution:

- Italy (6)
- Turkey (5)
- Egypt (3)
- Russian Federation (3)
- Canada (2)
- Korea (2)
- Finland (2)
- Thailand (2)
- Slovakia (2)
- Spain (1)
- Iran (1)

- Taiwan (Republic of China) (1)
- Germany (1)
- Greece (1)
- UK (1)
- USA (1)

References

1. Natalini, P.; Ricci, P.E. New Bell–Sheffer polynomial sets. *Axioms* **2018**, *7*, 71, doi:10.3390/axioms7040071. [CrossRef]
2. Agaoglou, M.; Fečkan, M.; Pospíšil, M.; Rothos, V.M.; Vakakis, A.F. Periodically forced nonlinear oscillatory acoustic vacuum. *Axioms* **2018**, *7*, 69, doi:10.3390/axioms7040069. [CrossRef]
3. Li, C.-K.; Humphries, T.; Plowman, H. Solutions to Abel's integral equations in distributions. *Axioms* **2018**, *7*, 66, doi:10.3390/axioms7030066. [CrossRef]
4. Djida, J.-D.; Fernandez, A. Interior regularity estimates for a degenerate elliptic equation with mixed boundary conditions. *Axioms* **2018**, *7*, 65, doi:10.3390/axioms7030065. [CrossRef]
5. Dattoli, G.; Germano, B.; Licciardi, S.; Martinelli, M.R. Umbral methods and harmonic numbers. *Axioms* **2018**, *7*, 62. [CrossRef]
6. Chaikham, N.; Sawangtong, W. Sub-Optimal control in the Zika virus epidemic model using differential evolution. *Axioms* **2018**, *7*, 61, doi:10.3390/axioms7030061. [CrossRef]
7. Kim, T.; Ryoo, C.S. Some identities for Euler and Bernoulli polynomials and their zeros. *Axioms* **2018**, *7*, 56. [CrossRef]
8. Gümüş, H.; Demir, N. A new type of generalization on W-asymptotically J_λ-statistical equivalence with the number of α. *Axioms* **2018**, *7*, 54. [CrossRef]
9. Zhukovsky, K.; Oskolkov, D.; Gubina, N. Some exact solutions to non-Fourier heat equations with substantial derivative. *Axioms* **2018**, *7*, 48. [CrossRef]
10. Masjed-Jamei, M.; Koepf, W. Some summation theorems for generalized hypergeometric functions. *Axioms* **2018**, *7*, 38. [CrossRef]
11. Wu, H.-C. Pre-Metric spaces along with different types of triangle inequalities. *Axioms* **2018**, *7*, 34. [CrossRef]
12. Rasila, A.; Sottinen, T. Yukawa potential, panharmonic measure and Brownian motion. *Axioms* **2018**, *7*, 28. [CrossRef]
13. Zayed, H.M.; Aouf, M.K.; Mostafa, A.O. Subordination properties for multivalent functions associated with a generalized fractional differintegral operator. *Axioms* **2018**, *7*, 27. [CrossRef]
14. Kişi, O.; Gümüş, H.; Savas, E. New definitions about AI-statistical convergence with respect to a sequence of modulus functions and Lacunary sequences. *Axioms* **2018**, *7*, 27, doi:10.3390/axioms7020027. [CrossRef]
15. Simsek, Y. Special numbers and polynomials including their generating functions in umbral analysis methods. *Axioms* **2018**, *7*, 22. [CrossRef]

axioms

MDPI

Article

New Bell–Sheffer Polynomial Sets

Pierpaolo Natalini [1,*] and Paolo Emilio Ricci [2]

1 Dipartimento di Matematica e Fisica, Università degli Studi Roma Tre, Largo San Leonardo Murialdo, 1, 00146 Roma, Italy
2 Sezione di Matematica, International Telematic University UniNettuno, Corso Vittorio Emanuele II, 39, 00186 Roma, Italy; paoloemilioricci@gmail.com
* Correspondence: natalini@mat.uniroma3.it

Received: 20 July 2018; Accepted: 2 October 2018; Published: 8 October 2018

Abstract: In recent papers, new sets of Sheffer and Brenke polynomials based on higher order Bell numbers, and several integer sequences related to them, have been studied. The method used in previous articles, and even in the present one, traces back to preceding results by Dattoli and Ben Cheikh on the monomiality principle, showing the possibility to derive explicitly the main properties of Sheffer polynomial families starting from the basic elements of their generating functions. The introduction of iterated exponential and logarithmic functions allows to construct new sets of Bell–Sheffer polynomials which exhibit an iterative character of the obtained shift operators and differential equations. In this context, it is possible, for every integer r, to define polynomials of higher type, which are linked to the higher order Bell-exponential and logarithmic numbers introduced in preceding papers. Connections with integer sequences appearing in Combinatorial analysis are also mentioned. Naturally, the considered technique can also be used in similar frameworks, where the iteration of exponential and logarithmic functions appear.

Keywords: Sheffer polynomials; generating functions; monomiality principle; shift operators; combinatorial analysis

1. Introduction

In recent articles [1,2], new sets of Sheffer [3] and Brenke [4] polynomials, based on higher order Bell numbers [2,5–7], have been studied. Furthermore, several integer sequences associated [8] with the considered polynomials sets both of exponential [9,10] and logarithmic type have been introduced [1].

It is worth noting that exponential and logarithmic polynomials have been recently studied in the multidimensional case [11–13].

In this article, new sets of Bell–Sheffer polynomials are considered and some particular cases are analyzed.

It is worth noting that the Sheffer A-type 0 polynomial sets have been also approached with elementary methods of linear algebra (see, e.g., [14–16] and the references therein).

Connection with umbral calculus has been recently emphasized (see, e.g., [17,18] and the references therein).

2. Sheffer Polynomials

The Sheffer polynomials $\{s_n(x)\}$ are introduced [3] by means of the exponential generating function [19] of the type:

$$A(t)\exp(xH(t)) = \sum_{n=0}^{\infty} s_n(x)\,\frac{t^n}{n!}\,, \tag{1}$$

where

$$A(t) = \sum_{n=0}^{\infty} a_n \frac{t^n}{n!}, \qquad (a_0 \neq 0),$$
$$H(t) = \sum_{n=0}^{\infty} h_n \frac{t^n}{n!}, \qquad (h_0 = 0). \tag{2}$$

According to a different characterization (see [20], p. 18), the same polynomial sequence can be defined by means of the pair $(g(t), f(t))$, where $g(t)$ is an invertible series and $f(t)$ is a delta series:

$$g(t) = \sum_{n=0}^{\infty} g_n \frac{t^n}{n!}, \qquad (g_0 \neq 0),$$
$$f(t) = \sum_{n=0}^{\infty} f_n \frac{t^n}{n!}, \qquad (f_0 = 0,\ f_1 \neq 0). \tag{3}$$

Denoting by $f^{-1}(t)$ the compositional inverse of $f(t)$ (i.e., such that $f\left(f^{-1}(t)\right) = f^{-1}\left(f(t)\right) = t$), the exponential generating function of the sequence $\{s_n(x)\}$ is given by

$$\frac{1}{g[f^{-1}(t)]} \exp\left(x f^{-1}(t)\right) = \sum_{n=0}^{\infty} s_n(x) \frac{t^n}{n!}, \tag{4}$$

so that

$$A(t) = \frac{1}{g[f^{-1}(t)]}, \qquad H(t) = f^{-1}(t). \tag{5}$$

When $g(t) \equiv 1$, the Sheffer sequence corresponding to the pair $(1, f(t))$ is called the associated Sheffer sequence $\{\sigma_n(x)\}$ for $f(t)$, and its exponential generating function is given by

$$\exp\left(x f^{-1}(t)\right) = \sum_{n=0}^{\infty} \sigma_n(x) \frac{t^n}{n!}. \tag{6}$$

A list of known Sheffer polynomial sequences and their associated ones can be found in [21].

3. New Bell–Sheffer Polynomial Sets

We introduce, for shortness, the following compact notation.

Put, by definition:

$$E_0(t) := \exp(t) - 1,$$
$$E_1(t) := E_0(E_0(t)) = \exp(\exp(t) - 1) - 1$$
$$\cdots\cdots\cdots$$
$$E_r(t) := E_0(E_{r-1}(t)) = \exp(\ldots \exp(\exp(t) - 1) - 1) \cdots - 1 \qquad [(r+1) - \text{times } \exp],$$

and in a similar way:

$$\Lambda_0(t) := \log(t+1)$$
$$\Lambda_1(t) := \Lambda_0(\Lambda_0(t)) = \log(\log(t+1) + 1)$$
$$\cdots\cdots\cdots$$
$$\Lambda_r(t) := \Lambda_0(\Lambda_{r-1}(t)) = \log\left(\log\left(\ldots(\log(t+1) + 1)\ldots\right) + 1\right), \qquad [(r+1) - \text{times } \log].$$

Remark 1. *Note that, for all integers* r, k, h,

$$E_r(\Lambda_r(t)) = t\,, \qquad \Lambda_r(E_r(t)) = t\,,$$

$$(\text{if } k > h) \qquad E_k(\Lambda_h(t)) = E_{k-h-1}(t)\,, \quad E_h(\Lambda_k(t)) = \Lambda_{k-h-1}(t)\,,$$

$$(\text{if } k > h) \qquad \Lambda_k(E_h(t)) = \Lambda_{k-h-1}(t)\,, \quad \Lambda_h(E_k(t)) = E_{k-h-1}(t)\,,$$

$$e^{E_r(t)} = E_{r+1}(t) + 1\,, \qquad e^{\Lambda_r(t)} = \Lambda_{r-1}(t) + 1\,.$$

Remark 2. *Note that the coefficients of the Taylor expansion of* $E_1(t)$ *are given by the Bell numbers* $b_n = b_n^{[1]}$

$$E_1(t) = \sum_{n=1}^{\infty} b_n^{[1]} \frac{t^n}{n!}\,,$$

and, in general, the coefficients of the Taylor expansion of $E_r(t)$ *are given by the higher order Bell numbers* $b_n^{[r]}$

$$E_r(t) = \sum_{n=1}^{\infty} b_n^{[r]} \frac{t^n}{n!}\,.$$

The higher order Bell numbers, also known as higher order exponential numbers, have been considered in [5,7,22], and used in [2] in the framework of Brenke and Sheffer polynomials.

Remark 3. *Note that the coefficients of the Taylor expansion of* $\Lambda_0(t)$ *are given by the logarithmic numbers* $l_n^{[1]} = (-1)^{n-1}(n-1)!$

$$\Lambda_0(t) = \sum_{n=1}^{\infty} l_n^{[1]} \frac{t^n}{n!} = \sum_{n=1}^{\infty} (-1)^{n-1}(n-1)! \frac{t^n}{n!}\,,$$

and, in general, the coefficients of the Taylor expansion of $\Lambda_{r-1}(t)$ *are given by the higher order logarithmic numbers* $l_n^{[r]}$

$$\Lambda_{r-1}(t) = \sum_{n=1}^{\infty} l_n^{[r]} \frac{t^n}{n!}\,.$$

The higher order logarithmic numbers, which are the counterpart of the higher order Bell (exponential) numbers, have been considered in [1], and used there in the framework of new sets of Sheffer polynomials.

3.1. The Polynomials $\mathcal{E}_k^{(1)}(x)$

Therefore, we consider the Sheffer polynomials, defined through their generating function, by putting

$$A(t) = E_1(t) + 1 = e^{E_0(t)}\,, \qquad H(t) = E_0(t)\,,$$
$$G(t,x) = \exp\left[(1+x)E_0(x)\right] = \sum_{k=0}^{\infty} \mathcal{E}_k^{(1)}(x) \frac{t^k}{k!}\,. \tag{7}$$

3.2. *Recurrence Relation for the* $\mathcal{E}_k^{(1)}(x)$

Theorem 1. *For any* $k \geq 0$*, the polynomials* $\mathcal{E}_k^{(1)}(x)$ *satisfy the following recurrence relation:*

$$\mathcal{E}_{k+1}^{(1)}(x) = \sum_{h=0}^{k} \binom{k}{h} (1+x)\, \mathcal{E}_h^{(1)}(x)\,. \tag{8}$$

Proof. Differentiating $G(t,x)$ with respect to t, we have

$$\frac{\partial G(t,x)}{\partial t} = G(t,x)\, e^t\, (1+x)\,, \tag{9}$$

and therefore

$$\sum_{k=0}^{\infty} (1+x)\, \mathcal{E}_k^{(1)}(x) \frac{t^k}{k!} \sum_{k=0}^{\infty} \frac{t^k}{k!} = \sum_{k=0}^{\infty} \mathcal{E}_{k+1}^{(1)}(x) \frac{t^k}{k!}\,,$$

i.e.,

$$\sum_{k=0}^{\infty} \sum_{h=0}^{k} \binom{k}{h} (1+x)\, \mathcal{E}_h^{(1)}(x) \frac{t^k}{k!} = \sum_{k=0}^{\infty} \mathcal{E}_{k+1}^{(1)}(x) \frac{t^k}{k!}$$

so that the recurrence relation (8) follows. □

3.3. *Generating Function's PDEs*

Theorem 2. *The generating function* $(7)_2$ *satisfies the homogeneous linear PDEs:*

$$(1-e^{-t}) \frac{\partial G(t,x)}{\partial t} = (1+x) \frac{\partial G(t,x)}{\partial x}\,, \tag{10}$$

$$\frac{\partial G(t,x)}{\partial t} = \frac{\partial G(t,x)}{\partial x} + G(t,x)(1+xe^t)\,, \tag{11}$$

$$\frac{\partial G(t,x)}{\partial t} = (1+x) \left[\frac{\partial G(t,x)}{\partial x} + G(t,x) \right]. \tag{12}$$

Proof. Differentiating $G(t,x)$ with respect to x, we have

$$\frac{\partial G(t,x)}{\partial x} = G(t,x)\,(e^t - 1)\,. \tag{13}$$

By taking the ratio between the members of Equations (9) and (13), we find Equation (10). The other ones easily follows by elementary algebraic manipulations. □

3.4. *Shift Operators*

We recall that a polynomial set $\{p_n(x)\}$ is called quasi-monomial if and only if there exist two operators $\hat{P}$ and $\hat{M}$ such that

$$\hat{P}\,(p_n(x)) = n p_{n-1}(x)\,, \qquad \hat{M}\,(p_n(x)) = p_{n+1}(x)\,, \qquad (n = 1, 2, \ldots). \tag{14}$$

$\hat{P}$ is called the *derivative* operator and $\hat{M}$ the *multiplication* operator, as they act in the same way of classical operators on monomials.

This definition traces back to a paper by Steffensen [23], recently improved by Dattoli [24] and widely used in several applications.

Ben Cheikh [25] proved that every polynomial set is quasi-monomial under the action of suitable derivative and multiplication operators. In particular, in the same article (Corollary 3.2), the following result is proved.

Theorem 3. *Let $(p_n(x))$ denote a Boas–Buck polynomial set, i.e., a set defined by the generating function*

$$A(t)\psi(xH(t)) = \sum_{n=0}^{\infty} p_n(x)\,\frac{t^n}{n!}\,, \tag{15}$$

where

$$\begin{aligned} A(t) &= \sum_{n=0}^{\infty} a_n t^n\,, \qquad (a_0 \neq 0)\,, \\ \psi(t) &= \sum_{n=0}^{\infty} \gamma_n t^n\,, \qquad (\gamma_n \neq 0 \quad \forall n)\,, \end{aligned} \tag{16}$$

with $\psi(t)$ not a polynomial, and lastly

$$H(t) = \sum_{n=0}^{\infty} h_n\, t^{n+1}\,, \qquad (h_0 \neq 0)\,. \tag{17}$$

Let $\sigma \in \Lambda^{(-)}$ the lowering operator defined by

$$\sigma(1) = 0, \qquad \sigma(x^n) = \frac{\gamma_{n-1}}{\gamma_n}\, x^{n-1}\,, \qquad (n = 1, 2, \ldots). \tag{18}$$

Put

$$\sigma^{-1}(x^n) = \frac{\gamma_{n+1}}{\gamma_n}\, x^{n+1} \quad (n = 0, 1, 2, \ldots). \tag{19}$$

Denoting, as before, by $f(t)$ the compositional inverse of $H(t)$, the Boas–Buck polynomial set $\{p_n(x)\}$ is quasi-monomial under the action of the operators

$$\hat{P} = f(\sigma)\,, \qquad \hat{M} = \frac{A'[f(\sigma)]}{A[f(\sigma)]} + xD_x H'[f(\sigma)]\sigma^{-1}\,, \tag{20}$$

where prime denotes the ordinary derivatives with respect to t.

Note that, in our case, we are dealing with a Sheffer polynomial set, so that since we have $\psi(t) = e^t$, the operator σ defined by Equation (16) simply reduces to the derivative operator D_x. Furthermore, we have:

$$\begin{aligned} &f(t) = H^{-1}(t) = \Lambda_0(t)\,, \\ &\frac{A'(t)}{A(t)} = e^t\,, \qquad H'(t) = e^t\,, \end{aligned}$$

and, consequently,

$$f(\sigma) = \Lambda_0(D_x)\,, \qquad \frac{A'[\Lambda_0(D_x)]}{A[\Lambda_0(D_x)]} = D_x + 1\,,$$

$$H'[f(\sigma)] = H'[\Lambda_0(D_x)] = D_x + 1\,.$$

Theorem 4. *The Bell–Sheffer polynomials* $\{\mathcal{E}_k^{(1)}(x)\}$ *are quasi-monomial under the action of the operators*

$$\hat{P} = \Lambda_0(D_x) = \sum_{k=0}^{\infty} (-1)^{k+1} \frac{D_x^k}{k}\,, \qquad \hat{M} = (1+x)(D_x+1)\,. \tag{21}$$

3.5. Differential Equation for the $\mathcal{E}_k^{(1)}(x)$

According to the results of monomiality principle [24], the quasi-monomial polynomials $\{p_n(x)\}$ satisfy the differential equation

$$\hat{M}\hat{P}\, p_n(x) = n\, p_n(x)\,. \tag{22}$$

In the present case, recalling Equation (22), we have

Theorem 5. *The Bell–Sheffer polynomials* $\{\mathcal{E}_k^{(1)}(x)\}$ *satisfy the differential equation*

$$(1+x) \sum_{k=1}^{n} (-1)^{k+1} \left[\frac{D_x^{k+1} + D_x^k}{k!} \right] \mathcal{E}_n^{(1)}(x) = n\, \mathcal{E}_n^{(1)}(x)\,. \tag{23}$$

Proof. Equation (22), by using Equation (21), becomes

$$\hat{M}\hat{P}\, \mathcal{E}_n^{(1)}(x) = (1+x)\,(D_x+1)\,\Lambda_0(D_x)\,\mathcal{E}_n^{(1)}(x) =$$

$$= (1+x)\,(D_x+1) \sum_{k=1}^{n} (-1)^{k+1} \frac{D_x^k}{k!}\, \mathcal{E}_n^{(1)}(x) = n\, \mathcal{E}_n^{(1)}(x)\,,$$

i.e.,

$$(1+x) \sum_{k=1}^{\infty} (-1)^{k+1} \left[\frac{D_x^{k+1} + D_x^k}{k!} \right] \mathcal{E}_n^{(1)}(x) = n\, \mathcal{E}_n^{(1)}(x)\,,$$

and, furthermore, for any fixed n, the last series expansion reduces to a finite sum, with upper limit $n-1$, when it is applied to a polynomial of degree n because the last not vanishing term (for $k = n-1$) contains the derivative of order n. □

3.6. First Few Values of the $\mathcal{E}_k^{(1)}(x)$

Here, we show the first few values for the Bell–Sheffer polynomials $\mathcal{E}_k^{(1)}(x)$, defined by the generating function (7)$_2$

$$\begin{aligned}
\mathcal{E}_0^{(1)}(x) &= 1, \\
\mathcal{E}_1^{(1)}(x) &= x+1, \\
\mathcal{E}_2^{(1)}(x) &= x^2+3x+2, \\
\mathcal{E}_3^{(1)}(x) &= x^3+6x^2+10x+5, \\
\mathcal{E}_4^{(1)}(x) &= x^4+10x^3+31x^2+37x+15, \\
\mathcal{E}_5^{(1)}(x) &= x^5+15x^4+75x^3+160x^2+151x+52.
\end{aligned}$$

Further values can be easily achieved by using Wolfram Alpha© (2009, Wolfram Research, Champaign, IL, USA).

Remark 4. *Note that the numerical values $\mathcal{E}_k^{(1)}(0)$ of the considered Bell–Sheffer polynomials*

$$(1,1,2,5,15,52,203,877,\dots)$$

appears in the Encyclopedia of Integer Sequences [8] under A000110: Bell or exponential numbers: number of ways to partition a set of n labeled elements.

The same sequence also appears under A164864, A164863, A276723, A276724, A276725, A276726, A287278, A287279, A287280.

4. Iterated Bell–Sheffer Polynomial Sets

Here, we iterate the procedure introduced in Section 3, by considering the Sheffer polynomial sets defined by putting

$$\begin{aligned}
&A(t) = E_2(t)+1 = e^{E_1(t)}, \qquad H(t) = E_1(t)\,, \\
&G(t,x) = \exp\left[(1+x)E_1(x)\right] = \sum_{k=0}^{\infty} \mathcal{E}_k^{(2)}(x)\,\frac{t^k}{k!}\,.
\end{aligned} \tag{24}$$

We find:

$$f(t) = H^{-1}(t) = \Lambda_1(t)\,,$$

$$\frac{A'(t)}{A(t)} = H'(t) = [E_1(t)+1]\,e^t = [E_1(t)+1]\,[E_0(t)+1]\,,$$

and, consequently,

$$\begin{aligned}
&f(\sigma) = \Lambda_1(D_x)\,, \qquad H'[f(\sigma)] = H'[\Lambda_1(D_x)] = [D_x+1]\;[\Lambda_0(D_x)+1]\;, \\
&\frac{A'[\Lambda_1(D_x)]}{A[\Lambda_1(D_x)]} = [E_1(\Lambda_1(D_x))+1]\,e^{\Lambda_1(D_x)} = (D_x+1)\;[\Lambda_0(D_x)+1]\;.
\end{aligned}$$

Theorem 6. *The Bell–Sheffer polynomials $\{\mathcal{E}_k^{(2)}(x)\}$ are quasi-monomial under the action of the operators*

$$\begin{aligned}
&\hat{P} = \Lambda_1(D_x)\,, \\
&\hat{M} = (1+x)\,(D_x+1)\;[\Lambda_0(D_x)+1]\;.
\end{aligned} \tag{25}$$

4.1. Differential Equation for the $\mathcal{E}_k^{(2)}(x)$

According to the results of monomiality principle [24,26], the quasi-monomial polynomials $\{p_n(x)\}$ satisfy the differential equation

$$\hat{M}\hat{P}\, p_n(x) = n\, p_n(x)\,. \tag{26}$$

In the present case, recalling Equation (19), we have

Theorem 7. *The Bell–Sheffer polynomials* $\{\mathcal{E}_k^{(2)}(x)\}$ *satisfy the differential equation*

$$(1+x)\,(D_x+1)\;[\Lambda_0(D_x)+1]\;\Lambda_1(D_x)\,\mathcal{E}_n^{(2)}(x) = n\,\mathcal{E}_n^{(2)}(x)\,. \tag{27}$$

4.2. First Few Values of the $\mathcal{E}_k^{(2)}(x)$

Here, we show the first few values for the Bell–Sheffer polynomials $\mathcal{E}_k^{(2)}(x)$, defined by the generating function $(7)_2$

$$\begin{aligned}
&\mathcal{E}_0^{(2)}(x) = 1,\\
&\mathcal{E}_1^{(2)}(x) = x+1,\\
&\mathcal{E}_2^{(2)}(x) = x^2+4x+3,\\
&\mathcal{E}_3^{(2)}(x) = x^3+9x^2+20x+12,\\
&\mathcal{E}_4^{(2)}(x)) = x^4+16x^3+74x^2+119x+60,\\
&\mathcal{E}_5^{(2)}(x) = x^5+25x^4+200x^3+635x^2+817x+358.
\end{aligned}$$

Further values can be easily achieved by using Wolfram Alpha©.

Remark 5. *Note that the numerical values* $\mathcal{E}_k^{(2)}(0)$ *of the considered Bell–Sheffer polynomials*

$$(1,1,3,12,60,358,2471,19302,\dots)$$

appear in the Encyclopedia of Integer Sequences under A000258: McLaurin coefficients of the function $E_2(x)$.

5. The General Case

In general, by putting

$$\begin{aligned}
&A(t) = E_r(t)+1 = e^{E_{r-1}(t)}, \qquad H(t) = E_{r-1}(t)\,,\\
&G(t,x) = \exp\left[(1+x)E_{r-1}(t)\right] = \sum_{k=0}^{\infty} \mathcal{E}_k^{(r)}(x)\,\frac{t^k}{k!}\,,
\end{aligned} \tag{28}$$

we find:

$$f(t) = H^{-1}(t) = \Lambda_{r-1}(t)\,,$$

$$\frac{A'(t)}{A(t)} = H'(t) = \prod_{\ell=1}^{r-1}\left[E_\ell(t)+1\right]e^t = \prod_{\ell=0}^{r-1}\left[E_\ell(t)+1\right],$$

and, consequently,

$$f(\sigma) = \Lambda_{r-1}(D_x), \qquad \frac{A'[\Lambda_{r-1}(D_x)]}{A[\Lambda_{r-1}(D_x)]} = \prod_{\ell=0}^{r-1} [E_\ell(\Lambda_{r-1}(D_x)) + 1],$$

$$H'[f(\sigma)] = H'[\Lambda_{r-1}(D_x)] = \prod_{\ell=0}^{r-1} [E_\ell(\Lambda_{r-1}(D_x)) + 1].$$

Recalling Remark 3.1, we find

$$E_\ell(\Lambda_{r-1}(D_x)) = \Lambda_{r-\ell-2}(D_x),$$

$$\prod_{\ell=0}^{r-1} [E_\ell(\Lambda_{r-1}(D_x)) + 1] = (D_x + 1) \prod_{k=0}^{r-2} [\Lambda_k(D_x) + 1],$$

so that we have the theorem:

Theorem 8. *The Bell–Sheffer polynomials* $\{\mathcal{E}_k^{(r)}(x)\}$ *are quasi-monomial under the action of the operators*

$$\hat{P} = \Lambda_{r-1}(D_x),$$
$$\hat{M} = (1+x)\ (D_x + 1) \prod_{k=0}^{r-2} [\Lambda_k(D_x) + 1]. \tag{29}$$

Differential Equation for the $\mathcal{E}_k^{(r)}(x)$

According to the results of monomiality principle [24], the quasi-monomial polynomials $\{p_n(x)\}$ satisfy the differential equation

$$\hat{M}\hat{P}\, p_n(x) = n\, p_n(x). \tag{30}$$

In the present case, recalling Equation (19), we have

Theorem 9. *The Bell–Sheffer polynomials* $\{\mathcal{E}_k^{(r)}(x)\}$ *satisfy the differential equation*

$$(1+x)\ (D_x + 1) \prod_{k=0}^{r-2} [\Lambda_k(D_x) + 1]\, \Lambda_{r-1}(D_x)\, \mathcal{E}_n^{(r)}(x) = n\, \mathcal{E}_n^{(r)}(x). \tag{31}$$

6. Conclusions

By introducing iterated exponential and logarithmic functions, we have shown how to construct new sets of Bell-Sheffer polynomials which exhibit an iterative character. We have found their main properties by using the monomiality property and a general result by Y. Ben Cheikh which gives explicitly the derivative and multiplication operators for polynomials of Sheffer type. The tools we used are internal to Sheffer's polynomial theory and do not use external techniques. In our opinion the demonstrated properties (in particular the differential equations for polynomials of higher order) could hardly be achieved by other methods.

Author Contributions: The authors declare to have both contributed to the final version of this article.

Funding: This research received no external funding.

Conflicts of Interest: The authors declare no conflict of interest.

References

1. Bretti, G.; Natalini, P.; Ricci, P.E. A new set of Sheffer-Bell polynomials and logarithmic numbers. *Georgian Math. J.* **2018**, in print.
2. Ricci, P.E.; Natalini, P.; Bretti, G. Sheffer and Brenke polynomials associated with generalized Bell numbers. *Jnanabha Vijnana Parishad India* **2017**, *47*, 337–352.
3. Sheffer, I.M. Some properties of polynomials sets of zero type. *Duke Math. J.* **1939**, *5*, 590–622. [CrossRef]
4. Brenke, W.C. On generating functions of polynomial systems. *Am. Math. Mon.* **1945**, *52*, 297–301. [CrossRef]
5. Natalini, P.; Ricci, P.E. Remarks on Bell and higher order Bell polynomials and numbers. *Cogent Math.* **2016**, *3*, 1–15. [CrossRef]
6. Bernardini, A.; Natalini, P.; Ricci, P.E. Multi-dimensional Bell polynomials of higher order. *Comput. Math. Appl.* **2005**, *50*, 1697–1708. [CrossRef]
7. Natalini, P.; Ricci, P.E. An extension of the Bell polynomials. *Comput. Math. Appl.* **2004**, *47*, 719–725. [CrossRef]
8. Sloane, N.J.A. The On-Line Encyclopedia of Integer Sequences. 2016. Available online: http://oeis.org (accessed on 24 May 2018).
9. Bell, E.T. Exponential polynomials. *Ann. Math.* **1934**, *35*, 258–277. [CrossRef]
10. Bell, E.T. The Iterated Exponential Integers. *Ann. Math.* **1938**, *39*, 539–557. [CrossRef]
11. Qi, F.; Niu, D.W.; Lim, D.; Guo, B.N. Some Properties and an Application of Multivariate Exponential Polynomials. HAL Archives. 2018. Available online: https://hal.archives-ouvertes.fr/hal-01745173 (accessed on 13 September 2018).
12. Qi, F. Integral representations for multivariate logarithmic potentials. *J. Comput. Appl. Math.* **2018**, *336*, 54–62. [CrossRef]
13. Qi, F. On multivariate logarithmic polynomials and their properties. *Indag. Math.* **2018**, *336*. [CrossRef]
14. Costabile, F.A.; Longo, E. A new recurrence relation and related determinantal form for Binomial type polynomial sequences. *Mediterr. J. Math.* **2016**, *13*, 4001–4017. [CrossRef]
15. Yang, Y.; Youn, H. Appell polynomial sequences: a linear algebraic approach. *JP J. Algebra Number Theory Appl.* **2009**, *13*, 65–98.
16. Yang, S.L. Recurrence relations for Sheffer sequences. *Linear Algebra Appl.* **2012**, *437*, 2986–2996. [CrossRef]
17. Kim, T.; Kim, D.S. On λ-Bell polynomials associated with umbral calculus. *Russ. J. Math. Phys.* **2017**, *24*, 69–78. [CrossRef]
18. Kim, T.; Kim, D.S.; Jang, G.-W. Some formulas of ordered Bell numbers and polynomials arising from umbral calculus. *Proc. Jangjeon Math. Soc.* **2017**, *20*, 659–670.
19. Srivastava, H.M.; Manocha, H.L. *A Treatise on Generating Functions*; Ellis Horwood: Chichester, UK, 1984.
20. Roman, S.M. *The Umbral Calculus*; Academic Press: New York, NY, USA, 1984.
21. Boas, R.P.; Buck, R.C. *Polynomial Expansions of Analytic Functions*; Springer: Berlin/Heidelberg, Germany; Gottingen, Germany; New York, NY, USA, 1958.
22. Natalini, P.; Ricci, P.E. Higher order Bell polynomials and the relevant integer sequences. *Appl. Anal. Discret. Math.* **2017**, *11*, 327–339. [CrossRef]
23. Steffensen, J.F. The poweroid, an extension of the mathematical notion of power. *Acta Math.* **1941**, *73*, 333–366. [CrossRef]
24. Dattoli, G. Hermite-Bessel and Laguerre-Bessel functions: A by-product of the monomiality principle. In *Advanced Special Functions and Applications, Proceedings of the Melfi School on Advanced Topics in Mathematics and Physics, Melfi, Italy, 9–12 May 1999*; Cocolicchio, D., Dattoli, G., Srivastava, H.M., Eds.; Aracne Editrice: Rome, Italy, 2000; pp. 147–164.
25. Cheikh, Y.B. Some results on quasi-monomiality. *Appl. Math. Comput.* **2003**, *141*, 63–76.
26. Dattoli, G.; Ricci, P.E.; Srivastava, H.M. (Eds.) Advanced Special Functions and Related Topics in Probability and in Differential Equations. In *Applied Mathematics and Computation, Proceedings of the Melfi School on Advanced Topics in Mathematics and Physics, Melfi, Italy, 24–29 June 2001*; Aracne Editrice: Rome, Italy, 2000; Volume 141, pp. 1–230.

axioms

MDPI

Article

Periodically Forced Nonlinear Oscillatory Acoustic Vacuum

Makrina Agaoglou [1,2], Michal Fečkan [1,3,*], Michal Pospíšil [1,3], Vassilis M. Rothos [2] and Alexander F. Vakakis [4]

1 Mathematical Institute of Slovak Academy of Sciences, Štefánikova 49, 814 73 Bratislava, Slovakia; makrina_agao@hotmail.com (M.A.); michal.pospisil@fmph.uniba.sk (M.P.)
2 Department of Mechanical Engineering, Faculty of Engineering, Aristotle University of Thessaloniki, 54124 Thessaloniki, Greece; rothos@auth.gr
3 Department of Mathematical Analysis and Numerical Mathematics, Comenius University in Bratislava, Mlynská dolina, 842 48 Bratislava, Slovakia
4 Department of Mechanical Science and Engineering, University of Illinois at Urbana-Champaign, Urbana, IL 61801, USA; avakakis@illinois.edu
* Correspondence: michal.feckan@fmph.uniba.sk

Received: 31 July 2018; Accepted: 19 September 2018; Published: 22 September 2018

Abstract: In this work, we study the in-plane oscillations of a finite lattice of particles coupled by linear springs under distributed harmonic excitation. Melnikov-type analysis is applied for the persistence of periodic oscillations of a reduced system.

Keywords: nonlinear oscillatory acoustic vacuum; periodic oscillations; Melnikov function; symmetry

1. Introduction

We analytically study the persistence of periodic oscillations for certain three-dimensional systems of ordinary differential equations (ODEs) with periodic perturbations and a slowly-varying variable. The considered ODEs are derived from a model of a finite lattice of particles coupled by linear springs under distributed harmonic excitation, which is described in detail in Section 2. This model presents a low-energy nonlinear acoustic vacuum. We refer the reader for more motivations, further details and applications to [1,2]. Following the computations of [1], we consider just two modes in Section 3, and we postpone higher modes investigation to our future paper, since another approach will be used. Melnikov analysis is demonstrated in Section 4 for finding conditions for the existence of periodic solutions for the perturbed ODEs corresponding to two modes. More precisely, following [1], we derive a three-dimensional periodically-perturbed system of ODEs with a slowly-varying variable. Then, we analyze an unperturbed autonomous system of ODEs to compute its family of periodic solutions by revising the results of [1] in more detail. Since we are interested in the persistence of periodic solutions for the perturbed ODEs, we compute the corresponding Melnikov functions by [3]. Due to the difficulty of finding simple roots of these Melnikov functions explicitly, we outline an asymptotic approach for the location of some of them. Note that the simple roots of Melnikov functions predict the persistence and location of periodic solutions for perturbed ODEs. This is a novelty and a contribution of our paper. Section 5 outlines possible future research along with summarizing our achievements in this paper.

2. The Model

We consider a lattice consisting of N identical particles coupled by identical linear springs (they are un-stretched when the lattice is in the horizontal position) and executing in-plane oscillations (see Figure 1). Fixed boundary conditions and dissipative terms are imposed, and the transverse harmonic forces are also applied. The equations of motion can be expressed as follows,

$$\begin{aligned} m\frac{d^2u_i}{dt^2} + \left(T_i - \xi\frac{d\epsilon_i}{dt}\right)\cos\phi_i - \left(T_{i+1} - \xi\frac{d\epsilon_{i+1}}{dt}\right)\cos\phi_{i+1} &= 0 \\ m\frac{d^2v_i}{dt^2} + \left(T_i - \xi\frac{d\epsilon_i}{dt}\right)\sin\phi_i - \left(T_{i+1} - \xi\frac{d\epsilon_{i+1}}{dt}\right)\sin\phi_{i+1} - F_i &= 0 \end{aligned} \quad (1)$$

with u_i, v_i being the longitudinal and transversal displacements of the i-th particle, respectively, ϕ_i the angle between the i-th spring and the horizontal direction, ξ the damping coefficient, $\epsilon_i = l_i' - l_i$ the deformation of the i-th spring, F_i the exciting transverse force and m the mass of each particle of the lattice. The tensile forces are proportional to the deformations of the springs, and considering the geometry of the deformed state of the lattice (see Figure 1), one may write:

$$\begin{aligned} T_i &= k(l_i' - l_i), \\ \epsilon_i = l_i' - l_i &= [(v_i - v_{i-1})^2 + (l_i + u_i - u_{i-1})^2]^{1/2} - l_i \end{aligned}$$

with l_i being the equilibrium length of the i-th spring (each spring has the same length) and k the linear stiffness coefficient of each coupling spring. Introducing $\delta_i = \epsilon_i/l_i$, $s_i = u_i/l_i$, $w_i = v_i/l_i$, $c = \xi/(km)^{1/2}$, where s_i and w_i are the normalized axial and transverse displacements, and the "slow" time scale $\tau = (\frac{k}{m})^{1/2}t$, Equation (1) can be rewritten in normalized form:

$$\begin{aligned} \frac{d^2s_i}{d\tau^2} &= \delta_i\cos\phi_i - \delta_{i-1}\cos\phi_{i-1} + c\delta_i'\cos\phi_i - c\delta_{i-1}'\cos\phi_{i-1} \\ \frac{d^2w_i}{d\tau^2} &= \delta_i\sin\phi_i - \delta_{i-1}\sin\phi_{i-1} + c\delta_i'\sin\phi_i - c\delta_{i-1}'\sin\phi_{i-1} + f_i, \end{aligned} \quad (2)$$

where:

$$\begin{aligned} \delta_i &= [(w_{i+1} - w_i)^2 + (1 + s_{i+1} - s_i)]^{1/2} - 1, \\ \delta_i' &= \frac{(w_{i+1}-w_i)(w_{i+1}'-w_i') + (1+s_{i+1}-s_i)(s_{i+1}'-s_i')}{[(w_{i+1}-w_i)^2+(1+s_{i+1}-s_i)^2]^{1/2}} \end{aligned}$$

and:

$$\begin{aligned} \cos\phi_i &= \frac{1+s_{i+1}-s_i}{[(w_{i+1}-w_i)^2+(1+s_{i+1}-s_i)^2]^{1/2}}, \\ \sin\phi_i &= \frac{w_{i+1}-w_i}{[(w_{i+1}-w_i)^2+(1+s_{i+1}-s_i)^2]^{1/2}}, \\ f_i &= \frac{F_i}{k}. \end{aligned}$$

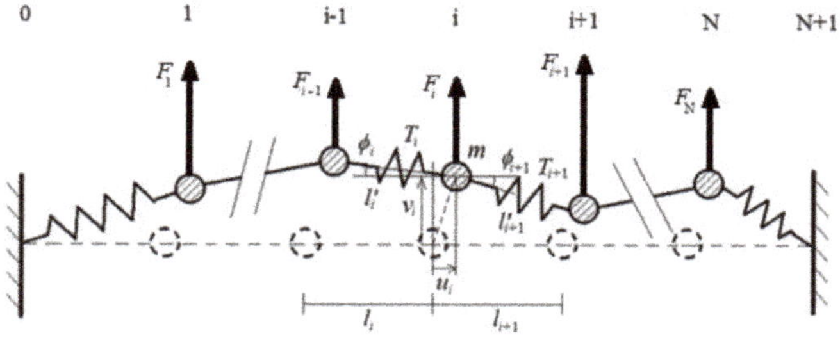

Figure 1. Forced and damped lattice oscillating in the plane (see [2]).

The normalized system (2) is referred to as the "exact lattice" in the following sections.

According to the previous research [1], when we introduce this system (2) without extra transverse force and damping terms, an interesting feature is that in the low energy limit and under the assumption that the axial displacements are assumed to be an order of magnitude smaller compared to the transverse ones, it was shown that, correct to the leading order of approximation, the transverse oscillations decouple from the axial ones and are governed by the following reduced system of equations for predominantly transverse oscillations of the particles:

$$\begin{aligned} &\frac{d^2w_i}{d\tau^2}+2^{-1}(N+1)^{-1}\sum_{q=0}^{N}(w_{q+1}-w_q)^2(2w_i-w_{i+1}-w_{i-1})=0,\\ &i=1,\cdots,N,\quad w_0(0)=w_{N+1}(0)=0. \end{aligned} \tag{3}$$

Then, the nearly linear axial oscillations are driven by the transverse responses (see [1,2] for more details). Therefore, we focus our analysis like in [1] just on Equation (3), which presents *a low-energy nonlinear acoustic vacuum*, because in the absence of linear terms, it possesses zero speed of sound in the context of classical linear acoustics. What is more, it is notable that the existence of the strongly nonlocal multiplier $2^{-1}(N+1)^{-1}\sum_{q=0}^{N}(w_{q+1}-w_q)^2$ indicates that the response of each particle is dependent (and hence, it is coupled) on the responses of all other particles. Equation (3) admits N exact nonlinear standing waves, or nonlinear normal modes (NNMs), in the form:

$$w_i(\tau)=A_p(\tau)\sin\frac{\pi p i}{N+1},\quad i=1,\cdots,N$$

for the p-th NNM, $1\le p\le N$, where $A_p(\tau)$ denotes the p-th modal amplitude. These, by construction, are mutually orthogonal, and there are no other NNMs in this system, nor any NNM bifurcations [1]. Substituting this NNM ansatz into Equation (3) yields a set of N uncoupled nonlinear equations governing the time-dependent amplitudes of the NNMs:

$$A_p''(\tau)+(1/4)\omega_p^4A_p^3(\tau)=0$$

with:

$$\omega_p=2\sin\frac{\pi p}{2(N+1)},$$

which is the p-th natural frequency of the corresponding linear system Equation (3) and the prime denoting differentiation with respect to τ. The exciting force, which is applied on each particle in the transverse direction, is expressed as:

$$f_i=F_p\cos\omega_p\tau\sin\frac{pi\pi}{n+1}$$

where $i=1,\cdots,N$, for the p-th NNM, $1\le p\le N$, and this exciting force includes NNMs in the form $\sin\frac{pi\pi}{n+1}$, $i=1,\cdots,N$, for the p-th NNM, $1\le p\le N$ and the p-th natural linear frequency ω_p.

The frequency of the p-th NNM is tunable with the force and energy, and it also paves the way for nonlinear resonances between NNMs widely separated in the nonlinear spectrum, given that their energies tune their frequencies to satisfy certain rational relationships.

Summarizing, we can write (3) as:

$$\frac{d^2w}{d\tau^2}+2^{-1}(N+1)^{-1}\langle Mw,w\rangle Mw=0,$$

where $w = [w_1, \cdots, w_N] \in \mathbb{R}^N$, M is a symmetric matrix given by

$$M = \begin{pmatrix} 2 & -1 & 0 & \cdots & 0 \\ -1 & 2 & -1 & \cdots & 0 \\ \vdots & \vdots & \vdots & \vdots & \vdots \\ 0 & \cdots & -1 & 2 & -1 \\ 0 & \cdots & 0 & -1 & 2 \end{pmatrix}$$

and $\langle \cdot, \cdot \rangle$ is the standard scalar product on $\mathbb{R}^N$. The eigenvectors of M are $\underline{\phi}_p = [\sin \frac{p\pi}{N+1}, \cdots, \sin \frac{pN\pi}{N+1}]$ with the corresponding eigenvalues ω_p^2, $1 \le p \le N$. Moreover, it holds (see [4], p. 37):

$$\begin{aligned} \langle \underline{\phi}_p, \underline{\phi}_p \rangle &= \sum_{i=1}^{N} \sin^2 \frac{pi\pi}{N+1} = \frac{N+1}{2}, \\ \langle \underline{\phi}_p, \underline{\phi}_k \rangle &= \sum_{i=1}^{N} \sin \frac{pi\pi}{N+1} \sin \frac{ki\pi}{N+1} \\ &= \frac{1}{2} \sum_{i=1}^{N} \left(\cos \frac{(p-k)i\pi}{N+1} - \cos \frac{(p+k)i\pi}{N+1} \right) = 0, \quad p \neq k. \end{aligned}$$

The forced (3) has the form:

$$\frac{d^2 w}{d\tau^2} + 2^{-1}(N+1)^{-1}\langle Mw, w \rangle Mw = \sum_{p=1}^{N} F_p \cos \omega_p \tau \underline{\phi}_p. \tag{4}$$

Therefore, considering the basis $\{\underline{\phi}_p\}_{p=1}^N$ of $\mathbb{R}^N$, we take $w(\tau) = \sum_{p=1}^N C_p(\tau)\underline{\phi}_p$ in (4) to get:

$$C_p''(\tau) + \frac{1}{4} \left(\sum_{i=1}^{N} C_i^2(\tau)\omega_i^2 \right) \omega_p^2 C_p(\tau) = F_p \cos \omega_p \tau, \quad 1 \le p \le N. \tag{5}$$

Next, applying the coordinate transformation to (5):

$$A_p(\tau) = \frac{\omega_p}{2} C_p(\tau), \quad 1 \le p \le N,$$

we get:

$$A_p''(\tau) + \left(\sum_{i=1}^{N} A_i^2(\tau) \right) \omega_p^2 A_p(\tau) = \frac{F_p \omega_p}{2} \cos \omega_p \tau, \quad 1 \le p \le N.$$

3. Two-Mode System

In this paper, we consider just two modes: k and p, so we study the system:

$$\begin{aligned} A_k''(\tau) + [A_k^2(\tau) + A_p^2(\tau)]\omega_k^2 A_k(\tau) + \epsilon\mu_1 \cos(\omega_k \tau) &= 0 \\ A_p''(\tau) + [A_k^2(\tau) + A_p^2(\tau)]\omega_p^2 A_p(\tau) + \epsilon\mu_2 \cos(\omega_p \tau) &= 0, \end{aligned} \tag{6}$$

for $\epsilon \neq 0$ small and parameters μ_1, μ_2. Using the transformation:

$$\begin{aligned} \psi_1(\tau) &= A_k'(\tau) + j\Omega A_k(\tau) \equiv \zeta_1(\tau) e^{j\Omega\tau} \\ \psi_2(\tau) &= A_p'(\tau) + j\Omega A_p(\tau) \equiv \zeta_2(\tau) e^{j\Omega\tau}, \end{aligned}$$

and performing an averaging approach like in [1], Equation (6) is modified to the form:

$$\begin{aligned}
\zeta_1' + \frac{j\Omega}{2}\zeta_1 - \frac{j\omega_k^2}{8\Omega^3}(\zeta_2^2\zeta_1^* + 2|\zeta_2|^2\zeta_1 + 3|\zeta_1|^2\zeta_1) + \epsilon\mu_1\cos(\omega_k\tau)e^{-j\Omega\tau} &= 0 \\
\zeta_2' + \frac{j\Omega}{2}\zeta_2 - \frac{j\omega_p^2}{8\Omega^3}(\zeta_1^2\zeta_2^* + 2|\zeta_1|^2\zeta_2 + 3|\zeta_2|^2\zeta_2) + \epsilon\mu_2\cos(\omega_p\tau)e^{-j\Omega\tau} &= 0.
\end{aligned}$$

Introducing $\zeta_i = a_i e^{j\beta_i}$ and $\Delta = \beta_2 - \beta_1$, we get:

$$\begin{aligned}
a_1' + \frac{\omega_k^2}{8\Omega^3}a_2^2 a_1 \sin 2\Delta + \epsilon\mu_1\cos(\omega_k\tau)\cos(\Omega\tau + \beta_1) &= 0 \\
a_2' - \frac{\omega_p^2}{8\Omega^3}a_1^2 a_2 \sin 2\Delta + \epsilon\mu_2\cos(\omega_p\tau)\cos(\Omega\tau + \beta_1 + \Delta) &= 0 \\
\Delta' - \frac{\omega_p^2}{8\Omega^3}(3a_2^2 + a_1^2\cos 2\Delta + 2a_1^2) + \frac{\omega_k^2}{8\Omega^3}(3a_1^2 + a_2^2\cos 2\Delta + 2a_2^2) & \\
-\frac{\epsilon\mu_2}{a_2}\cos(\omega_p\tau)\sin(\Omega\tau + \beta_1 + \Delta) + \frac{\epsilon\mu_1}{a_1}\cos(\omega_k\tau)\sin(\Omega\tau + \beta_1) &= 0
\end{aligned} \tag{7}$$

where we consider β_1 as a constant parameter. Now, by introducing the coordinate transformations $a_1 = (\frac{\rho}{\omega_p})\sin\theta$ and $a_2 = (\frac{\rho}{\omega_k})\cos\theta$ into Equation (7), we get:

$$\begin{aligned}
\rho' = -\epsilon\mu_1\omega_p\sin\theta\cos(\omega_k\tau)\cos(\Omega\tau + \beta_1) - \epsilon\mu_2\omega_k\cos\theta\cos(\omega_p\tau)\cos(\Omega\tau + \Delta + \beta_1) \\
\theta' + \frac{\rho^2}{16\Omega^3}\sin 2\theta\sin 2\Delta + \epsilon\mu_1\omega_p\frac{\cos\theta}{\rho}\cos(\omega_k\tau)\cos(\Omega\tau + \beta_1) \\
-\epsilon\mu_2\omega_k\frac{\sin\theta}{\rho}\cos(\omega_p\tau)\cos(\Omega\tau + \Delta + \beta_1) = 0 \\
\Delta' - \frac{\rho^2}{8\Omega^3}\left[\frac{3\omega_p^2}{\omega_k^2}\cos^2\theta - \frac{3\omega_k^2}{\omega_p^2}\sin^2\theta - \cos 2\theta(2 + \cos 2\Delta)\right] \\
-\frac{\epsilon\mu_2\omega_k}{\rho\cos\theta}\cos(\omega_p\tau)\sin(\Omega\tau + \Delta + \beta_1) + \frac{\epsilon\mu_1\omega_p}{\rho\sin\theta}\cos(\omega_k\tau)\sin(\Omega\tau + \beta_1) = 0.
\end{aligned}$$

In the rest of the paper, we assume $\omega_p = \omega_k = P$ and $\Omega = kP$ for a natural number k, so we study the periodically-perturbed system:

$$\begin{aligned}
\rho' = -\epsilon\mu_1\sin\theta\cos(P\tau)\cos(kP\tau + \beta_1) - \epsilon\mu_2\cos\theta\cos(P\tau)\cos(kP\tau + \Delta + \beta_1) \\
\theta' + \frac{\rho^2}{16k^3P^3}\sin 2\theta\sin 2\Delta + \epsilon\mu_1\frac{\cos\theta}{\rho}\cos(P\tau)\cos(kP\tau + \beta_1) \\
-\epsilon\mu_2\frac{\sin\theta}{\rho}\cos(P\tau)\cos(kP\tau + \Delta + \beta_1) = 0 \\
\Delta' - \frac{\rho^2}{4k^3P^3}\cos 2\theta\sin^2\Delta + \frac{\epsilon\mu_1}{\rho\sin\theta}\cos(P\tau)\sin(kP\tau + \beta_1) \\
-\frac{\epsilon\mu_2}{\rho\cos\theta}\cos(P\tau)\sin(kP\tau + \Delta + \beta_1) = 0
\end{aligned} \tag{8}$$

where we scaled $P\mu_i \leftrightarrow \mu_i$, $i = 1, 2$. We may suppose:

$$\mu_1^2 + \mu_2^2 = 1.$$

First, consider the unperturbed case where $\epsilon = 0$, so the system:

$$\begin{aligned} \rho' &= 0 \\ \theta' + \frac{\rho^2}{16k^3P^3}\sin 2\theta \sin 2\Delta &= 0 \\ \Delta' - \frac{\rho^2}{4k^3P^3}\cos 2\theta \sin^2\Delta &= 0. \end{aligned} \tag{9}$$

By introducing the temporal variable $\tau_2 = \frac{\rho^2}{8k^3P^3}\tau$ in Equation (9), we get:

$$\begin{aligned} \frac{d\theta}{d\tau_2} &= -\tfrac{1}{2}\sin 2\theta \sin 2\Delta \\ \frac{d\Delta}{d\tau_2} &= 2\cos 2\theta \sin^2\Delta \end{aligned} \tag{10}$$

which is fully integrable and gives us the first integral $I = \sin 2\theta \sin \Delta = K = const.$ of the degenerate slow flow. If we consider the following initial conditions: $\theta(0) = \theta_0$, where $0 < \theta_0 < \frac{\pi}{4}$ and $\Delta(0) = \pi/2$, then we get $K = \sin 2\theta_0 \in (0, 1)$. To find exact solutions of (10), we first derive:

$$\begin{aligned} \frac{d\tau_2}{d\theta} &= -\frac{2}{\sin 2\theta \sin 2\Delta} = -\frac{1}{K\cos\Delta} = \frac{1}{K\sqrt{1-\sin^2\Delta}} \\ &= \frac{1}{K\sqrt{1-\frac{K^2}{\sin^2 2\theta}}} = \frac{1}{K}\frac{\sin 2\theta}{\sqrt{\sin^2 2\theta - K^2}}, \end{aligned}$$

for $\tau_2 > 0$ small, since (10) gives $0 < \theta < \frac{\pi}{2}$, $0 < \Delta < \pi$ (see Figure 2) and $\frac{d\Delta}{d\tau_2}(0) = 2\cos 2\theta_0 > 0$, so for $\tau_2 > 0$ small, we have $\Delta(\tau_2) > \frac{\pi}{2}$.

By using the formula in [4] ((2.599.4) p. 205), we obtain:

$$\tau_2 = \int_{\theta_0}^{\theta} \frac{1}{K}\frac{\sin 2\theta}{\sqrt{\sin^2 2\theta - K^2}} d\theta = \frac{1}{2K}\left(\frac{\pi}{2} - \sin^{-1}\left(\frac{\cos 2\theta}{\sqrt{1-K^2}}\right)\right),$$

which gives:

$$\theta(\tau_2) = \frac{1}{2}\cos^{-1}\left(\sqrt{1-K^2}\cos(2K\tau_2)\right), \tag{11}$$

recalling $0 < \theta < \frac{\pi}{2}$. Of course, Formula (11) holds for all τ_2, not just for small positive ones. Next, using (10) and (11), we have:

$$\frac{d\Delta}{d\tau_2} = 2\sqrt{1-K^2}\cos(2K\tau_2)\sin^2\Delta,$$

which can be easily solved to arrive at:

$$\Delta(\tau_2) = \pi - \cot^{-1}\left(\frac{\sqrt{1-K^2}}{K}\sin(2K\tau_2)\right).$$

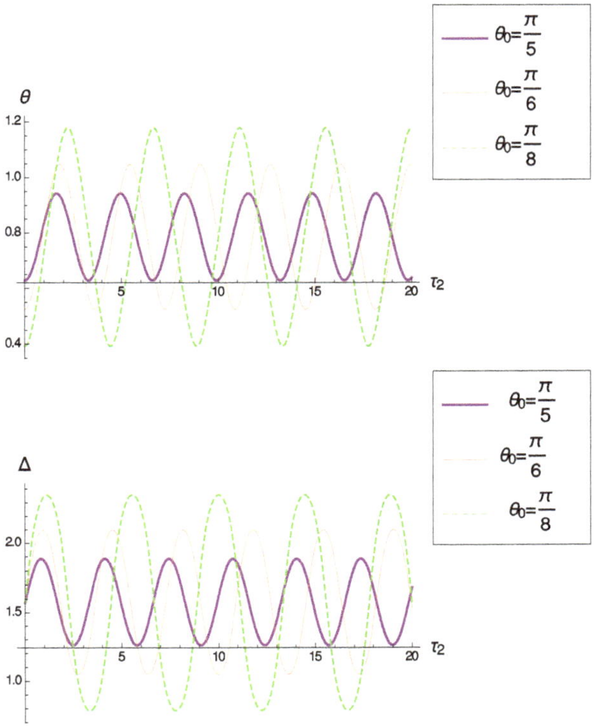

Figure 2. Top panel: Graph for θ for different initial values of θ_0. Bottom panel: Graph for Δ for different initial values of θ_0.

Hence, the exact solution of the system (10) is given as:

$$\theta(\tau_2) = \frac{1}{2}\cos^{-1}\left(\sqrt{1-K^2}\cos(2K\tau_2)\right)$$
$$\Delta(\tau_2) = \pi - \cot^{-1}\left(\frac{\sqrt{1-K^2}}{K}\sin(2K\tau_2)\right)$$

where the period is:

$$T(\theta_0) = \frac{2}{K}\int_{\theta_0}^{\frac{\pi}{2}-\theta_0} \frac{d\theta}{\left(1-\frac{\sin^2 2\theta_0}{\sin^2 2\theta}\right)^{1/2}} = \frac{\pi}{K}.$$

It is easy to verify the following symmetry property (see Figure 3):

$$\Delta\left(\tau_2+\frac{T}{2}\right) = \pi - \Delta(\tau_2), \quad \theta\left(\tau_2+\frac{T}{2}\right) = \frac{\pi}{2} - \theta(\tau_2).$$

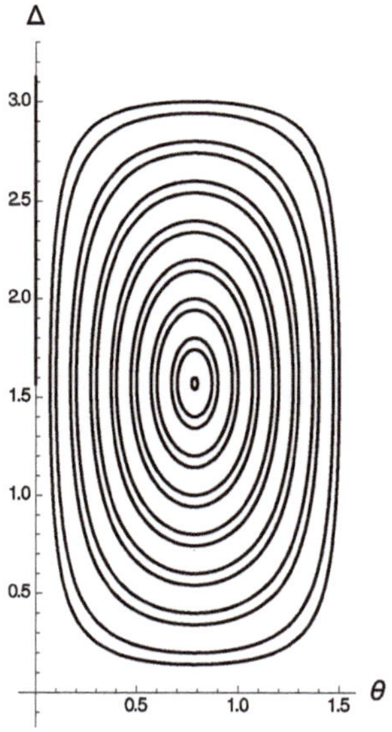

Figure 3. Orbits in the phase portrait of (10), where $(\theta, \Delta) \in [0, \frac{\pi}{2}] \times [0, \pi]$.

Summarizing, the exact solution of the unperturbed (9) is the following:

$$\begin{aligned}\rho(\tau) &= const. \\ \theta(\tau) &= \frac{1}{2}\cos^{-1}\left(\sqrt{1-K^2}\cos\left(\frac{K\rho^2}{4k^3P^3}\tau\right)\right) \\ \Delta(\tau) &= \pi - \cot^{-1}\left(\frac{\sqrt{1-K^2}}{K}\sin\left(\frac{K\rho^2}{4k^3P^3}\tau\right)\right)\end{aligned}$$

with the period:

$$T = \frac{8k^3P^3\pi}{K\rho^2}.$$

Consequently, for any:

$$\rho > 2kP^2\sqrt{k},$$

taking:

$$K(\rho) = \frac{4k^3P^4}{\rho^2} \in (0,1),$$

Equation (9) has the $T = 2\pi/P$-periodic solution:

$$\begin{aligned}\theta(\rho,\tau) &= \frac{1}{2}\cos^{-1}\left(\frac{\sqrt{\rho^4 - 16k^6P^8}}{\rho^2}\cos(P\tau)\right) \\ \Delta(\rho,\tau) &= \pi - \cot^{-1}\left(\frac{\sqrt{\rho^4 - 16k^6P^8}}{4k^3P^4}\sin(P\tau)\right).\end{aligned} \tag{12}$$

4. Melnikov Analysis for Periodic Oscillations

Writing (8) as:

$$\begin{aligned}
\theta' + \frac{\rho^2}{16k^3P^3}\sin 2\theta \sin 2\Delta &= \epsilon\rho^{-1}g_1(\rho,\theta,\Delta,\tau)\\
\Delta' - \frac{\rho^2}{4k^3P^3}\cos 2\theta \sin^2\Delta &= \epsilon\rho^{-1}g_2(\rho,\theta,\Delta,\tau)\\
\rho' &= \epsilon g_3(\rho,\theta,\Delta,\tau)
\end{aligned}$$

and using [5], (3.5.11), p. 111, with $\alpha = 0$, and [6], Lemma 2.5, p. 283, we compute the Melnikov function:

$$M(\beta_1,\rho) = (M_1(\beta_1,\rho), M_2(\beta_1,\rho))$$

as:

$$\begin{aligned}
M_1(\beta_1,\rho) &= \int_0^T \left(\frac{\partial I}{\partial\theta}\rho^{-1}g_1 + \frac{\partial I}{\partial\Delta}\rho^{-1}g_2 - \frac{\partial I}{\partial\theta}\frac{\partial\theta}{\partial\rho}g_3 - \frac{\partial I}{\partial\Delta}\frac{\partial\Delta}{\partial\rho}g_3\right)d\tau\\
&= \int_0^T \left(\frac{\partial I}{\partial\theta}\rho^{-1}g_1 + \frac{\partial I}{\partial\Delta}\rho^{-1}g_2\right)d\tau - \frac{dK}{d\rho}\int_0^T g_3 d\tau
\end{aligned} \tag{13}$$

since differentiating by ρ the identity:

$$I(\theta(\rho,\tau),\Delta(\rho,\tau)) = K(\rho),$$

we get:

$$\frac{\partial I}{\partial\theta}\frac{\partial\theta}{\partial\rho} + \frac{\partial I}{\partial\Delta}\frac{\partial\Delta}{\partial\rho} = \frac{dK}{d\rho},$$

which is independent of τ, and:

$$M_2(\beta_1,\rho) = \int_0^T g_3 d\tau. \tag{14}$$

Formulas (13) and (14) are similar to [3], (2.7). We are looking for a simple zero of M, which is equivalent to considering:

$$\begin{aligned}
&\bar{M}(\beta_1,\rho) = (\bar{M}_1(\beta_1,\rho), \bar{M}_2(\beta_1,\rho))\\
&\bar{M}_1(\beta_1,\rho) = \int_0^T \left(\frac{\partial I}{\partial\theta}g_1 + \frac{\partial I}{\partial\Delta}g_2\right)d\tau, \quad \bar{M}_2(\beta_1,\rho) = \int_0^T g_3 d\tau.
\end{aligned}$$

Since:

$$g_i = \mu_1 g_{i1} + \mu_2 g_{i2}, \quad i = 1,2,3$$

for:

$$\begin{aligned}
g_{11}(\rho,\theta,\Delta,\tau) &= -\cos\theta\cos(P\tau)\cos(kP\tau+\beta_1)\\
g_{12}(\rho,\theta,\Delta,\tau) &= \sin\theta\cos(P\tau)\cos(kP\tau+\Delta+\beta_1)\\
g_{21}(\rho,\theta,\Delta,\tau) &= -\frac{\cos(P\tau)}{\sin\theta}\sin(kP\tau+\beta_1)\\
g_{22}(\rho,\theta,\Delta,\tau) &= \frac{\cos(P\tau)}{\cos\theta}\sin(kP\tau+\Delta+\beta_1)\\
g_{31}(\rho,\theta,\Delta,\tau) &= -\sin\theta\cos(P\tau)\cos(kP\tau+\beta_1)\\
g_{32}(\rho,\theta,\Delta,\tau) &= -\cos\theta\cos(P\tau)\cos(kP\tau+\Delta+\beta_1)
\end{aligned}$$

we get:

$$\bar{M}_i(\beta_1,\rho) = \mu_1\bar{M}_{i1}(\beta_1,\rho) + \mu_2\bar{M}_{i2}(\beta_1,\rho), \quad i = 1,2,$$

for

$$\bar{M}_{1j}(\beta_1,\rho) = \int_0^T \left(\frac{\partial I}{\partial \theta} g_{1j} + \frac{\partial I}{\partial \Delta} g_{2j}\right) d\tau, \quad \bar{M}_{2j}(\beta_1,\rho) = \int_0^T g_{3j} d\tau, \quad j = 1,2.$$

To solve:

$$\mu_1 \bar{M}_{i1}(\beta_1,\rho) + \mu_2 \bar{M}_{i2}(\beta_1,\rho) = 0, \quad i = 1,2,$$

we first solve the scalar equation:

$$\bar{M}(\beta_1,\rho) = \bar{M}_{11}(\beta_1,\rho)\bar{M}_{22}(\beta_1,\rho) - \bar{M}_{12}(\beta_1,\rho)\bar{M}_{21}(\beta_1,\rho) = 0 \tag{15}$$

to get its root $\beta_{1,0}$ and ρ_0. Then, we look for $\mu_{1,0}$ and $\mu_{2,0}$ with $\mu_{1,0}^2 + \mu_{2,0}^2 = 1$ such that:

$$\begin{gathered} \mu_{1,0}\bar{M}_{i1}(\beta_{1,0},\rho_0) + \mu_{2,0}\bar{M}_{i2}(\beta_{1,0},\rho_0) = 0, \quad i = 1,2 \\ \det\begin{pmatrix} \mu_{1,0}\nabla\bar{M}_{11}(\beta_{1,0},\rho_0)^\top + \mu_{2,0}\nabla\bar{M}_{12}(\beta_{1,0},\rho_0)^\top \\ \mu_{1,0}\nabla\bar{M}_{21}(\beta_{1,0},\rho_0)^\top + \mu_{2,0}\nabla\bar{M}_{22}(\beta_{1,0},\rho_0)^\top \end{pmatrix} \neq 0. \end{gathered} \tag{16}$$

Summarizing, we have the following result.

Theorem 1. *If there are $\beta_{1,0} \in [0,2\pi)$, ρ_0 satisfying (15), $\mu_{1,0}$ and $\mu_{2,0}$ with $\mu_{1,0}^2 + \mu_{2,0}^2 = 1$ solving (16), then for any μ_1 near $\mu_{1,0}$ and μ_2 near $\mu_{2,0}$ with $\mu_1^2 + \mu_2^2 = 1$ and $\epsilon \neq 0$ small, there are $\beta_1(\epsilon)$ near $\beta_{1,0}$ and $\rho(\epsilon)$ near ρ_0 such that (8) with $\beta_1 = \beta_1(\epsilon)$ and $\rho = \rho(\epsilon)$ has a $T = 2\pi/P$-periodic solution near (12) with $\rho = \rho(\epsilon)$.*

Note:

$$\begin{aligned} \bar{M}_{11}(\beta_1,\rho) &= -\int_0^T \Big(2\cos(2\theta(\rho,\tau))\sin\Delta(\rho,\tau)\cos\theta(\rho,\tau)\cos(P\tau)\cos(kP\tau+\beta_1) \\ &\quad +2\cos\theta(\rho,\tau)\cos\Delta(\rho,\tau)\cos(P\tau)\sin(kP\tau+\beta_1))\Big)d\tau \\ \bar{M}_{12}(\beta_1,\rho) &= \int_0^T \Big(2\cos(2\theta(\rho,\tau))\sin\Delta(\rho,\tau)\sin\theta(\rho,\tau)\cos(P\tau)\cos(kP\tau+\Delta(\rho,\tau)+\beta_1) \\ &\quad +2\sin\theta(\rho,\tau)\cos\Delta(\rho,\tau)\cos(P\tau)\sin(kP\tau+\Delta(\rho,\tau)+\beta_1)\Big)d\tau \\ \bar{M}_{21}(\beta_1,\rho) &= -\int_0^T \sin\theta(\rho,\tau)\cos(P\tau)\cos(kP\tau+\beta_1)d\tau \\ \bar{M}_{22}(\beta_1,\rho) &= -\int_0^T \cos\theta(\rho,\tau)\cos(P\tau)\cos(kP\tau+\Delta(\rho,\tau)+\beta_1)d\tau. \end{aligned}$$

Next, taking $\rho \to \infty$ in (12), we obtain:

$$\begin{gathered} \theta(\infty,\tau) = \begin{cases} \dfrac{P\tau}{2} & \tau \in \left[0,\dfrac{\pi}{P}\right], \\ \pi - \dfrac{P\tau}{2} & \tau \in \left[\dfrac{\pi}{P},\dfrac{2\pi}{P}\right], \end{cases} \\ \Delta(\infty,\tau) = \begin{cases} \pi & \tau \in \left(0,\dfrac{\pi}{P}\right), \\ 0 & \tau \in \left(\dfrac{\pi}{P},\dfrac{2\pi}{P}\right), \\ \dfrac{\pi}{2} & \tau \in \left\{0,\dfrac{\pi}{P},\dfrac{2\pi}{P}\right\}, \end{cases} \\ \frac{\partial\theta}{\partial\rho}(\infty,\tau) = 0, \quad \frac{\partial\Delta}{\partial\rho}(\infty,\tau) = 0. \end{gathered} \tag{17}$$

Hence:

$$\begin{aligned}
\bar{M}_{11}(\beta_1,\infty) &= \int_0^{\frac{2\pi}{P}} 2\cos\frac{P\tau}{2}\cos(P\tau)\sin(kP\tau+\beta_1)d\tau \\
&= \frac{16k(4k^2-5)\cos\beta_1}{(16k^4-40k^2+9)P} \\
\bar{M}_{12}(\beta_1,\infty) &= \int_0^{\frac{2\pi}{P}} 2\sin\frac{P\tau}{2}\cos(P\tau)\sin(kP\tau+\beta_1)\Big)d\tau \\
&= -\frac{8(4k^2+3)\sin\beta_1}{(16k^4-40k^2+9)P} \\
\bar{M}_{21}(\beta_1,\infty) &= -\int_0^{\frac{2\pi}{P}} \sin\frac{P\tau}{2}\cos(P\tau)\cos(kP\tau+\beta_1)d\tau \\
&= \frac{4(4k^2+3)\cos\beta_1}{(16k^4-40k^2+9)P} \\
\bar{M}_{22}(\beta_1,\infty) &= \int_0^{\frac{2\pi}{P}} \cos\frac{P\tau}{2}\cos(P\tau)\cos(kP\tau+\beta_1)d\tau \\
&= -\frac{8k(4k^2-5)\sin\beta_1}{(16k^4-40k^2+9)P}.
\end{aligned}$$

Then, (15) gives as $\rho\to\infty$,

$$\tilde{M}(\beta_1,\infty) = -\frac{16\sin 2\beta_1}{(4k^2-9)P^2}$$

with asymptotic solutions $\beta_{1,0}^{\infty}$ satisfying either $\sin\beta_{1,0}^{\infty}=0$ or $\cos\beta_{1,0}^{\infty}=0$. The asymptotic equation of (16) is as follows:

$$\begin{aligned}
0 &= \mu_{1,0}\bar{M}_{11}(\beta_{1,0}^{\infty},\infty)+\mu_{2,0}\bar{M}_{12}(\beta_{1,0}^{\infty},\infty) \\
&= \mu_{1,0}\frac{16k(4k^2-5)\cos\beta_{1,0}^{\infty}}{(16k^4-40k^2+9)P} - \mu_{2,0}\frac{8(4k^2+3)\sin\beta_{1,0}^{\infty}}{(16k^4-40k^2+9)P},
\end{aligned}$$

which has solutions: $\mu_{1,0}^{\infty}=0$ and $\mu_{2,0}^{\infty}=\pm1$ when $\sin\beta_{1,0}^{\infty}=0$ or $\mu_{1,0}^{\infty}=\pm1$ and $\mu_{2,0}^{\infty}=0$ when $\cos\beta_{1,0}^{\infty}=0$. However, $\frac{\partial\bar{M}_{ij}}{\partial\rho}(\beta_1,\infty)=0$ for $i,j\in\{1,2\}$, so Theorem 1 cannot be applied directly. Note we consider just the first asymptotic equation of (16), since the second one is a scalar multiple of the first one.

On the other hand, following the method of [5], p. 111, we get the following result.

Corollary 1. *For any $\rho>0$ sufficiently large and $\epsilon\neq0$ sufficiently small, there is $\mu_1(\rho,\epsilon)$, $\mu_2(\rho,\epsilon)$ with $\mu_1^2(\rho,\epsilon)+\mu_2^2(\rho,\epsilon)=1$ and $\beta_1(\rho,\epsilon)$ such that (8) has a T-periodic solution near (17) for $\epsilon\neq0$ small and either $\mu_1(\infty,0)=0$, $\mu_2(\infty,0)=\pm1$, $\sin\beta_1(\infty,0)=0$ or $\mu_1(\infty,0)=\pm1$, $\mu_2(\infty,0)=0$, $\cos\beta_1(\infty,0)=0$.*

Proof. The bifurcation equation has the form (see [5], p. 111):

$$\mu_1\bar{M}_{i1}(\beta_1,\rho)+\mu_2\bar{M}_{i2}(\beta_1,\rho)=O(\epsilon),\quad i=1,2,\quad \mu_1^2+\mu_2^2=1$$

i.e.,

$$\sin\Gamma\bar{M}_{i1}(\beta_1,\rho)+\cos\Gamma\bar{M}_{i2}(\beta_1,\rho)=O(\epsilon),\quad i=1,2 \tag{18}$$

for $\mu_1=\sin\Gamma$ and $\mu_2=\cos\Gamma$. For $\rho=\infty$ and $\epsilon=0$, (18) takes the form:

$$\begin{aligned}
&\frac{16k(4k^2-5)}{(16k^4-40k^2+9)P}\sin\Gamma\cos\beta_1-\frac{8(4k^2+3)}{(16k^4-40k^2+9)P}\cos\Gamma\sin\beta_1=0 \\
&\frac{4(4k^2+3)\cos\beta_1}{(16k^4-40k^2+9)P}\sin\Gamma\cos\beta_1-\frac{8k(4k^2-5)}{(16k^4-40k^2+9)P}\cos\Gamma\sin\beta_1=0.
\end{aligned} \tag{19}$$

The determinant of (19) is $-\frac{32}{(4k^2-9)P^2} \neq 0$, so (19) has the only solutions:

$$\sin\Gamma_\infty = \sin\beta_{1,0}^\infty = 0, \quad \cos\Gamma_\infty = \cos\beta_{1,0}^\infty = 0.$$

The determinants of Jacobians of (19) at these zeros are as follows:

$$\det\begin{pmatrix} \frac{16k(4k^2-5)}{(16k^4-40k^2+9)P}\cos\Gamma_\infty\cos\beta_{1,0}^\infty & -\frac{8(4k^2+3)}{(16k^4-40k^2+9)P}\cos\Gamma_\infty\cos\beta_{1,0}^\infty \\ \frac{4(4k^2+3)\cos\beta_{1,0}^\infty}{(16k^4-40k^2+9)P}\cos\Gamma_\infty\cos\beta_{1,0}^\infty & -\frac{8k(4k^2-5)}{(16k^4-40k^2+9)P}\cos\Gamma_\infty\cos\beta_{1,0}^\infty \end{pmatrix} = -\frac{32\cos^2\Gamma_\infty\cos^2\beta_{1,0}^\infty}{(4k^2-9)P^2} \neq 0$$

and:

$$\det\begin{pmatrix} \frac{8(4k^2+3)}{(16k^4-40k^2+9)P}\sin\Gamma_\infty\sin\beta_{1,0}^\infty & -\frac{16k(4k^2-5)}{(16k^4-40k^2+9)P}\sin\Gamma_\infty\sin\beta_{1,0}^\infty \\ \frac{8k(4k^2-5)}{(16k^4-40k^2+9)P}\sin\Gamma_\infty\sin\beta_{1,0}^\infty & -\frac{4(4k^2+3)\cos\beta_{1,0}^\infty}{(16k^4-40k^2+9)P}\sin\Gamma_\infty\sin\beta_{1,0}^\infty \end{pmatrix} = \frac{32\sin^2\Gamma_\infty\sin^2\beta_{1,0}^\infty}{(4k^2-9)P^2} \neq 0,$$

respectively. Hence, the zeroes Γ_∞ and $\beta_{1,0}^\infty$ are simple, so we can apply the implicit function theorem to get the result. The proof is complete. □

5. Discussion

Melnikov analysis is applied for the persistence of periodic oscillations for periodically-perturbed systems of ODEs with a slowly-varying variable. The ODEs are obtained from a model of a finite lattice of particles coupled by linear springs under distributed harmonic excitation, which presents a low-energy nonlinear acoustic vacuum (see [1]), but we consider just two modes. We extend the study of [1] to a problem with small exciting harmonic forces. We apply an analytical method based on derivation of Melnikov functions and then on the location of their simple roots. Melnikov functions are derived by using the approach of [3,5] developed for slowly-varying ODEs. Since the Melnikov functions are rather complicated, we follow an asymptotic way for solving the corresponding Melnikov equations. It would be nice to solve these Melnikov equations numerically for finding other simple roots, which is postponed to our next research. These roots determine and locate periodic solutions of the periodically-perturbed systems of ODEs derived from the two-mode low-energy nonlinear acoustic vacuum system. Moreover, our next investigation will be also to consider higher numbers of modes represented by a system of ODEs in (5). The method used will be different from that in this paper, since it will be based on the results of Section 3.3 of [5]. Note that higher modes of (2) were numerically studied in [2], which is another challenge for our study.

Author Contributions: The authors contributed equally to this work.

Funding: M.A. is supported by the National Scholarship Programme of the Slovak Republic for the Support of Mobility of Students, Ph.D. Students, University Teachers, Researchers and Artists. M.F. and M.P. are supported by the Slovak Research and Development Agency (Grant Number APVV-14-0378) and the Slovak Grant Agency VEGA(Grant Numbers 2/0153/16 and 1/0078/17).

Conflicts of Interest: The authors declare no conflict of interest.

References

1. Manevich, L.I.; Vakakis, A.F. Nonlinear oscillatory acoustic vacuum. *SIAM J. Appl. Math.* **2014**, *74*, 1742–1762. [CrossRef]

2. Zhang, Z.; Manevitch, L.I.; Smirnov, V.; Bergman, L.; Vakakis, A.F. Extreme nonlinear energy exchanges in a geometrically nonlinear lattice oscillating in the plane. *J. Mech. Phys. Solids* **2018**, *110*, 1–20. [CrossRef]
3. Wiggins, S.; Holmes, P. Periodic orbits in slowly varying oscillators. *SIAM J. Math. Anal.* **1987**, *18*, 592–611. [CrossRef]
4. Gradshteyn, I.S.; Ryzhik, I.M. *Table of Integrals, Series, and Products*, 7th ed.; Academic Press: Cambridge, MA, USA, 2007.
5. Fečkan, M. *Topological Degree Approach to Bifurcation Problems*; Springer: Berlin, Germany, 2008.
6. Hale, J.K. *Ordinary Differential Equations*, 2nd ed.; Robert E. Krieger Publishing Company, Inc.: Malabar, FL, USA, 1980.

axioms

MDPI

Article

Solutions to Abel's Integral Equations in Distributions

Chenkuan Li *, Thomas Humphries and Hunter Plowman

Department of Mathematics and Computer Science, Brandon University, Brandon, MB R7A 6A9, Canada; humphrte65@brandonu.ca (T.H.); plowmahh10@brandonu.ca (H.P.)

* Correspondence: lic@brandonu.ca; Tel.: +1-204-571-8549

Received: 10 August 2018; Accepted: 31 August 2018; Published: 2 September 2018

Abstract: The goal of this paper is to study fractional calculus of distributions, the generalized Abel's integral equations, as well as fractional differential equations in the distributional space $\mathcal{D}'(R^+)$ based on inverse convolutional operators and Babenko's approach. Furthermore, we provide interesting applications of Abel's integral equations in viscoelastic systems, as well as solving other integral equations, such as $\int_\theta^{\pi/2} \frac{y(\varphi)}{\cos^\beta \varphi(\cos\theta - \cos\varphi)^\alpha} d\varphi = f(\theta)$, and $\int_0^\infty x^{1/2} g(x) y(x+t) dx = f(t)$.

Keywords: distribution; fractional calculus; Mittag–Leffler function; Abel's integral equation; convolution

MSC: 46F10; 26A33

---

1. Introduction

Fractional modeling is an emergent tool which uses fractional differential and integral equations to describe non-local dynamic processes associated with complex systems [1–8]. Integral and fractional differential equations arise in numerous physical problems [9–12], in the fields of chemistry, biology, electronics, noncommutative quantum field theories [13], and quantum mechanics [14]. Mathematical models of systems and processes in the mentioned areas of engineering [15] and scientific disciplines involve integrals of unknown functions and derivatives of fractional order. As far as we know, fractional calculus provides an excellent tool to construct certain electro-chemical problems and characterizes long-term behaviors [16,17], allometric scaling laws, hereditary properties of various materials and so on [18]. This is the main advantage of fractional differential equations, in comparison with classical integer-order models in practice. Recently, Srivastava et al. presented the model under-actuated mechanical system with fractional order derivative [19]. Many initial and boundary value problems associated with ordinary (or partial) differential equations, can be converted into Volterra integral equations [1,20]. The Volterra's population growth model, biological species living together, and the heat change can all be characterized by integral equations. For example, Gorenflo and Mainardi [21] provided applications of Abel's integral equations, of the first and second kind, in solving the partial differential equation which describes the problem of the heating (or cooling) of a semi-infinite rod by influx (or efflux) of heat across the boundary into (or from) it's interior. In 1985, Hatcher [22] worked on a nonlinear Hilbert problem of a power type, solved in closed form by representing a sectionally holomorphic function by means of an integral with power kernel, and transformed the problem to one of solving a generalized Abel's integral equation. The development of integral equations has led to the construction of many real world problems, such as mathematical physics models [23,24], scattering in quantum mechanics and water waves. There have been lots of techniques, such as numerical analysis and integral transforms [25–27], thus far to studying fractional differential and integral equations, including Abel's equations, with many applications [1,20,28–42].

Kilbas et al. [43] presented a solution in a closed form of multi-dimensional integral equations of the first kind with the Gauss hypergeometric function in the kernel over special pyramidal domains.

Raina et al. [44] later on investigated the solvability of the one-dimensional Abel-type hypergeometric integral equation, given by

$$\frac{(x-a)^{-\alpha}}{\Gamma(\gamma)} \int_a^x (x-t)^{\gamma-1} F\left(\alpha, \beta; \gamma; \frac{x-t}{x-a}\right) \phi(t)dt = f(x)$$

where $x > a$ with $\alpha, \beta \in R$ and $0 < \gamma < 1$, as well as the multidimensional Abel-type hypergeometric integral equation over a pyramidal domain in R^n. The generalized fractional integral and differential operators are introduced and their properties are investigated systematically based on the results obtained.

Srivastava and Buschman [20] presented the comprehensive theory and numerous applications of the integral equations of convolution type, and of certain classes of integro-differential and non-linear integral equations, including Abel's integral equations, in the classical sense.

We start with the necessary concepts and definitions of fractional calculus of distributions in $\mathcal{D}'(R^+)$ based on the generalized convolution in the Schwartz space. Using inverse convolutional operators and Babenko's approach, we study and solve several Abel's integral (for all $\alpha \in R$) and fractional differential equations, as convergent series or in terms of the Mittag–Leffler functions. Many of the results derived can not be archived in the classical sense including numerical analysis methods, or by the Laplace transform. Applications are presented at the end in viscoelastic systems, and for solving other types of integral equations which can be converted into Abel's ones.

2. Abel's Integral Equations in Distribution

In order to study Abel's integral and fractional differential equations distributionally, we briefly introduce the following basic concepts in distribution. Let $\mathcal{D}(R)$ be the Schwartz space (testing function space) [45] of infinitely differentiable functions with compact support in R, and $\mathcal{D}'(R)$ the (dual) space of distributions defined on $\mathcal{D}(R)$. A sequence $\phi_1, \phi_2, \cdots, \phi_n, \cdots$ goes to zero in $\mathcal{D}(R)$ if and only if these functions vanish outside a certain fixed bounded set, and converge to zero uniformly together with their derivatives of any order. We further assume that $\mathcal{D}'(R^+)$ is the subspace of $\mathcal{D}'(R)$ with support contained in R^+.

The functional $\delta^{(n)}(x-x_0)$ for $x_0 \in R$ is defined as

$$(\delta^{(n)}(x-x_0), \phi(x)) = (-1)^n \phi^{(n)}(x_0)$$

where $\phi \in \mathcal{D}(R)$. Clearly, $\delta^{(n)}(x-x_0)$ is a linear and continuous functional on $\mathcal{D}(R)$, and hence $\delta^{(n)}(x-x_0) \in \mathcal{D}'(R)$.

Define

$$f(x) = \begin{cases} \sin x & \text{if } 0 < x < 1, \\ 0 & \text{otherwise.} \end{cases}$$

Then, $f(x)$ is a locally integrable function on R (clearly not continuous) and

$$(f(x), \phi(x)) = \int_0^1 \sin x \, \phi(x) dx \quad \text{for} \quad \phi \in \mathcal{D}(R), \tag{1}$$

defines a regular distribution $f(x) \in \mathcal{D}'(R^+)$.

Let $f \in \mathcal{D}'(R)$. The distributional derivative of f, denoted by f' or df/dx, is defined as

$$(f', \phi) = -(f, \phi')$$

for $\phi \in \mathcal{D}(R)$.

Assume f is a distribution in $\mathcal{D}'(R)$ and g is a function in $C^\infty(R)$. Then the product fg is well defined by

$$(fg, \phi) = (f, g\phi)$$

for all functions $\phi \in \mathcal{D}(R)$ as $g\phi \in \mathcal{D}(R)$.

Clearly, $f' \in \mathcal{D}'(R)$ and every distribution has a derivative.

It can be shown that the ordinary rules of differentiation also apply to distributions. For instance, the derivative of a sum, is the sum of the derivatives, and a constant can be commuted with the derivative operator.

It follows from [45] that $\Phi_\lambda = \frac{x_+^{\lambda-1}}{\Gamma(\lambda)} \in \mathcal{D}'(R^+)$ is an entire function of λ on the complex plane, and

$$\left.\frac{x_+^{\lambda-1}}{\Gamma(\lambda)}\right|_{\lambda=-n} = \delta^{(n)}(x), \quad \text{for} \quad n = 0, 1, 2, \cdots. \tag{2}$$

Clearly, the Laplace transform of Φ_λ is given by

$$\mathcal{L}\{\Phi_\lambda(x)\} = \int_0^\infty e^{-sx}\Phi_\lambda(x)dx = \frac{1}{s^\lambda}, \quad \text{Re}\lambda > 0, \quad \text{Re}s > 0$$

which plays an important role in solving integral equations [46].

For the functional $\Phi_\lambda = \frac{x_+^{\lambda-1}}{\Gamma(\lambda)}$, the (distributional) derivative formula is simpler than that for x_+^λ. In fact,

$$\frac{d}{dx}\Phi_\lambda = \frac{d}{dx}\frac{x_+^{\lambda-1}}{\Gamma(\lambda)} = \frac{(\lambda-1)x_+^{\lambda-2}}{\Gamma(\lambda)} = \frac{x_+^{\lambda-2}}{\Gamma(\lambda-1)} = \Phi_{\lambda-1}. \tag{3}$$

The convolution of certain pairs of distributions is usually defined as follows, see Gel'fand and Shilov [45] for example.

Definition 1. *Let f and g be distributions in $\mathcal{D}'(R)$ satisfying either of the following conditions:*

(a) either f or g has bounded support (set of all essential points), or
(b) the supports of f and g are bounded on the same side.

Then the convolution $f * g$ is defined by the equation

$$((f*g)(x), \phi(x)) = (g(x), (f(y), \phi(x+y)))$$

for $\phi \in \mathcal{D}(R)$.

The classical definition of the convolution is as follows:

Definition 2. *If f and g are locally integrable functions, then the convolution $f * g$ is defined by*

$$(f*g)(x) = \int_{-\infty}^{\infty} f(t)g(x-t)dt = \int_{-\infty}^{\infty} f(x-t)g(t)dt$$

for all x for which the integrals exist.

Note that if f and g are locally integrable functions satisfying either of the conditions in (a) or (b) in Definition 1, then Definition 1 is in agreement with Definition 2. It also follows that if the convolution $f * g$ exists by Definitions 1 or 2, then the following equations hold:

$$f * g = g * f \tag{4}$$

$$(f*g)' = f*g' = f'*g \tag{5}$$

where all the derivatives above are in the distributional sense.

Let λ and μ be arbitrary complex numbers. Then it is easy to show

$$\Phi_\lambda * \Phi_\mu = \Phi_{\lambda+\mu} \tag{6}$$

by Equation (3), without any help of analytic continuation mentioned in all current books.

Let λ be an arbitrary complex number and $g(x)$ be the distribution concentrated on $x \geq 0$. We define the primitive of order λ of g as convolution in the distributional sense

$$g_\lambda(x) = g(x) * \frac{x_+^{\lambda-1}}{\Gamma(\lambda)} = g(x) * \Phi_\lambda. \tag{7}$$

Note that the convolution on the right-hand side is well defined since supports of g and Φ_λ are bounded on the same side.

Thus Equation (7) with various λ will not only give the fractional derivatives, but also the fractional integrals of $g(x) \in \mathcal{D}'(R^+)$ when $\lambda \notin Z$, and it reduces to integer-order derivatives or integrals when $\lambda \in Z$. We shall define the convolution

$$g_{-\lambda} = g(x) * \Phi_{-\lambda}$$

as the fractional derivative of the distribution $g(x)$ with order λ, writing it as

$$g_{-\lambda} = \frac{d^\lambda}{dx^\lambda} g$$

for $\mathrm{Re}\lambda \geq 0$. Similarly, $\frac{d^\lambda}{dx^\lambda} g$ is interpreted as the fractional integral if $\mathrm{Re}\lambda < 0$.

In 1996, Matignon [47] also studied fractional derivatives in the distributional sense using the kernel distribution $\Phi_{-\lambda}$, and defined the fractional derivative of order λ of a continuous (in the normal sense) causal (zero for $t < 0$) function g, as $g_{-\lambda} = g(x) * \Phi_{-\lambda}$, and further obtained a relation between the distributional derivative and the classical one for a smooth function. Mainardi [42] extended Matignon's work and formally defined the fractional derivative of order $\lambda > 0$ of a causal function (not necessarily continuous) as

$$\frac{d^\lambda}{dx^\lambda} g(x) = \Phi_{-\lambda} * g = \frac{1}{\Gamma(-\lambda)} \int_0^x \frac{g(\zeta) d\zeta}{(x-\zeta)^{1+\lambda}}, \quad \lambda \in R^+.$$

The limit case $\lambda = 0$ is defined as

$$\frac{d^0}{dx^0} g(x) = \Phi_0 * g = \delta * g = g.$$

In addition, Podlubny [46] investigated fractional calculus of generalized functions by the distributional convolution and derived the following identities of fractional derivatives and integrals

$$\begin{aligned} \frac{d\lambda}{dx^\lambda} \theta(x-a) &= \frac{(x-a)_+^{-\lambda}}{\Gamma(-\lambda+1)}, \\ \frac{d\lambda}{dx^\lambda} \delta^{(k)}(x-a) &= \frac{(x-a)_+^{-k-\lambda-1}}{\Gamma(-k-\lambda)} \end{aligned}$$

where $\lambda \in C$ and k is a nonnegative integer.

The following theorem can be obtained from [41] with a minor change in the proof.

Theorem 1. *Let $G(x)$ be a given distribution and y be an unknown distribution in $\mathcal{D}'(R^+)$. Then the generalized Abel's integral equation of the first kind*

$$G(x) = \frac{1}{\Gamma(\alpha)} \int_0^x (x-\zeta)^{\alpha-1} y(\zeta) d\zeta \tag{8}$$

has the solution

$$y(x) = G(x) * \Phi_{-\alpha}$$

where α is any real number in R. In particular, if $G(x) = \theta(x - x_0)g(x)$, where $g(x)$ is an infinitely differentiable function on $[0, \infty]$ and $x_0 \geq 0$, then we have four different cases depending on the value of α.

(i) *If $m < \alpha < m+1$ for $m = 0, 1, \ldots,$ then*

$$\begin{aligned} y(x) &= \frac{d^{m+1}}{dx^{m+1}} \theta(x-x_0)g(x) * \frac{x_+^{-\alpha+m}}{\Gamma(-\alpha+m+1)} \\ &= g(x_0)\frac{(x-x_0)_+^{-\alpha}}{\Gamma(-\alpha+1)} + \cdots + g^{(m)}(x_0)\frac{(x-x_0)_+^{-\alpha+m}}{\Gamma(-\alpha+m+1)} \\ &\quad + \frac{1}{\Gamma(-\alpha+m+1)} \int_{x_0}^x g^{(m+1)}(\zeta)(x-\zeta)^{-\alpha+m} d\zeta \end{aligned}$$

for $x \geq x_0$.

(ii) *If $\alpha = 1, 2, \ldots,$ then*

$$y(x) = g(x_0)\delta^{(\alpha-1)}(x-x_0) + \cdots + g^{(\alpha-1)}(x_0)\delta(x-x_0) + \theta(x-x_0)g^{(\alpha)}(x).$$

(iii) *If $\alpha = 0$, then $y(x) = \theta(x-x_0)g(x)$.*

(iv) *If $\alpha < 0$, then for $x \geq x_0$*

$$y(x) = \frac{1}{\Gamma(-\alpha)} \int_{x_0}^x g(\zeta)(x-\zeta)^{-\alpha-1} d\zeta$$

which is well defined.

Example 1. *Let k be a nonnegative integer, and $n \in N$, $s \in R$ with $s \neq -1, -2, \cdots$. Then, the integral equation*

$$x_+^s + \delta^{(k)}(x) = \int_0^x y(\tau)(x-\tau)^{n/2-1} d\tau \tag{9}$$

has the solution in the space $\mathcal{D}'(R^+)$

$$y(x) = \frac{2^{(n-1)/2}}{(n-2)!!\sqrt{\pi}} \left(\Gamma(s+1)\Phi_{-n/2+s+1}(x) + \Phi_{-n/2-k}(x) \right).$$

Proof. Equation (9) is equivalent to

$$x_+^s + \delta^{(k)}(x) = \frac{\Gamma(n/2)}{\Gamma(n/2)} \int_0^x y(\tau)(x-\tau)^{n/2-1} d\tau$$

which gives

$$x_+^s + \delta^{(k)}(x) = \Gamma(n/2) \left(y(x) * \Phi_{n/2}(x) \right).$$

Theorem 1 implies

$$y(x) = \frac{1}{\Gamma(n/2)} \left(\Phi_{-n/2}(x) * (x_+^s + \delta^{(k)}(x)) \right)$$

which simplifies to

$$\begin{aligned} y(x) &= \frac{1}{\Gamma(n/2)}((\Phi_{-n/2}(x) * x_+^s) + (\Phi_{-n/2}(x) * \delta^{(k)}(x))) \\ &= \frac{1}{\Gamma(n/2)}((\Phi_{-n/2} * \frac{\Gamma(s+1)}{\Gamma(s+1)} x_+^s) + (\Phi_{-n/2}(x) * \Phi_{-k}(x))) \\ &= \frac{1}{\Gamma(n/2)}((\Phi_{-n/2} * \Gamma(s+1)\Phi_{s+1}(x)) + \Phi_{-n/2-k}(x)) \\ &= \frac{1}{\Gamma(n/2)}((\Gamma(s+1)\Phi_{-n/2+s+1}) + \Phi_{-n/2-k}(x)). \end{aligned}$$

Using the formula

$$\Gamma(n/2) = \frac{(n-2)!!\sqrt{\pi}}{2^{(n-1)/2}},$$

we infer that

$$y(x) = \frac{2^{(n-1)/2}}{(n-2)!!\sqrt{\pi}}(\Gamma(s+1)\Phi_{-n/2+s+1}(x) + \Phi_{-n/2-k}(x)).$$

This completes the proof of Example 1. □

In particular, the integral equation

$$x_+^{-3/2} + \delta'(x) = \int_0^x y(\tau)(x-\tau)^{-1/2} d\tau \tag{10}$$

has the solution in the space $\mathcal{D}'(R^+)$

$$y(x) = -2\delta'(x) + \frac{1}{\sqrt{\pi}}\Phi_{-3/2}(x)$$

using

$$\Gamma(-1/2) = -2\sqrt{\pi}.$$

Remark 1. *We must mention that Equation (10) cannot be solved by the Laplace transform since the distribution $x_+^{-3/2}$ is not locally integrable and its Laplace transform does not exist.*

Similarly, the integral equation

$$x_+ + \delta(x) = \int_0^x y(\tau)(x-\tau) d\tau$$

has the solution in the space $\mathcal{D}'(R^+)$

$$y(x) = \delta(x) + \delta''(x)$$

by Equation (2).

Example 2. *Let $s, \alpha \in R$, and $\alpha \neq 0, -1, -2 \cdots$. Then, the integral equation*

$$\sin x_+ + \Phi_s(x) = \int_0^x y(\tau)(x-\tau)^{\alpha-1} d\tau \tag{11}$$

has the solution in the space $\mathcal{D}'(R^+)$

$$y(x) = \frac{1}{\Gamma(\alpha)}\left(\sum_{k=0}^{\infty}(-1)^k \Phi_{2k-\alpha+2}(x) + \Phi_{s-\alpha}(x)\right),$$

where

$$\sin x_+ = \begin{cases} \sin x & \text{if } x \geq 0, \\ 0 & \text{otherwise.} \end{cases}$$

Proof. Equation (11) can be written as

$$\sin x_+ + \Phi_s(x) = \frac{\Gamma(\alpha)}{\Gamma(\alpha)} \int_0^x y(\tau)(x-\tau)^{\alpha-1} d\tau$$

which is equal to

$$\sin x_+ + \Phi_s(x) = \Gamma(\alpha)(\Phi_\alpha * y)(x).$$

Applying Theorem 1, we get

$$y(x) = \frac{1}{\Gamma(\alpha)}(\Phi_{-\alpha}(x) * (\sin x_+ + \Phi_s(x)))$$

which distributes to

$$y(x) = \frac{1}{\Gamma(\alpha)}(\Phi_{-\alpha}(x) * \sin x_+ + \Phi_{-\alpha}(x) * \Phi_s(x)).$$

The Taylor expansion

$$\sin x_+ = \sum_{k=0}^{\infty} \frac{(-1)^k x_+^{2k+1}}{(2k+1)!} = \sum_{k=0}^{\infty} (-1)^k \Phi_{2k+2}(x)$$

gives

$$y(x) = \frac{1}{\Gamma(\alpha)} \left(\sum_{k=0}^{\infty} (-1)^k \Phi_{2k-\alpha+2}(x) + \Phi_{s-\alpha}(x) \right).$$

We note that the series

$$\sum_{k=0}^{\infty} (-1)^k \Phi_{2k-\alpha+2}$$

is absolutely convergent by the ratio test. Indeed,

$$\begin{aligned} \lim_{k\to\infty} \frac{\Phi_{2(k+1)-\alpha+2}(x)}{\Phi_{2k-\alpha+2}(x)} &= \lim_{k\to\infty} \frac{\dfrac{x_+^{2k-\alpha+3}}{\Gamma(2k-\alpha+4)}}{\dfrac{x_+^{2k-\alpha+1}}{\Gamma(2k-\alpha+2)}} \\ &= \lim_{k\to\infty} \frac{x_+^2}{(2k-\alpha+3)(2k-\alpha+2)} = 0. \end{aligned}$$

This completes the proof of Example 2. □

Remark 2. *We should point out that the series* $\sum_{k=0}^{\infty} (-1)^k \Phi_{2k-\alpha+2}(x)$ *is the sum of singular and regular distributions. Indeed, let* j *be the largest non-negative integer, such that* $2j - \alpha + 2 \leq 0$. *Then*

$$\sum_{k=0}^{\infty} (-1)^k \Phi_{2k-\alpha+2}(x) = \sum_{k=0}^{j} (-1)^k \Phi_{2k-\alpha+2}(x) + \sum_{k=j+1}^{\infty} (-1)^k \Phi_{2k-\alpha+2}(x)$$

where the term

$$\sum_{k=0}^{j}(-1)^k\Phi_{2k-\alpha+2}(x)$$

is a singular distribution, while

$$\sum_{k=j+1}^{\infty}(-1)^k\Phi_{2k-\alpha+2}(x)$$

is regular.

In the special case of $\alpha < 2$, we get

$$\begin{aligned} y(x) &= \frac{1}{\Gamma(\alpha)}\left(\sum_{k=0}^{\infty}\frac{(-1)^k x_+^{2k-\alpha+1}}{\Gamma(2k-\alpha+2)} + \Phi_{s-\alpha}(x)\right) \\ &= \frac{1}{\Gamma(\alpha)}\left(x_+^{-\alpha+1}\sum_{k=0}^{\infty}\frac{(-x_+^2)^k}{\Gamma(2k-\alpha+2)} + \Phi_{s-\alpha}(x)\right) \\ &= \frac{1}{\Gamma(\alpha)}\left(x_+^{-\alpha+1}E_{2,-\alpha+2}(-x_+^2) + \Phi_{s-\alpha}(x)\right). \end{aligned}$$

Note that the function $x_+^{-\alpha+1}$ is locally integrable on R.

In particular, we have for $\alpha = 1$ that

$$y(x) = \theta(x)\cosh(ix) + \Phi_{s-1}(x) = \theta(x)\cos x + \Phi_{s-1}(x)$$

using

$$E_{2,1}(z) = \cosh\sqrt{z}.$$

This also can be derived directly from Equation (11). In fact, it becomes for $\alpha = 1$

$$\sin x_+ + \Phi_s(x) = \int_0^x y(\tau)d\tau$$

which claims that

$$y(x) = \frac{d}{dx}(\theta(x)\sin x + \Phi_s(x)) = \theta(x)\cos x + \Phi_{s-1}(x)$$

by noting that

$$\frac{d}{dx}\theta(x)\sin x = \delta(x)\sin x + \theta(x)\cos x = \theta(x)\cos x.$$

We further note that Equation (11) becomes

$$\sin x_+ + \delta^{(-s)}(x) = \int_0^x y(\tau)(x-\tau)^{\alpha-1}d\tau.$$

for $s = 0, -1, -2, \cdots$.

Similarly, the integral equation

$$\cos x_+ + e_+^x = \int_0^x y(\tau)(x-\tau)^{\alpha-1}d\tau \tag{12}$$

has the solution in the space $\mathcal{D}'(R^+)$

$$y(x) = \frac{1}{\Gamma(\alpha)}\left(\sum_{k=0}^{\infty}(-1)^k\Phi_{2k-\alpha+1}(x) + \sum_{k=0}^{\infty}\Phi_{k-\alpha+1}(x)\right),$$

where $\alpha \neq 0, -1, \cdots$ and

$$e_+^x = \begin{cases} e^x & \text{if } x \geq 0, \\ 0 & \text{otherwise.} \end{cases}$$

Indeed, we apply Theorem 1 to get

$$y(x) = \frac{1}{\Gamma(\alpha)}(\Phi_{-\alpha}(x) * (\cos x_+ + e_+^x)).$$

Applying the following Taylor's expansions

$$\cos x_+ = \sum_{k=0}^{\infty} \frac{(-1)^k}{(2k)!} x_+^{2k} = \sum_{k=0}^{\infty} (-1)^k \Phi_{2k+1}(x),$$
$$e_+^x = \sum_{k=0}^{\infty} \frac{x_+^k}{k!} = \sum_{k=0}^{\infty} \Phi_{k+1}(x)$$

we arrive at

$$y(x) = \frac{1}{\Gamma(\alpha)} \left(\sum_{k=0}^{\infty} (-1)^k \Phi_{2k+1-\alpha}(x) + \sum_{k=0}^{\infty} \Phi_{k+1-\alpha}(x) \right).$$

This completes the proof by noting that both $\sum_{k=0}^{\infty} (-1)^k \Phi_{2i-\alpha+1}(x)$ and $\sum_{k=0}^{\infty} \Phi_{k-\alpha+1}(x)$ are absolutely convergent by the ratio test.

In the case of $\alpha < 1$ we get

$$y(x) = \frac{x_+^{-\alpha}}{\Gamma(\alpha)}(E_{2,-\alpha+1}(-x_+^2) + E_{1,-\alpha+1}(x_+)).$$

In particular when $\alpha = 1/2$

$$f(x) = \frac{x_+^{-1/2}}{\sqrt{\pi}} \left(E_{2,1/2}(-x_+^2) + E_{1,1/2}(x_+) \right).$$

We shall extend the techniques used by Yu. I. Babenko in his book [48], for solving various types of fractional differential and integral equations in the classical sense, to generalized functions. The method itself is close to the Laplace transform method in the ordinary sense, but it can be used in more cases [46], such as solving integral or fractional differential equations with distributions whose Laplace transforms do not exist in the classical sense as indicated below. Clearly, it is always necessary to show convergence of the series obtained as solutions. In [46], Podlubny also provided interesting applications to solving certain partial differential equations for heat and mass transfer by Babenko's method.

To illustrate Babenko's approach in detail, we solve the following Abel's integral equation of the second kind in the space $\mathcal{D}'(R^+)$

$$\Phi_{-1/2} = y(x) + \int_0^x (x-\zeta)^\alpha y(\zeta) d\zeta$$

for $\alpha > 0$. Note that the Laplace transform does not work for this equation, since the Laplace transform of $\Phi_{-1/2}$ does not exist. However, this equation can be converted into

$$\Phi_{-1/2} = y(x) + \frac{\Gamma(\alpha+1)}{\Gamma(\alpha+1)} \int_0^x (x-\zeta)^\alpha y(\zeta) d\zeta = (\delta + \Gamma(\alpha+1)\Phi_{\alpha+1}) * y(x)$$

This implies by Babenko's method that

$$\begin{aligned}
y(x) &= (\delta + \Gamma(\alpha+1)\Phi_{\alpha+1})^{-1} * \Phi_{-1/2} \\
&= \sum_{n=0}^{\infty} (-1)^n \Gamma^n(\alpha+1)\Phi_{\alpha+1}^n * \Phi_{-1/2} \\
&= \sum_{n=0}^{\infty} (-1)^n \Gamma^n(\alpha+1)\Phi_{(\alpha+1)n} * \Phi_{-1/2} \\
&= \sum_{n=0}^{\infty} (-1)^n \Gamma^n(\alpha+1)\Phi_{(\alpha+1)n-1/2} \\
&= \Phi_{-1/2} + \sum_{n=1}^{\infty} (-1)^n \Gamma^n(\alpha+1)\Phi_{(\alpha+1)n-1/2} \\
&= \Phi_{-1/2} + \sum_{n=0}^{\infty} (-1)^{n+1} \Gamma^{n+1}(\alpha+1)\Phi_{(\alpha+1)n+\alpha+1/2} \\
&= \Phi_{-1/2} - \Gamma(\alpha+1)x_+^{\alpha-1} E_{\alpha+1,\alpha+1/2}\left(-\Gamma(\alpha+1)x_+^{\alpha+1}\right)
\end{aligned}$$

using

$$\Phi_{\alpha+1}^n = \Phi_{(\alpha+1)n}.$$

Example 3. *Let $\alpha > \beta \geq 0$, and $\gamma \geq 0$. Then the fractional differential equation*

$$ay^{(\alpha)}(x) + by^{(\beta)}(x) = c\Phi_\gamma(x) + x_+^m \tag{13}$$

has the solution in the space $\mathcal{D}'(R^+)$

$$y(x) = \frac{cx_+^{\alpha+\gamma-1}}{a} E_{\alpha-\beta,\alpha+\gamma}\left(-\frac{b}{a}x_+^{\alpha-\beta}\right) + \frac{m!\,x_+^{\alpha+m}}{a} E_{\alpha-\beta,\alpha+m+1}\left(-\frac{b}{a}x_+^{\alpha-\beta}\right)$$

where $m = 0, 1, 2 \ldots$, and $a, b, c \in R$ with $a \neq 0$.

Proof. We see that Equation (13) is equivalent to

$$ay(x) * \Phi_{-\alpha}(x) + by(x) * \Phi_{-\beta}(x) = c\Phi_\gamma(x) + x_+^m.$$

Applying Φ_α to both sides, we get

$$\begin{aligned}
ay(x) + by(x) * \Phi_{\alpha-\beta}(x) &= ay(x) * \left(\delta(x) + \frac{b}{a}\Phi_{\alpha-\beta}(x)\right) \\
&= c\Phi_{\alpha+\gamma}(x) + x_+^m * \Phi_\alpha(x).
\end{aligned}$$

This implies, by Babenko's approach

$$\begin{aligned}
y(x) &= \frac{1}{a}\left(\delta(x) + \frac{b}{a}\Phi_{\alpha-\beta}(x)\right)^{-1} (c\Phi_{\alpha+\gamma}(x) + x_+^m * \Phi_\alpha(x)) \\
&= \frac{1}{a}\sum_{k=0}^{\infty}(-1)^k \left(\frac{b}{a}\Phi_{\alpha-\beta}(x)\right)^k * (c\Phi_{\alpha+\gamma}(x) + x_+^m * \Phi_\alpha(x)) \\
&= \frac{c}{a}\sum_{k=0}^{\infty}\frac{(-b)^k}{a^k}\Phi_{k(\alpha-\beta)+\alpha+\gamma}(x) + \frac{m!}{a}\sum_{k=0}^{\infty}\frac{(-b)^k}{a^k}\Phi_{k(\alpha-\beta)+\alpha+m+1}(x) \\
&= \frac{c}{a}\sum_{k=0}^{\infty}\frac{(-b)^k x_+^{k(\alpha-\beta)+\alpha+\gamma-1}}{a^k\Gamma(k(\alpha-\beta)+\alpha+\gamma)} + \frac{m!}{a}\sum_{k=0}^{\infty}\frac{(-b)^k x_+^{k(\alpha-\beta)+\alpha+m}}{a^k\Gamma(k(\alpha-\beta)+\alpha+m+1)} \\
&= \frac{cx_+^{\alpha+\gamma-1}}{a} E_{\alpha-\beta,\alpha+\gamma}\left(-\frac{b}{a}x_+^{\alpha-\beta}\right) + \frac{m!\,x_+^{\alpha+m}}{a} E_{\alpha-\beta,\alpha+m+1}\left(-\frac{b}{a}x_+^{\alpha-\beta}\right).
\end{aligned}$$

This completes the proof of Example 3. □

In particular, the ordinary differential equation

$$ay'(x) + by(x) = c\Phi_2(x) + x_+,$$

has the solution in the space $\mathcal{D}'(R^+)$

$$y(x) = \frac{(c+1)x_+^2}{a} E_{1,3}\left(-\frac{bx_+}{a}\right)$$

where

$$E_{1,3}(z) = \frac{e^z - z - 1}{z^2}.$$

On the other hand, we derive that the fractional differential equation

$$ay'(x) + by^{(1/2)}(x) = c\delta(x) + x_+^m, \quad a \neq 0$$

has the solution in the space $\mathcal{D}'(R^+)$

$$y(x) = \frac{c\theta(x)}{a} e^{\frac{b^2}{a^2}x_+} \operatorname{erfc}\left(\frac{b}{a}\sqrt{x_+}\right) + \frac{x_+^{1+m} m!}{a} E_{0.5,2+m}\left(-\frac{b}{a}\sqrt{x_+}\right).$$

by using

$$E_{0.5,1}(z) = e^{z^2}\operatorname{erfc}(-z),$$

where erfc is the complement to the error function (erf),

$$\operatorname{erfc}(z) = \frac{2}{\sqrt{\pi}} \int_z^\infty e^{-u^2} du = 1 - \operatorname{erf}(z), \quad z \in C.$$

Clearly, this example can also be solved using the Laplace transform. Applying the Laplace transform to the equation

$$ay^{(\alpha)}(x) + by^{(\beta)}(x) = c\Phi_\gamma(x) + x_+^m,$$

we come to

$$y^*(s) = \frac{cs^{m+1} + m!s^\gamma}{as^{\alpha+\gamma+m+1} + bs^{\beta+\gamma+m+1}} = \left(\frac{c}{a}\right)\left(\frac{s^{-\gamma-\beta}}{s^{\alpha-\beta} + \frac{b}{a}}\right) + \left(\frac{m!}{a}\right)\left(\frac{s^{-\beta-m-1}}{s^{\alpha-\beta} + \frac{b}{a}}\right).$$

Using the inverse transform, we have

$$y(x) = \frac{cx_+^{\alpha+\gamma-1}}{a} E_{\alpha-\beta,\alpha+\gamma}\left(-\frac{b}{a}x_+^{\alpha-\beta}\right) + \frac{m!\, x_+^{\alpha+m}}{a} E_{\alpha-\beta,\alpha+m+1}\left(-\frac{b}{a}x_+^{\alpha-\beta}\right)$$

by the formula [46]

$$\int_0^\infty e^{-sx} x^{\beta-1} E_{\alpha,\beta}(-ax^\alpha) dx = \frac{s^{\alpha-\beta}}{s^\alpha + a}, \quad \operatorname{Re}(s) > |a|^{1/\alpha}.$$

Remark 3. *We must add that the following fractional differential equation*

$$ay^{(\alpha)}(x) + by^{(\beta)}(x) = c\Phi_\gamma(x) + x_+^{-3/2}$$

can also be solved by the same technique used in Example 3. Though it fails to do so by the Laplace transform, as the distribution $x_+^{-3/2}$ *is singular.*

Many applied problems from physical, engineering and chemical processes lead to integral equations, which at first glance have nothing in common with Abel's integral equations. Due to this perception, additional efforts are undertaken for the development of analytical or numerical procedure for solving these equations. However, their transformations to the form of Abel's integral equations will speed up the solution process [20], or, more significantly, lead to distributional solutions in cases where classical ones do not exist [40,41].

Example 4. *Let $0 \le \theta < \pi/2$ and $\alpha < 1$. Then the following integral equation*

$$\int_\theta^{\pi/2} \frac{\sin\varphi\, y(\varphi)}{(\cos\theta - \cos\varphi)^\alpha} d\varphi = f(\theta) \tag{14}$$

has the solution

$$y(\arccos x) = \frac{1}{\Gamma(1-\alpha)} \frac{d^{1-\alpha}}{dx^{1-\alpha}} f(\arccos x)$$

where f is a differential function in $\mathcal{D}'(R^+)$.

Proof. Making the variable changes $\tau = \cos\varphi$ and $x = \cos\theta$. Then Equation (14) becomes

$$\int_0^x \frac{1}{(x-\tau)^\alpha} y(\arccos\tau)d\tau = \frac{\Gamma(1-\alpha)}{\Gamma(1-\alpha)} \int_0^x \frac{1}{(x-\tau)^\alpha} y(\arccos\tau)d\tau = f(\arccos x)$$

which is Abel's integral equation of the first kind. Therefore, we arrive at

$$\Phi_{1-\alpha}(\tau) * y(\arccos\tau) = \frac{1}{\Gamma(1-\alpha)} f(\arccos x)$$

which implies that

$$y(\arccos x) = \frac{1}{\Gamma(1-\alpha)} \Phi_{\alpha-1}(\tau) * f(\arccos\tau) = \frac{1}{\Gamma(1-\alpha)} \frac{d^{1-\alpha}}{dx^{1-\alpha}} f(\arccos x).$$

This completes the proof of Example 4. □

In particular, we have that for $\alpha = 1/2$ and $f(\theta) = \theta$

$$y(\arccos x) = \frac{1}{\Gamma(1/2)} \frac{d^{1/2}}{dx^{1/2}} \arccos x = \frac{1}{\sqrt{\pi}} \frac{d^{1/2}}{dx^{1/2}} \arccos x.$$

Using the Taylor series

$$\arccos x = \frac{1}{2}\pi - \sum_{n=0}^\infty \frac{(2n)!}{2^{2n}(n!)^2(2n+1)} x^{2n+1}$$

if $0 < x \le 1$, and

$$\frac{d^{1/2}}{dx^{1/2}} x_+^{2n+1} = (2n+1)!\Phi_{-1/2}(x) * \frac{x_+^{2n+1}}{\Gamma(2n+2)} = (2n+1)!\Phi_{2n+3/2}(x)$$

we come to

$$y(\arccos x) = -\frac{1}{\sqrt{\pi}} \sum_{n=0}^\infty \frac{[(2n)!]^2}{2^{2n}(n!)^2} \Phi_{2n+3/2}(x)$$

which is obviously convergent. Furthermore, setting $t = \arccos x$ we finally infer that

$$y(t) = -\frac{1}{\sqrt{\pi}} \sum_{n=0}^{\infty} \frac{[(2n)!]^2}{2^{2n}(n!)^2} \Phi_{2n+3/2}(\cos t).$$

Remark 4. *Clearly, Equation (14) can be converted into*

$$\begin{aligned}
\int_{\theta}^{\pi/2} \frac{\sin\varphi\, y(\varphi)}{(\cos\theta - \cos\varphi)^{\alpha}} d\varphi &= \int_{\pi/2}^{\theta} \frac{y(\varphi)}{(\cos\theta - \cos\varphi)^{\alpha}} d\cos\varphi \\
&= \int_{0}^{\cos\theta} \frac{y(\arccos\tau)}{(\cos\theta - \tau)^{\alpha}} d\tau \\
&= \int_{0}^{x} \frac{y(\arccos\tau)}{(x - \tau)^{\alpha}} d\tau \\
&= f(\arccos x).
\end{aligned}$$

Setting

$$y(\varphi) = \frac{1}{\sin\varphi} Y(\varphi),$$

then the integral equation

$$\int_{\theta}^{\pi/2} \frac{Y(\varphi)}{(\cos\theta - \cos\varphi)^{\alpha}} d\varphi = f(\theta)$$

has the solution

$$Y(\arccos x) = \frac{\sqrt{1-x^2}}{\Gamma(1-\alpha)} \frac{d^{1-\alpha}}{dx^{1-\alpha}} f(\arccos x)$$

since

$$\sin(\arccos x) = \sqrt{1-x^2}.$$

Further, setting

$$y(\varphi) = \frac{1}{\sin\varphi \cos^{\beta}\varphi} Y(\varphi) \quad \textit{for } \beta < 1,$$

then the integral equation

$$\int_{\theta}^{\pi/2} \frac{Y(\varphi)}{\cos^{\beta}\varphi(\cos\theta - \cos\varphi)^{\alpha}} d\varphi = f(\theta)$$

has the solution

$$Y(\arccos x) = \frac{x^{\beta}\sqrt{1-x^2}}{\Gamma(1-\alpha)} \frac{d^{1-\alpha}}{dx^{1-\alpha}} f(\arccos x).$$

Example 5. *Assume that the functions g and f are given and g is a nonzero function satisfying the condition*

$$g(x+t) = g(x)g(t)$$

for all $x,\ t \in R$. Then the integral equation

$$\int_{0}^{\infty} x^{1/2} g(x) y(x+t) dx = f(t) \tag{15}$$

has the solution

$$y(1/s) = \frac{2s^{2.5}}{\sqrt{\pi}\, g\left(\frac{1}{s}\right)} \left(\frac{d^{1.5}}{ds^{1.5}} \frac{f(1/s)s^{0.5}}{g(-1/s)} \right).$$

Proof. Making the substitution

$$\tau = \frac{1}{x+t},$$

Equation (15) becomes

$$\int_0^{1/t} \left(\frac{1}{\tau} - t\right)^{1/2} g\left(\frac{1}{\tau} - t\right) \frac{y\left(\frac{1}{\tau}\right)}{\tau^2} d\tau = f(t)$$

which infers that

$$\int_0^{1/t} \left(\frac{1}{\tau} - t\right)^{1/2} g\left(\frac{1}{\tau}\right) \frac{y\left(\frac{1}{\tau}\right)}{\tau^2} d\tau = \frac{f(t)}{g(-t)}$$

since g is a nonzero function. Further, setting $s = 1/t$ we come to

$$\frac{\Gamma(1.5)}{\Gamma(1.5)} \int_0^s (s-\tau)^{1/2} \frac{g\left(\frac{1}{\tau}\right) y\left(\frac{1}{\tau}\right)}{\tau^{2.5}} d\tau = \frac{s^{0.5} f(1/s)}{g(-1/s)}$$

which is Abel's integral equation. Hence, we get the solution

$$y(1/s) = \frac{2s^{2.5}}{\sqrt{\pi}\, g\left(\frac{1}{s}\right)} \left(\frac{d^{1.5}}{ds^{1.5}} \frac{f(1/s)s^{0.5}}{g(-1/s)}\right)$$

using

$$\Gamma(1.5) = \sqrt{\pi}/2.$$

This completes the proof of Example 5. □

A particular example can be derived from setting $g(x) = e^{-x}$. We leave this to interested readers. We should point out that the term

$$\frac{d^{1.5}}{ds^{1.5}} \left(\frac{f(1/s)s^{0.5}}{g(-1/s)}\right)$$

is in the distributional sense. Otherwise, it is undefined if we let $g(x) \equiv 1$ and f be chosen in $\mathcal{D}'(R^+)$ such that $f(1/s)s^{0.5} = s_+^{-1.4}$.

3. The Applications in Viscoelastic Systems

A modeling is a cognitive activity which we use to describe how devices, or objects of interest, behave.

Elasticity is the ability of a material to resist a distortion or a deforming force and return to its original form when the force is removed. According to the classical theory in the infinitesimal deformation, the most elastic materials, based on Hooke's Law, can be described by a linear relation between the strain ϵ and stress σ and

$$\epsilon(t) = \frac{1}{E}\sigma(t)$$

where E is a constant, known as the elastic or Young's modulus.

However, in a more complicated fractional viscoelastic model, one [49,50] constructs the following integral equation

$$\epsilon(t) = \sigma(t)J(0^+) + \frac{1}{E\tau^\alpha}\left[\frac{1}{\Gamma(\alpha)}\int_0^t (t-\zeta)^{\alpha-1}\sigma(\zeta)d\zeta\right] \tag{16}$$

where

$$J\alpha(t) = \frac{1}{E}\frac{(t/\tau)^{\alpha}}{\Gamma(1+\alpha)},$$

and $\tau = E/\eta$, η being the shear modulus.

According to the kernel function $t^{\alpha-1}/\Gamma(\alpha)$ in Equation (16), $\alpha = 0$ and $\alpha = 1$ are equal with a memory-less system and full-memory system in creeping state respectively owing to $\Gamma(0) = \infty$. Clearly, we can derive that for $\alpha = 0$

$$\sigma(t) = \frac{\epsilon(t)}{J(0^+) + \frac{1}{E}} = \frac{E\epsilon(t)}{1 + EJ(0^+)}$$

using Equation (2) in distribution.

When $0 < \alpha \leq 1$, we convert Equation (16) into

$$\frac{\epsilon(t)}{J(0^+)} = \sigma(t) + \frac{1}{E\tau^{\alpha}J(0^+)}(\Phi_\alpha * \sigma)(t) = \left(\delta + \frac{1}{E\tau^{\alpha}J(0^+)}\Phi_\alpha\right) * \sigma(t).$$

By Babenko's approach, we imply that

$$\begin{aligned} \sigma(t) &= \frac{1}{J(0^+)}\left(\delta + \frac{1}{E\tau^{\alpha}J(0^+)}\Phi_\alpha\right)^{-1} * \epsilon(t) \\ &= \frac{1}{J(0^+)}\sum_{n=0}^{\infty}\frac{(-1)^n}{E^n\tau^{\alpha n}J^n(0^+)}\Phi_{\alpha n} * \epsilon(t) \end{aligned}$$

which is the relation between the stress $\sigma(t)$ and strain $\epsilon(t)$.

In particular, we derive that

$$\sigma(t) = \frac{1}{J(0^+)}\sum_{n=0}^{\infty}\frac{(-1)^n}{E^n\tau^{\alpha n}J^n(0^+)}\Phi_{\alpha n+1}$$

if the strain $\epsilon(t) = \theta(t)$.

4. Conclusions

With Babenko's approach, we have studied and solved several Abel's integral and fractional differential equations based on fractional calculus and convolutions of distributions in the space $\mathcal{D}'(R^+)$. Some of the results obtained are not achievable in the classical sense, such as numerical analysis methods or the Laplace transform, since the equations involve generalized functions which are not locally integrable, and undefined at points in R. Generally speaking, these equations can be expressed in terms of series or the Mittag–Leffer functions using inverse convolution operators in distribution. At the end, we demonstrate applications of Abel's integral equations in viscoelastic systems, and for solving other different types of integral equations with potential demands in physical problems.

Author Contributions: The order of the author list reflects contributions to the paper.

Funding: This research was funded by NSERC (Canada) under grant number 2017-00001.

Acknowledgments: The authors are grateful to the reviewers and academic editor for their careful reading of the paper with very productive suggestions and corrections, which certainly improved its quality.

Conflicts of Interest: The authors declare no conflict of interest.

References

1. Kilbas, A.A.; Srivastava, H.M.; Trujillo, J.J. *Theory and Applications of Fractional Differential Equations*; Elsevier: Amsterdam, The Netherlands, 2006.
2. Ding, J.F.; Zhang, Y. Noether symmetries for the El-Nabulsi-Pfaff variational problem from extended exponentially fractional integral. *Acta Sci. Nat. Univ. Sunyatseni* **2014**, *53*, 150–154.
3. El-Wakil, S.A.; Abulwafa, E.M.; El-Shewy, E.K.; Mahmoud, A.A. Time-fractional KdV equation for plasma of two different temperature electrons and stationary ion. *Phys. Plasmas* **2011**, *18*, 092116. [CrossRef]
4. Neirameh, A. Soliton solutions of the time fractional generalized Hirota Satsuma coupled KdV equations. *Appl. Math. Inf. Sci.* **2015**, *9*, 1847–1853.
5. Zhang, X.; Zhang, Y. Lie symmetry and conserved quantity based on El-Nabulsi models in phase space. *J. Jiangxi Norm. Univ. Nat. Sci.* **2016**, *1*, 65–70.
6. Chen, J.; Zhang, Y. Perturbation to Noether symmetries and adiabatic invariants for disturbed Hamiltonian systems based on El-Nabulsi nonconservative dynamics model. *Nonlinear Dyn.* **2014**, *77*, 353–360. [CrossRef]
7. Golmankhaneh, A.K.; Lambert, L. *Investigations in Dynamics: With Focus on Fractional Dynamics*; Academic Publishing: Cambridge, MA, USA, 2012.
8. Monje, C.A.; Chen, Y.Q.; Vinagre, B.M.; Xue, D.; Feliu, V. *Fractional-Order Systems and Controls*; Series: Advances in Industrial Control; Springer: Berlin, Germany, 2010.
9. El-Nabulsi, R.A. The fractional Boltzmann transport equation. *Comp. Math. Appl.* **2011**, *62*, 1568–1575. [CrossRef]
10. Pu, Y.; Wang, W.X.; Zhou, J.L.; Wand, Y.Y.; Jia, H.D. Fractional differential approach to detecting textural features of digital image and its fractional differential filter implementation. *Sci. China Ser. F Inf. Sci.* **2008**, *51*, 1319–1339. [CrossRef]
11. El-Nabulsi, R.A. Spectrum of Schrodinger Hamiltonian operator with singular inverted complex and Kratzer's molecular potentials in fractional dimensions. *Eur. Phys. J. Plus* **2018**, *133*, 277. [CrossRef]
12. Chuanjing, S.; Zhang, Y. Conserved quantities and adiabatic invariants for El-Nabulsi's fractional Birkhoff system. *Int. J. Theor. Phys.* **2015**, *54*, 2481–2493.
13. El-Nabulsi, R.A. Glaeske-Kilbas-Saigo fractional integration and fractional Dixmier traces. *Acta Math. Vietnam.* **2012**, *37*, 149–160.
14. El-Nabulsi, R.A. Path integral formulation of fractionally perturbed Lagrangian oscillators on fractal. *J. Stat. Phys.* **2018**. [CrossRef]
15. Magin, R.L. *Fractional Calculus in Bioengineering*; Begell House: Redding, CT, USA, 2006.
16. West, B.J.; Bologna, M.; Grigolini, P. *Physics of Fractal Operators, Institute for Nonlinear Science*; Springer: New York, NY, USA, 2003.
17. Cafagna, D. Fractional calculus: A mathematical toll from the past for present engineering. *IEEE Ind. Electron. Mag.* **2007**, 1, 35–40. [CrossRef]
18. El-Nabulsi, R.A.; Wu, G.C. Fractional complexified field theory from Saxena-Kumbhat fraction integral, fractional derivative of order (α, β) and dynamical fractional integral exponent. *Afr. Diaspora J. Math. New Ser.* **2012**, *13*, 45–61.
19. Srivastava, T.; Singh, A.P.; Agarwal, H. Modeling the Under-Actuated Mechanical System with Fractional Order Derivative. *Progr. Fract. Differ. Appl.* **2015**, *1*, 57–64.
20. Srivastava, H.M.; Buschman, R.G. *Theory and Applications of Convolution Integral Equations*; Kluwer Academic Publishers: Dordrecht, The Netherlands; Boston, MA, USA; London, UK, 1992.
21. Gorenflo, R.; Mainardi, F. *Fractional Calculus: Integral and Differential Equations of Fractional Order*; Carpinteri, A., Mainardi, F., Eds.; Fractals and Fractional Calculus in Continuum Mechanics; Springer: New York, NY, USA, 1997; pp. 223–276.
22. Hatcher, J.R. A nonlinear boundary problem. *Proc. Am. Math. Soc.* **1985**, *95*, 441–448. [CrossRef]
23. Hilfer, R. *Applications of Fractional Calculus in Physics*; World Scientific Publishing: River Edge, NJ, USA, 2000.
24. Caponetto, R.; Dongola, G.; Fortuna, L.; Petras, I. *Fractional Order Systems: Modeling and Control Applications*; World Scientific: Singapore, 2010.
25. Yang, X.-J.; Baleanu, D.; Srivastava, H.M. *Local Fractional Integral Transforms and Their Applications*; Elsevier: Amsterdam, The Netherlands; Heidelberg, Germany; London, UK; New York, NY, USA, 2016.

26. Herrmann, R. *Fractional Calculus: An Introduction for Physicists*; World Scientific Publishing Company: Singapore, 2011.
27. El-Nabulsi, R.A. Fractional elliptic operator of order 2/3 from Glaeske-Kilbas-Saigo fractional integral transform. *Funct. Anal. Approx. Comput.* **2015**, *7*, 29–33.
28. Wang, J.R.; Zhu, C.; Fečkan, M. Analysis of Abel-type nonlinear integral equations with weakly singular kernels. *Bound. Value Probl.* **2014**, *2014*, *20*. [CrossRef]
29. Atkinson, K.E. An existence theorem for Abel integral equations. *SIAM J. Math. Anal.* **1974**, *5*, 729–736. [CrossRef]
30. Bushell, P.J.; Okrasinski, W. Nonlinear Volterra integral equations with convolution kernel. *J. Lond. Math. Soc.* **1990**, *41*, 503–510. [CrossRef]
31. Gorenflo, R.; Vessella, S. *Abel Integral Equations: Analysis and Applications*; Lect. Notes Math 1461; Springer: Berlin, Germany, 1991.
32. Okrasinski, W. Nontrivial solutions to nonlinear Volterra integral equations. *SIAM J. Math. Anal.* **1991**, *22*, 1007–1015. [CrossRef]
33. Gripenberg, G. On the uniqueness of solutions of Volterra equations. *J. Integral Equ. Appl.* **1990**, *2*, 421–430. [CrossRef]
34. Mydlarczyk, W. The existence of nontrivial solutions of Volterra equations. *Math. Scand.* **1991**, *68*, 83–88. [CrossRef]
35. Kilbas, A.A.; Saigo, M. On solution of nonlinear Abel-Volterra integral equation. *J. Math. Anal. Appl.* **1999**, *229*, 41–60. [CrossRef]
36. Karapetyants, N.K.; Kilbas, A.A.; Saigo, M. Upper and lower bounds for solutions of nonlinear Volterra convolution integral equations with power nonlinearity. *J. Integral Equ. Appl.* **2000**, *12*, 421–448. [CrossRef]
37. Lima, P.; Diogo, T. Numerical solution of nonuniquely solvable Volterra integral equation using extrapolation methods. *J. Comput. App. Math.* **2002**, *140*, 537–557. [CrossRef]
38. Rahimy, M. Applications of fractional differential equations. *Appl. Math. Sci.* **2010**, *4*, 2453–2461.
39. Li, C.; Li, C.P.; Kacsmar, B.; Lacroix, R.; Tilbury, K. The Abel integral equations in distribution. *Adv. Anal.* **2017**, *2*, 88–104. [CrossRef]
40. Li, C.; Clarkson, K. Babenko's approach to Abel's integral equations. *Mathematics* **2018**, *6*, 32. [CrossRef]
41. Li, C.; Li, C.P.; Clarkson, K. Several results of fractional differential and integral equations in distribution. *Mathematics* **2018**, *6*, 97. [CrossRef]
42. Mainardi, F. Fractional relaxation-oscillation and fractional diffusion-wave phenomena. *Chaos Solitons Fractals* **1996**, *7*, 1461–1477. [CrossRef]
43. Kilbas, A.A.; Raina, R.K.; Saigo, M.; Srivastava, H.M. Solution of multidimensional hypergeometric integral equations of the Abel type. *Dokl. Natl. Acad. Sci. Belarus* **1999**, *43*, 23–26. (In Russian)
44. Raina, R.K.; Srivastava, H.M.; Kilbas, A.A.; Saigo, M. Solvability of some Abel-type integral equations involving the Gauss hypergeometric function as kernels in the spaces of summable functions. *ANZJAM J.* **2001**, *43*, 291–320
45. Gel'fand, I.M.; Shilov, G.E. *Generalized Functions*; Academic Press: New York, NY, USA, 1964; Volume I.
46. Podlubny, I. *Fractional Differential Equations*; Academic Press: New York, NY, USA, 1999.
47. Matignon, D. Stability results for fractional differential equations with applications to control processing. In Proceedings of the IEEE-SMC Computational Engineering in System Applications, Lille, France, 10 February 1996; Volume 2, pp. 963–968.
48. Babenko, Y.I. *Heat and Mass Transfer*; Khimiya: Leningrad, Russian, 1986. (In Russian)
49. Amirian, M.; Jamali, Y. The Concepts and Applications of Fractional Order Differential Calculus in Modelling of Viscoelastic Systems: A primer. *arXiv* **2017**, arXiv:1706.06446.
50. Stiassnie, M. On the application of fractional calculus for the formulation of viscoelastic models. *Appl. Math. Model.* **1979**, *3*, 300–302. [CrossRef]

axioms

MDPI

Article

Interior Regularity Estimates for a Degenerate Elliptic Equation with Mixed Boundary Conditions

Jean-Daniel Djida [1,*] and Arran Fernandez [2]

1 Departamento de Análise Matemática, Universidade de Santiago de Compostela, 15782 Santiago de Compostela, Spain
2 Department of Applied Mathematics & Theoretical Physics, University of Cambridge, Cambridge CB3 0WA, UK; af454@hermes.cam.ac.uk
* Correspondence: jeandaniel.djida@usc.es

Received: 9 August 2018; Accepted: 27 August 2018; Published: 1 September 2018

Abstract: The Marchaud fractional derivative can be obtained as a Dirichlet-to-Neumann map via an extension problem to the upper half space. In this paper we prove interior Schauder regularity estimates for a degenerate elliptic equation with mixed Dirichlet–Neumann boundary conditions. The degenerate elliptic equation arises from the Bernardis–Reyes–Stinga–Torrea extension of the Dirichlet problem for the Marchaud fractional derivative.

Keywords: marchaud fractional derivative; interior regularity; schauder estimate; extension problem; fractional order weighted Sobolev spaces

1. Introduction

In the last years there has been a growing interest in the study of fractional elliptic equations involving the right fractional Marchaud derivative $(\mathscr{D}_{right})^{\alpha}$, such as equations of the form

$$(\mathscr{D}_{right})^{\alpha} v = f \quad \text{in}\, \Omega, \quad v = 0 \quad \text{in} \quad [b, \infty), \tag{1}$$

where without loss of generality $\Omega := [a, b) \subset \mathbb{R}$, with $a < b$ and $0 < \alpha < 1$.

Fractional diffusion problems of type (1) arise for example in the modelling of neuronal transmission in Purkinje cells, whose malfunctioning is known to be related to the lack of voluntary coordination and the appearance of tremors [1]. Further motivation comes from various experimental results which showed anomalous diffusion of fractional type, see for example [2,3] and references therein.

The right fractional Marchaud derivative of a function $w : \mathbb{R} \to \mathbb{R}$ is defined via Fourier transforms as

$$\widehat{(\mathscr{D}_{right})^{\alpha} w}(\xi) = (\pm i\xi)^{\alpha}\, \widehat{w}(\xi), \tag{2}$$

and it can also be expressed by the pointwise formula

$$(\mathscr{D}_{right})^{\alpha} v(x) = \frac{c_{\alpha}}{\Gamma(-\alpha)} \int_x^{\infty} \frac{v(y) - v(x)}{(y-x)^{1+\alpha}}\, dy, \tag{3}$$

where c_{α} is a positive normalization constant. We observe from (3) that the right fractional Marchaud derivative is a nonlocal operator. Nonlocal operators have the peculiarity of taking memory effects into account and capturing long-range interactions, i.e., events that happen far away in time or space. Further discussion of the difference between local integro-differential operators and nonlocal or fractional ones can be found in [4] and references therein. In this context, the nonlocality of the fractional Marchaud derivative prevents us from applying local PDE techniques to treat nonlinear problems for $(\mathscr{D}_{right})^{\alpha}$. To overcome this difficulty, Bernardis, Reyes, Stinga and Torrea showed in [5] that the right fractional Marchaud derivative can be determined as an operator that maps a Dirichlet boundary condition to a Neumann-type condition via an extension problem. Similar extension properties have been found for the fractional Laplacian by Caffarelli and Silvestre in [6].

To be more precise, consider the function $\mathcal{U} : \mathbb{R}^2_+ := \mathbb{R} \times [0, \infty) \to \mathbb{R}$ that solves the boundary value problem

$$\begin{cases} \mathcal{M}_\alpha \mathcal{U}(t,x) = 0 & \text{for } (t,x) \in \mathbb{R}^2_+, \\ \lim_{t\to 0} \mathcal{N}_\alpha \mathcal{U}(t,x) = f(x) & \text{for } x \in \Omega, \\ \mathcal{U}(t,x) = 0 & \text{for } x \in \mathbb{R} \setminus \Omega, \end{cases} \tag{4}$$

Then we have [5]:

$$\lim_{t\to 0} \mathcal{N}_\alpha \mathcal{U}(t,x) = c_\alpha (\mathscr{D}_{right})^\alpha v(x),$$

where $c_\alpha := \dfrac{4^{\alpha-1/2}\Gamma(\alpha)}{\Gamma(1-\alpha)}$ is a positive multiplicative constant depending only on $\alpha \in (0,1)$. Here the differential operators $\mathcal{M}_\alpha$ and $\mathcal{N}_\alpha$ are given respectively by:

$$\mathcal{M}_\alpha \mathcal{U} := -(\mathscr{D}_{right})\mathcal{U} + \frac{1-2\alpha}{t}\mathcal{U}_t + \mathcal{U}_{tt}; \tag{5}$$

$$\mathcal{N}_\alpha \mathcal{U} := -t^{1-2\alpha}\mathcal{U}_t. \tag{6}$$

We use the notation $(\mathscr{D}_{right})$ for the derivative *from the right* at the point $x \in \mathbb{R}$, that is:

$$(\mathscr{D}_{right})v(x) = \lim_{t\to 0^+} \frac{v(x) - v(x+t)}{t}, \tag{7}$$

for good enough functions v. Observe that $(\mathscr{D}_{right})$ equals the negative of the lateral derivative $\left(\frac{d}{dx^+}\right)$ as usually defined in calculus [5].

This characterization of $(\mathscr{D}_{right})^\alpha v$ via the local (degenerate) PDE (5) was used for the first time in [5] to get maximum principles. To solve (4), Stinga and Torrea noted that (5) can be thought of as the harmonic extension of v into $2-2\alpha$ extra dimensions (see [5]). From there, they established the fundamental solution and, using a conjugate equation, a Poisson formula for $\mathcal{U}$. Furthermore, taking advantage of the general theory of degenerate elliptic equations developed by Fabes, Jerison, Kenig and Serapioni in 1982–83, they proved comparison principles for $\mathcal{U}$ (and thus for v).

The aim of this paper is to prove an interior Schauder estimate for the problem (4), involving any fractional power of the derivative $(\mathscr{D}_{right})^\alpha$ as an operator that maps a Dirichlet condition to a Neumann-type condition via an extension problem as in [5].

A significant contribution of the above extension problem is to provide a way of applying classical analysis methods to partial differential equations containing one-sided Marchaud derivative operators. By means of such extension techniques, a series of important results, such as comparison principles, Harnack inequalities, and regularity estimates for solutions to degenerate elliptic equations involving the fractional Laplacian, have been studied by many authors, for example [6–17]. The same analysis was done for the one-sided fractional derivative operator in the sense of Marchaud ([5] Theorem 1.1 and Corollary 1.2).

In view of these results, we immediately observe that interior regularity and boundary regularity for the degenerate elliptic equation with mixed boundary conditions involving the one-sided Marchaud derivative is missing in the literature. Indeed, from the pioneering work of [5,7,18] on the analogue extension problem for nonlocal operators that map Dirichlet to Neumann, one can reduce a nonlocal problem involving fractional derivatives to a local one by keeping their qualitative properties. Using this technique, one can study interior and boundary regularity. Hence the raison d'être for this work.

In order to underline what makes the difference between the extension problems introduced by Bernardis, Reyes, Stinga and Torrea [5], and the one introduced by Bucur and Ferrari [18], we point out that the extension problem established in [18] is based on a time-dependent initial condition, which leads to a heat conduction problem. Indeed, considering the function $\varphi : \mathbb{R} \to \mathbb{R}$ of one variable, formally representing the time variable, their approach relies on constructing a parabolic local operator by adding an extra variable, say the space variable, on the positive half-line, and working on the following problem in the half-plane $[0, \infty) \times \mathbb{R}$:

$$\begin{cases} \partial_t \mathcal{U} &= \Delta \mathcal{U} \qquad (t,x) \in (0,\infty) \times \mathbb{R}, \\ \mathcal{U}(0,t) &= \varphi(t) \qquad t \in \mathbb{R}. \end{cases} \tag{8}$$

The problem (8) is not the usual Cauchy problem for the heat operator, but a heat conduction problem.

In view of the type of problem we are interested in here, we choose to deal with the Bernardis–Reyes–Stinga–Torrea extension problem [5]. Our main result, which will be proved in Section 3 below, is as follows. We note that this result can be proved only using extension techniques.

Theorem 1. *Let $\alpha \in (0,1)$ and let $\mathcal{U} \in L^\infty(\mathbb{R}^+_2) \cap H^1(t^\mu; \mathcal{B}^+_2)$ be a weak solution to*

$$\begin{cases} \mathcal{M}_\alpha \mathcal{U} = 0 & \text{in } \mathcal{B}^+_2, \\ \lim_{t\to 0} \mathcal{N}_\alpha \mathcal{U}(t,\cdot) = f & \text{on } B_2, \\ \mathcal{U} = 0 & \text{on } \mathbb{R} \setminus \Omega. \end{cases}$$

(a) *For $1 < p < \infty$, if $f \in L^p(B_1, w)$ and $\gamma \in (0, \min(1,\alpha))$ is such that $0 < \alpha - \gamma - \frac{1}{p} < 1$, then $\mathcal{U} \in C^{0,\alpha-\gamma-\frac{1}{p}}\left(\overline{\mathcal{B}^+_1}, w\right)$. Moreover,*

$$\| \mathcal{U} \|_{C^{\alpha-\gamma-\frac{1}{p}}(\overline{\mathcal{B}^+_1}, w)} \leq C \left(\| \mathcal{U} \|_{L^p(\mathbb{R}^+_2, w)} + \| f \|_{L^p(\mathbb{R}, w)} \right),$$

where C is a positive constant depending only on α, γ, and p.

(b) *If $f \in L^\infty(B_1)$ and $\gamma \in (0, \min(1,\alpha))$, then $\mathcal{U} \in C^{\alpha-\gamma}\left(\overline{\mathcal{B}^+_1}\right)$. Moreover,*

$$\|\mathcal{U}\|_{C^{\alpha-\gamma}\left(\overline{\mathcal{B}^+_1}\right)} \leq C \left(\|\mathcal{U}\|_{\mathbb{R}^2_+} + \|f\|_{L^\infty(\mathbb{R})} \right),$$

where C is a positive constant depending only on α and γ.

The paper is organised as follows. In Section 2, we give some notations and definitions of function spaces and their associated norms which will be needed in this work. We also provide some preliminary results and finally state our main result. In Section 3, we prove an intermediate result and provide the proof of the regularity estimate up to the boundary for the degenerate Equation (4) with the Neumann boundary condition stated in Theorem 1. Finally we end with the conclusion in Section 4.

2. Notations and Preliminary Results

In this section we introduce some notations, definitions, and preliminary results used throughout the paper.

Here and in the following, we consider $\alpha \in (0,1)$, $\mathbb{R}^2_+ := \mathbb{R}_+ \times \mathbb{R} = \{z = (t,x) : t > 0\}$ and $\Omega \subset \mathbb{R}$ a bounded Lipschitz domain. For an open set Ω, an integer $k \geq 1$, and a real number $\lambda \in (0,1]$, the Hölder spaces $\mathcal{C}^{k,\lambda}(\Omega)$ are defined as the subspaces of $\mathcal{C}^k(\Omega)$ consisting of functions whose k-th order derivatives are uniformly Hölder continuous with exponent λ in Ω.

Furthermore, we introduce the following notation for intervals, boxes, and balls:

$$\begin{aligned} B_r(x_0) &:= \{x \in \mathbb{R} : |x - x_0| < r\}, \\ \mathcal{B}_r^+(x_0) &:= [0, r) \times B_r(x_0), \\ \mathcal{B}_r(z_0) &:= \{z = (t, x) \in \mathbb{R} \times \mathbb{R} : |z - z_0| < r\}, \end{aligned} \tag{9}$$

We consider the function space

$$\mathcal{L}_\alpha^1 := \left\{ \varphi : \mathbb{R} \to \mathbb{R} \quad : \quad \|\varphi\|_{\mathcal{L}_\alpha^1} := \int_{\mathbb{R}} \frac{|\varphi(x)|}{1 + |x|^{1+\alpha}} \, dx < \infty \right\}.$$

For $\Omega \subset \mathbb{R}$ an open set, we say $v : \Omega \to \mathbb{R}$ is in $C^{0,\gamma}(\Omega)$, i.e., Hölder continuous with exponent $\gamma \in (0, 1)$, if

$$\|v\|_{C^{0,\gamma}(\Omega)} := \sup_{x \neq y} \frac{|v(x) - v(y)|}{|x - y|^\gamma} < +\infty.$$

We recall the following definition of Sobolev spaces.

Definition 1 (Sobolev spaces). *For any real number α, the αth Sobolev space on $\mathbb{R}$ is defined to be*

$$H^\alpha(\mathbb{R}) := \{u \in \mathcal{S}'(\mathbb{R}) : \hat{u} \in L^2_{loc}(\mathbb{R}), ||u||_{H^\alpha} < \infty\},$$

where the Sobolev norm $|| \cdot ||_{H^\alpha}$ is defined by

$$||u||_{H^\alpha} := \left(\int_{\mathbb{R}} |\hat{u}(\lambda)|^2 \left(1 + |\lambda|^2\right)^\alpha \mathrm{d}\lambda \right)^{1/2}.$$

For a general domain $X \subset \mathbb{R}$, the αth Sobolev space on X is defined to be

$$H^\alpha_{loc}(X) := \{u \in \mathcal{D}'(X) : u\phi \in H^\alpha(\mathbb{R}) \text{ for all } \phi \in \mathcal{D}(X)\}.$$

Let $a \in \mathbb{R}$ and $\alpha \in (0, 1)$ be two arbitrary parameters. We define the functional space

$$C_a^{1,\alpha} := \left\{ f : \mathbb{R} \to \mathbb{R} \quad : \quad \text{for any } x > a,\ f \in AC([a, x]) \text{ and } f'(\cdot)(x - \cdot)^{-\alpha} \in L^1((a, x)) \right\}. \tag{10}$$

We denote here by $AC(I)$ the space of absolutely continuous functions on I.

Definition 2 (Caputo derivative). *The Caputo derivative of $v \in C_a^{1,\alpha}$ with initial point $a \in \mathbb{R}$ at the point $x > a$ is given by*

$$D_a^\alpha v(x) := \frac{1}{\Gamma(1 - \alpha)} \int_a^x v'(y)(x - y)^{-\alpha} \, dy. \tag{11}$$

Definition 3. *The right Marchaud derivative of a well defined function v is given by*

$$(\mathscr{D}_{right})^\alpha v(x) = \lim_{\delta \to 0^+} \frac{C}{\Gamma(-\alpha)} \int_{x+\delta}^\infty \frac{v(y) - v(x)}{(y - x)^{1+\alpha}} \, dy, \tag{12}$$

with C_α a positive normalisation constant.

Remark 1. *Notice that the one-sided nonlocal derivative in the sense of Marchaud can also be obtained by extending the Caputo derivative. Indeed, by making an integration by parts of Equation (11), we obtain an equivalent definition [19,20] as follows:*

$$D^{\alpha}v(x) = C(\alpha)\int_{x}^{\infty}\frac{v(y)-v(x)}{(y-x)^{1+\alpha}}\,dy,$$

for all $x < 0$, so that $v(x) = v(0)$, where $C(\alpha)$ is a constant depending on α. Indeed, for sufficiently regular functions v, we have:

$$\begin{aligned} D^{\alpha}v(x) &= \int_0^x \frac{v'(y)}{(x-y)^{\alpha}}\,dy = \int_0^x \left(\frac{d}{dy}\frac{(v(y)-v(x))}{(x-y)^{\alpha}} - \alpha\,\frac{(v(y)-v(x))}{(x-y)^{1+\alpha}}\right)dy \\ &= \frac{v(x)-v(0)}{x^{\alpha}} - \lim_{y\to x}\frac{v(x)-v(y)}{(x-y)^{\alpha}} - \alpha\int_0^x\frac{(v(y)-v(x))}{(x-y)^{1+\alpha}}\,dy \\ &= \frac{v(x)-u(0)}{x^{\alpha}} - v'(y)\lim_{y\to x}(x-y)^{1-\alpha} - \alpha\int_0^x\frac{v(y)-v(x)}{(x-y)^{1+\alpha}}\,dy \\ &= \frac{v(x)-v(0)}{x^{\alpha}} + \alpha\int_0^x\frac{v(x)-v(y)}{(x-y)^{1+\alpha}}\,dy. \end{aligned} \tag{13}$$

Hence, we take the convention that $v(x) = v(0)$ for any $x \leq 0$. With this extension, one has that, for any $x > 0$,

$$\alpha\int_{-\infty}^{0}\frac{v(x)-v(y)}{(x-y)^{1+\alpha}}\,dy = \alpha\int_{-\infty}^{0}\frac{v(x)-u(0)}{(x-y)^{1+\alpha}}\,dy = \frac{v(x)-v(0)}{x^{\alpha}}.$$

So one can write (13) as

$$D^{\alpha}v(x) = C(\alpha)\int_{-\infty}^{x}\frac{v(x)-v(y)}{(x-y)^{1+\alpha}}\,dy.$$

This type of formula also relates the Caputo derivative to the so-called Marchaud derivative [20,21]. Therefore the results obtained in this paper could also be applied for the extended Caputo derivative.

Note that the integral in (12) is absolutely convergent for functions in the Schwartz class $\mathcal{S}$. Furthermore one should notice that the nonlocal operators $(\mathscr{D}_{right})^{\alpha}$ and $(\mathscr{D}_{right})^{-\alpha}$ depend on the values of v on the whole half line (x, ∞).

We recall that the inverse of the right fractional Marchaud derivative $(\mathscr{D}_{right})^{-\alpha}$ is defined as

$$(\mathscr{D}_{right})^{-\alpha}v(x) := \int_{\mathbb{R}}\frac{v(y)}{|x-y|^{1-\alpha}}\,dy = \mathcal{I}_{\alpha} * v(x) \tag{14}$$

where the Riesz potential (see [7,21]) is defined as

$$\mathcal{I}_{\alpha} = C_{\alpha}|x-y|^{\alpha-1} \quad \text{for } \alpha < 1, \tag{15}$$

with the constant $C_{\alpha} = \frac{1}{\pi}\Gamma(1-\alpha)\sin\frac{\pi\alpha}{2}$.

From [5], we have that for $u \in \mathcal{S}$, $(\mathscr{D}_{right})^{\alpha}u \in \mathcal{S}_{\alpha}$, where

$$\mathcal{S}_{\alpha} := \left\{ f \in C^{\infty}(\mathbb{R}) : (1+|x|^{1+\alpha})f^{k}(x) \in L^{\infty}(\mathbb{R}), \text{ for each } k \geq 0 \right\}.$$

The topology in $\mathcal{S}_{\alpha}$ is given by the family of seminorms $[f]_k := \sup_{x\in\mathbb{R}}\left|(1+|x|^{1+\alpha})f^{(k)}(x)\right|$, for $k \geq 0$. Let $\mathcal{S}'_{\alpha}$ be the dual space of $\mathcal{S}_{\alpha}$; then $(\mathscr{D}_{right})^{\alpha}$ defines a continuous operator from $\mathcal{S}'_{\alpha}$ into $\mathcal{S}'$.

2.1. Weighted Spaces

Weighted spaces of smooth functions play an important role in the context of partial differential equations (PDEs). They are widely used, for instance, to treat PDEs with degenerate coefficients or domains with a nonsmooth geometry (see e.g., [22–25]), as is the case here. For evolution equations, power weights in time play an important role in order to obtain results for rough initial data (see [26,27]). This subsection dedicated to weighted spaces is motivated by the appearance of the Muckenhoupt weight $w := t^{1-2\alpha}$ which appears in (5) and (6). For general literature on weighted function spaces we refer to [23–25,28–31] and references therein.

In a general framework, a function $w : \mathbb{R}^d \to [0, \infty)$, for an integer $d \geq 1$, is called a *weight* if w is locally integrable and the zero set $\{x : w(x) = 0\}$ has Lebesgue measure zero. For $p \in [1, \infty]$ we denote by A_p the *Muckenhoupt class* of weights. In the case $p \in (1, \infty)$, we say that $w \in A_p$ if

$$\sup_{\mathcal{B} \text{ cubes in } \mathbb{R}^d} \left(\frac{1}{|\mathcal{B}|}\int_{\mathcal{B}} w(x)\,dx\right)\left(\frac{1}{|\mathcal{B}|}\int_{\mathcal{B}} w(x)^{-\frac{1}{p-1}}\,dx\right)^{p-1} < \infty.$$

In the case $p = 1$, we say that $w : \mathbb{R} \to [0, \infty)$ belongs to A_1 if there exists some constant C such that

$$\frac{1}{|\mathcal{B}|}\int_{\mathcal{B}} w(y)\,dy \leq Cw(x)$$

for all $x \in \mathcal{B}$ and all balls $\mathcal{B} \subset \mathbb{R}^d$. In the case $p = \infty$, we define $A_\infty = \bigcup_{1 \leq p < \infty} A_p$. Note that, for functions with support contained in $(-\infty, 0)$ or $(0, \infty)$, the class of weights is denoted by A_p^+ or A_p^- respectively. We refer to [27,30,32] for the general properties of these classes.

Example 1. *Problem (4) is a weighted—singular or degenerate, depending on the value of $\alpha \in (0,1)$—elliptic equation on $\mathbb{R}^2_+$ with mixed boundary conditions. The weight $w := t^{1-2\alpha}$ belongs to the Muckenhoupt class A_2^+, i.e., there exists a constant C such that for any $\mathcal{B} \subset \mathbb{R}^2_+$,*

$$\left(\frac{1}{|\mathcal{B}|}\int_{\mathcal{B}} |t|^{1-2\alpha} dt\,dx\right)\left(\frac{1}{|\mathcal{B}|}\int_{\mathcal{B}} |t|^{2\alpha-1} dt\,dx\right) \leq C.$$

For this reason, when working with one-sided weights, we can assume without loss of generality that $\Omega := [a, b) = \mathbb{R}$ (see e.g., [30] for more details).

Next, for a strongly measurable function f and a number $p \in [1, \infty)$, we define the weighted L^p norm by

$$\|f\|_{L^p(\mathbb{R}^d, w)} := \left(\int_{\mathbb{R}^d} \|f(x)\|^p w(x)\,dx\right)^{1/p},$$

and we define the weighted L^p space to be the following Banach space:

$$L^p(\mathbb{R}^d, w) := \{f \text{ strongly measurable} : \|f\|_{L^p(\mathbb{R}^d, w)} < \infty\}.$$

Definition 4 (see [8])**.** *Given $\alpha \in (0,1)$, $\mu = 1 - 2\alpha \in (-1, 1)$, and an open set $\mathcal{B} \subset \overline{\mathbb{R}^2_+}$, we denote*

$$L^2(t^\mu; \mathcal{B}) := \left\{\mathcal{U} : \mathbb{R}^2_+ \to \mathbb{R},\ \int_{\mathcal{B}} t^\mu |\mathcal{U}|^2 dt\,dx < +\infty\right\},$$

endowed with the norm

$$\|\mathcal{U}\|_{L^2(t^\mu; \mathcal{B})} := \left(\int_{\mathcal{B}} t^\mu |\mathcal{U}|^2 dt\,dx\right)^{1/2}.$$

We also denote

$$H^1(t^\mu; \mathcal{B}) := \left\{\mathcal{U} \in L^2(t^\mu; \mathcal{B}) : \nabla\mathcal{U} \in L^2(t^\mu; \mathcal{B})\right\},$$

with the induced norm

$$\|\mathcal{U}\|_{H^1(t^\mu; \mathcal{B})} := \left(\int_{\mathcal{B}} t^\mu \left(|\mathcal{U}|^2 + |\nabla\mathcal{U}|^2\right) dt\,dx\right)^{1/2}.$$

Using the variable $(t,x) \in \mathbb{R}^2_+$, the space $H^\alpha(\mathbb{R})$ coincides with the trace on $\partial\mathbb{R}^2_+$ of

$$\dot{H}^1(t^\mu;\mathcal{B}) := \left\{ \mathcal{U} \in L^2_{loc}(\mathbb{R}^2_+) : \int_{\mathbb{R}^2_+} t^\mu \left(\mathcal{U}^2 + |\nabla \mathcal{U}|^2 \right) dt\, dx < +\infty \right\}.$$

In other words [8,9], for any given function $\mathcal{U} \in \dot{H}^1(t^\mu;\mathcal{B}) \cap \mathcal{C}\left(\overline{\mathbb{R}^2_+}\right)$, we have $v := \mathcal{U}|_{\partial\mathbb{R}^2_+} \in H^\alpha(\mathbb{R})$, and there exists a constant $C = C(\alpha) > 0$ such that

$$\| v \|_{H^\alpha(\mathbb{R})} \leq C \| \mathcal{U} \|_{H^1(t^\mu;\mathcal{B})} \,.$$

So by a density argument, every $\mathcal{U} \in \dot{H}^1(t^\mu;\mathcal{B}) \cap \mathcal{C}\left(\overline{\mathbb{R}^2_+}\right)$ has a well defined trace $v \in H^\alpha(\mathbb{R})$. Conversely, any $v \in H^\alpha(\mathbb{R})$ is the trace (restriction to $t = 0$) of a function $\mathcal{U} \in \dot{H}^1(t^\mu;\mathcal{B}) \cap \mathcal{C}\left(\overline{\mathbb{R}^2_+}\right)$.

Definition 5. *We say that a function $\mathcal{U} \in \dot{H}^1(t^\mu;\mathcal{B})$ is a weak solution of (4) if*

$$\int_{\mathcal{B}} t^\mu \nabla \mathcal{U}(t,x) \nabla \Psi(t,x) dt\, dx - c_\alpha^{-1} \int_\Omega f(x) \mathrm{Tr}(\Psi)(x)\, dx = 0, \tag{16}$$

where f is as in (1), $\mathrm{Tr}(\Psi)$ denotes the trace $\Psi|_{\{0\}\times\mathbb{R}}$, and $\Psi \in \dot{H}^1(t^\mu;\mathcal{B}) \cap \mathcal{C}\left(\overline{\mathbb{R}^2_+}\right)$ is an arbitrary test function.

2.2. The Extension Problem

In the next statement we recall the results obtained from [5] which show that the fractional derivatives on the line are Dirichlet-to-Neumann operators for an extension degenerate PDE problem in $\mathbb{R} \times (0,\infty)$, where the data f have been taken in the more general setting: more precisely a weighted $L^p(w)$ space, where w satisfies the one-sided version A_p^+ (see [30]) of the familiar A_p condition of Muckenhoupt.

Fix $0 < \alpha < 1$. Given a semigroup $\{T_t\}_{t\geq 0}$ acting on real functions, the *generalized Poisson integral* of f is given by

$$P_t^\alpha f(x) = \frac{t^{2\alpha}}{4^\alpha \Gamma(\alpha)} \int_0^\infty e^{-t^2/(4s)} T_s f(x) \frac{ds}{s^{1+\alpha}}, \quad x \in \mathbb{R}; \tag{17}$$

see ([5] (1.9)) for more details.

By considering the semigroup of translations $T_s f(x) = f(x+s)$, $s \geq 0$, we find

$$P_t^\alpha f(x) = f * P_t^\alpha(x) := \int_{\mathbb{R}} f(s) P_t^\alpha(x-s)\, ds,$$

where

$$P_t^\alpha(x) := \frac{t^{2\alpha} e^{t^2/4x}}{4^\alpha \Gamma(\alpha)(-x)^{1+\alpha}} \chi_{(-\infty,0)}(x). \tag{18}$$

Since the kernel P_t^α is increasing and integrable in $(-\infty,0)$, it is well known that the function

$$P_*^\alpha f(x) := \sup_{t>0} |f| * P_t^\alpha(x) = \int_{\mathbb{R}} |f(t)| P_t^\alpha(x-t)\, dt,$$

is pointwise controlled by the usual Hardy–Littlewood maximal operator. However, since the support of P_t^α is $(-\infty,0)$, a sharper control can be obtained by using the one-sided Hardy–Littlewood maximal operator. This control and the behavior of P_*^α in weighted L^p-spaces will be used in the results of this paper. We revise briefly recall the two fundamental theorems from [5].

Theorem 2 ([5]). *Consider the semigroup of translations $T_t f(x) = f(x+t)$, $t \geq 0$, initially acting on functions $f \in \mathcal{S}$. Let $P_t^\alpha f$, $0 < \alpha < 1$, be as in (17). Then:*

1. *For $1 \leq p \leq \infty$, P_t^α is a bounded linear operator from $L^p(\mathbb{R})$ into itself and $\|P_t^\alpha f\|_{L^p(\mathbb{R})} \leq \|f\|_{L^p(\mathbb{R})}$.*
2. *When $f \in \mathcal{S}$, the Fourier transform of $P_t^\alpha f$ is given by*

$$\widehat{P_t^\alpha f}(\xi) = \frac{2^{1-\alpha}}{\Gamma(\alpha)}(-it\xi^{1/2})^\alpha \|_\alpha(-it\xi^{1/2})\widehat{f}(\xi), \quad \xi \in \mathbb{R},$$

where $\|_\nu(z)$ is the modified Bessel function of the third kind or Macdonald's function, which is defined for arbitrary ν and $z \in \mathbb{C}$, see ([33] Chapter 5). In particular,

$$\widehat{P_t f}(\xi) = e^{-t(-i\xi)^{1/2}}\widehat{f}(\xi).$$

3. *The maximal operator P_*^α defined by $P_*^\alpha f(x) = \sup_{t>0}|P_t^\alpha f(x)|$ is bounded from $L^p(\mathbb{R}, w)$ into itself, for $w \in A_p^+, 1 < p < \infty$, and from $L^1(\mathbb{R}, w)$ into weak-$L^1(\mathbb{R}, w)$, for $w \in A_1^+$.*
4. *Let $f \in L^p(w)$, for $w \in A_p^+$, $1 \leq p < \infty$. The function $\mathcal{U}(x,t) \equiv P_t^\alpha f(x)$ is a classical solution to the extension problem (4).*

Theorem 3 (Extension problem). *Let $f \in L^p(w)$, $w \in A_p^+$, $1 < p < \infty$. Then the function*

$$\mathcal{U}(x,t) := \frac{t^{2\alpha}}{4^\alpha\Gamma(\alpha)}\int_0^\infty e^{-t^2/(4\tau)}T_\tau f(x)\frac{d\tau}{\tau^{1+\alpha}}, \quad x \in \mathbb{R},\ t > 0,$$

is a classical solution to the extension problem

$$\begin{cases} -(\mathscr{D}_{right})\mathcal{U} + \frac{1-2\alpha}{t}\mathcal{U}_t + \mathcal{U}_{tt} = 0, & \text{in } \mathbb{R} \times (0,\infty), \\ \lim_{t\to 0^+}\mathcal{U}(x,t) = f(x), & \text{a.e. and in } L^p(w). \end{cases}$$

Moreover, for $c_\alpha := \dfrac{4^{\alpha-1/2}\Gamma(\alpha)}{\Gamma(1-\alpha)} > 0$, *we have*

$$-c_\alpha \lim_{t\to 0^+} t^{1-2\alpha}\mathcal{U}_t(x,t) = (\mathscr{D}_{right})^\alpha f(x) \quad \text{in the distributional sense.}$$

Remark 2. *This parallel result regarding the extension problem in the case of the Marchaud fractional time derivative has been derived as well in [18,20].*

3. Regularity Estimate up to the Boundary for the Degenerate Equation with the Neumann Boundary Condition

In this section, we prove the interior regularity estimate up to the boundary for the degenerate equation with the Neumann boundary condition associated to problem (4). Namely we provide the proof of Theorem 1. But before we get into that, it is necessary to explain the main ideas in the proof of interior regularity provided by Theorem 1. The proof of Theorem 1 is inspired by [5,7,8,34]. The method for this proof differs substantially from interior regularity methods for second-order equations, but is similar to the proof for the fractional Laplacian. Recall that for second-order equations, one first shows that D^2u is bounded, and then the estimate for equations with bounded measurable coefficients implies a $C^{2,\sigma}$ estimate for $\sigma \in (0, \min(1,\alpha))$. This is also true for the boundary regularity for solutions to fully nonlinear equations [35].

We shall start by the regularity property of the problem (1). We show in Proposition 1 that the solution of the problem (1) is of class $C^{0,\sigma}$. To the best of the authors' knowledge, the proofs available in the literature are those dealing with the case of the fractional Laplacian (see for instance [7,36] (Proposition 2.1.9)). With this result in hand, and by making an appropriate change of variables, we will use this result and estimate to prove our main theorem.

We start by recalling the following lemma from [37], which gives a Liouville-type theorem for (1) in the case $f = 0$.

Lemma 1. *Let $u \in C(\mathbb{R})$ be a function satisfying $(\mathscr{D}_{right})^\alpha u = 0$ in $\mathbb{R}_+$, $u = 0$ in $\mathbb{R}_-$, and $|u(x)| \leq C(1+|x|^\gamma)$ for some $\gamma < \alpha$. Then $u(x) = kx^\alpha$.*

The proof of this lemma relies on similar reasoning as the proof of ([37] (Theorem 2.2.3)) for the Caputo density function.

In the case where we have a non-vanishing right hand side ($f \neq 0$) as in (1), we state the following Liouville-type theorem for the one-sided Marchaud derivative.

Proposition 1. *Let $\alpha \in (0,1)$ and let $u \in \mathcal{L}^1_\alpha \cap L^\infty_{loc}(B_1)$ be the solution to*

$$(\mathscr{D}_{right})^\alpha u = f \quad in \quad B_1.$$

(a) For $1 < p < \infty$, if $f \in L^p(B_1, w)$ and $r \in (0,1)$ and $\gamma \in (0, \min(1,\alpha))$ is such that $0 < \gamma - \frac{1}{p} < 1$, then $u \in \mathcal{C}^{0,\gamma-\frac{1}{p}}(B_1, w)$ and there exists a constant $C := C(\alpha, r, \gamma, p) > 0$ such that

$$\|u\|_{\mathcal{C}^{0,\gamma-\frac{1}{p}}(B_r,w)} \leq C\left(\|u\|_{L^p(B_r,w)} + \|f\|_{L^p(B_1,w)}\right). \tag{19}$$

(b) If $f \in L^\infty(B_1)$ and $r \in (0,1)$ and $\gamma \in (0, \min(1,\alpha))$, then $u \in \mathcal{C}^{0,\gamma}(B_1)$ and there exists a constant $C := C(\alpha, r, \gamma) > 0$ such that

$$\|u\|_{\mathcal{C}^{0,\gamma}(B_r)} \leq C\left(\|u\|_{L^\infty(B_r)} + \|f\|_{L^\infty(B_1)}\right). \tag{20}$$

Proof. We will show that u has the corresponding regularity in a neighbourhood of the origin. We split the proof into two parts, as follows.

Proof of (a): $f \in L^p(B_1, w)$. Let $\eta \in C_c^\infty(\mathbb{R})$ be a smooth cutoff function such that $\eta = 1$ on B_r, $\eta = 0$ on $\mathbb{R} \setminus B_1$, and $0 \leq \eta \leq 1$ on $\mathbb{R}$. Consider the Riesz potential as defined in (15). Then the function

$$v(x) := \int_{\mathbb{R}} \mathcal{I}_\alpha(x,y)(\eta f)(y)dy, \quad \text{for all} \quad x \in \mathbb{R},$$

satisfies

$$(\mathscr{D}_{right})^\alpha v(x) = \eta(x) f(x) \quad \text{for all } x \in \mathbb{R}_+. \tag{21}$$

We first estimate the L^p norm of v for $\alpha < 1$. Since the kernel $(\mathscr{D}_{right})^{-\alpha}$ is positive and $\eta \geq 0$ is a smooth function with compact support in B_r, we write $v = (\mathscr{D}_{right})^{-\alpha}(\eta f) = (\mathscr{D}_{right})^{1-\alpha} \circ (\mathscr{D}_{right})^{-1}(\eta f)$. We note that, by using a similar argument as for the Poisson equation for the fractional Laplacian, we find that $(\mathscr{D}_{right})^{-1}(\eta f)$ is an element of $\mathcal{C}^{1,\gamma}$ with norm depending only on $\|f\|_{\mathcal{C}^{0,\gamma}}$.

Since ηf is compactly supported, we get

$$\|v\|_{\mathcal{C}^{0,\gamma}(\mathbb{R})} \leq C_{\alpha,\gamma}\|\eta f\|_{L^p(B,w^p)} + C\|v\|_{L^p(\mathbb{R},w^p)} \leq C_{\alpha,\gamma,B'}\|f\|_{L^p(B,w^p)}.$$

For $\alpha < 1$ and $\gamma \in (0, \min(\alpha, 1))$ and $x, y \in B_r$, we have

$$\begin{aligned} v(x) - v(y) &= \int_{\mathbb{R}} \left(\mathcal{I}_\alpha(x,z) - \mathcal{I}_\alpha(y,z)\right)(\eta f)(z)dz \\ &= C_{1,\alpha}\int_{\mathbb{R}} \left(|x-z|^{\alpha-1} - |y-z|^{\alpha-1}\right)(\eta f)(z)\,dz. \end{aligned}$$

Next we consider the following inequalities [38], valid for $\gamma \in (0, \min(1, \alpha))$ and $m \in \mathbb{R}$ with $m + \gamma > 0$ and for every $x, y, z \in B_r$:

$$\left|\left(|x-z|^{-m} - |y-z|^{-m}\right)\right| \leq \frac{|m|}{m+\gamma}|x-y|^{\gamma}\left(|x-z|^{-(m+\gamma)} + |y-z|^{-(m+\gamma)}\right). \tag{22}$$

For $m = 1 - \alpha$, and for $1 < p < \infty$, we can write

$$\begin{aligned}|v(x) - v(y)| &< \frac{(1-\alpha)}{(1-\alpha+\gamma)}\int_{\mathbb{R}} |x-y|^{\gamma}\left(|x-z|^{\alpha-1-\gamma} + |y-z|^{\alpha-1-\gamma}\right)|(\eta f)(z)|dz \\ &\leq C_{\alpha,\gamma}|x-y|^{\gamma-\frac{1}{p}}\int_{\mathbb{R}} |x-y|^{\frac{1}{p}}|x-z|^{\alpha-1-\gamma}|(\eta f)(z)|dz,\end{aligned}$$

since $\frac{|y - z|}{|x - z|} > 1$. Using the fact that the support of η is always contained in the ball of radius 2 centred at $x \in B_r$, we have that

$$\begin{aligned}|v(x) - v(y)| &\leq C(1,\alpha,\gamma)|x-y|^{\gamma}\int_{\mathbb{R}} |x-y|^{\alpha-1-\gamma+\frac{1}{p}}|(\eta f)(y)|dy, \\ &\leq C(1,\alpha,\gamma)|x-y|^{\gamma-\frac{1}{p}}\left(\int_{\mathbb{R}} w^{-1}|x-y|^{p(\alpha-1-\gamma)+1} w\,|(\eta f)(y)|^p\,dy\right)^{1/p} \\ &\leq C(1,\alpha,\gamma)|x-y|^{\gamma-\frac{1}{p}}\left(\int_{B(x,2)} w\,|(\eta f)(y)|^p\,dy\right)^{\frac{1}{p}}\left(\int_{B(x,2)} \frac{w^{-\frac{1}{p-1}}}{|x-y|^{\frac{p(-\alpha+1+\gamma)-1}{p-1}}}dy\right)^{\frac{p-1}{p}} \\ &\leq C(1,\alpha,\gamma)|x-y|^{\gamma-\frac{1}{p}}\left(\int_{B_1} w\,|f(y)|^p\,dy\right)^{\frac{1}{p}}\left(\int_{B_1} w^{-\frac{1}{p-1}}|y|^{\frac{p(\alpha-1-\gamma)+1}{p-1}}dy\right)^{\frac{p-1}{p}} \\ &\leq 2C(1,\alpha,\gamma)|x-y|^{\gamma-\frac{1}{p}}\|f\|_{L^p(B_1,w)}\left(\int_{B_1} w^{-\frac{1}{p-1}}|y|^{\frac{p(\alpha-1-\gamma)+1}{p-1}}dy\right)^{\frac{p-1}{p}},\end{aligned}$$

up to relabelling of the positive constant $C(\alpha, \gamma)$ that depends on α and γ. Replacing $w(y)$ by its value $|y|^{1-2\alpha}$ and using the polar coordinates $y = rx$, $r > 0$, we get that

$$\begin{aligned}\int_{B_1} w(y)^{-\frac{1}{p-1}}|y|^{\frac{p(\alpha-1-\gamma)+1}{p-1}}dy &:= \int_{S_1}\int_0^1 |r|^{\frac{1}{p-1}(2\alpha+p(\alpha-1-\gamma))}\,dr\,d\sigma(x) \leq \frac{(p-1)}{p(\alpha-1-\gamma)+p-1)}|S_1| \\ &\leq C(p,\gamma,\alpha).\end{aligned}$$

Then,

$$|v(x) - v(y)| \leq C(p,\gamma,\alpha)|x-y|^{\gamma-\frac{1}{p}}\|f\|_{L^p(B_1,w)}.$$

Hence, we conclude that

$$\|v\|_{C^{0,\gamma-\frac{1}{p}}(B_r)} \leq C_{\alpha}\|f\|_{L^p(B_1,w)}, \tag{23}$$

for every $\gamma \in (0, \min(1, \alpha))$.

Next, by change of variables, the function $\xi := u - v$ satisfies $(\mathscr{D}_{right})^{\alpha}\xi = 0$ in B_r by (21). Then, thanks to the derivative estimate, for every $r' \in (0, r)$,

$$\|\nabla\xi\|_{L^p(B_{r'},w)} \leq C_{\alpha,r'}\|\xi\|_{L^p(B_1,w)} \leq C_{\alpha,r'}\left(\|u\|_{L^p(B_r,w)} + \|v\|_{L^p(B_r,w)}\right).$$

The difference function $\xi = u - v$ is smooth in B_1 and is bounded. From this observation, together with (23), we have that

$$\|u\|_{C^{0,\gamma-\frac{1}{p}}(B_{r'})} = \|\xi + v\|_{C^{0,\gamma}(B_{r'})} \leq C_{\alpha,r'}\left(\|u\|_{L^p(B_r,w)} + \|f\|_{L^p(B_1,w)}\right),$$

for every $\gamma \in (0, \min(1, \alpha))$ with $0 < \gamma - \frac{1}{p} < 1$, as required.

Proof of (b): $f \in L^\infty(B_1)$. The proof in this case is similar to the previous one. We consider as above a smooth cutoff function $\eta \in C_c^\infty(\mathbb{R})$ such that $\eta = 1$ on B_r, $\eta = 0$ on $\mathbb{R} \setminus B_1$, and $0 \leq \eta \leq 1$ on $\mathbb{R}$. Then we consider the Riesz potential as defined in (15), so that we can estimate the L^∞ norm of v for $\alpha < 1$. Since the kernel $\mathcal{I}_\alpha$ is positive and $\eta \geq 0$ is a smooth function with compact support in B_r, we get

$$\|v\|_{C^{0,\gamma}(\mathbb{R})} \leq C_{\alpha,\gamma} \|\eta f\|_{L^\infty(\mathbb{R})} + C\|v\|_{L^\infty(\mathbb{R})} \leq C_{\alpha,\gamma,B'} \|f\|_{L^\infty(B)}.$$

Next, by using the inequality stated in (22), we get that for $\gamma \in (0, \min(\alpha, 1))$ and $x, y \in B_r$,

$$\begin{aligned}
|v(x) - v(y)| &\leq C_{\alpha,\gamma} |x-y|^\gamma \int_{\mathbb{R}} |x-y|^{\alpha-1-\gamma} \eta(y) |f(y)| dy \\
&\leq C_{\alpha,\gamma} |x-y|^\gamma \|f\|_{L^\infty(B_1)} \int_{B_2} |x-y|^{\alpha-1-\gamma} dy \\
&\leq C_{\alpha,\gamma} \|f\|_{L^\infty(B_1)} |x-y|^\gamma.
\end{aligned}$$

Hence, we conclude that

$$\|v\|_{C^{0,\gamma}(B_r)} \leq C_\alpha \|f\|_{L^\infty(B_1)}, \tag{24}$$

for every $\gamma \in (0, \min(1, \alpha))$.

Next, by change of variables, the function $\xi := u - v$ satisfies $(\mathscr{D}_{right})^\alpha \xi = 0$ in B_r by (21). Therefore, thanks to ([7] (Corollary 1.13)), we have the derivative estimate for every $r' \in (0, r)$:

$$\|\nabla w\|_{L^\infty(B_r')} \leq C_{\alpha,r'} \|w\|_{L^\infty(B_1)} \leq C_{\alpha,r'} (\|u\|_{L^\infty(B_r)} + \|v\|_{L^\infty(B_r)}).$$

The difference function $\xi = u - v$ is smooth in B_1 and is bounded. From this, together with (24), we conclude that

$$\|u\|_{C^{0,\gamma}(B_{r'})} = \|w + v\|_{C^{0,\gamma}(B_{r'})} \leq C_{\alpha,r'} \left(\|u\|_{L^\infty(B_r)} + \|f\|_{L^\infty(B_1)} \right),$$

for every $\gamma \in (0, \min(1, \alpha))$. □

Now we are in a position to state and prove our main result on the interior Schauder estimate for the solution function $\mathcal{U}$ on the set $\overline{\mathcal{B}_1^+}$.

Proof of the Main Result: Theorem 1

Proof. Again, we present the two parts of the proof separately.

Proof of (a): $f \in L^p(\mathbb{R}, w)$. We choose a cut-off function $\eta \in C_c^\infty(B_2)$ such that $\eta \equiv 1$ on B_2 and $0 \leq \eta \leq 1$ on $\mathbb{R}$. Let $\overline{v}$ be the unique solution to the equation

$$(\mathscr{D}_{right})^\alpha \overline{v} = \overline{f} \qquad \text{in} \quad \mathbb{R},$$

where $\overline{f} := \eta f$. Making use of the previous result Proposition 1, we know that $\overline{v} \in C^{\alpha-\gamma-\frac{1}{p}}(\mathbb{R}, w)$ and

$$\|\overline{v}\|_{C^{\alpha-\gamma-\frac{1}{p}}(\mathbb{R},w)} \leq C \left(\|\overline{v}\|_{L^p(\mathbb{R},w)} + \left\|\overline{f}\right\|_{L^p(\mathbb{R},w)} \right),$$

where $C > 0$ is a constant that depends only on α, γ, and p.

The next step is to consider the Bernardis–Reyes–Stinga–Torrea extension $\overline{\mathcal{U}}$ of $\overline{v}$, i.e., the function

$$\overline{\mathcal{U}}(t, \cdot) = P_t^\alpha(t, \cdot) \star \overline{v},$$

which satisfies the equations

$$\begin{cases} \mathcal{M}_\alpha \, \overline{\mathcal{U}}(t,x) = 0 & \text{in } \mathbb{R}^2_+, \\ \lim_{t\to 0} \mathcal{N}_\alpha \, \overline{\mathcal{U}}(t,x) = (\mathscr{D}_{right})^\alpha \overline{v}(x) = \overline{f}(x) & \text{on } \mathbb{R}. \end{cases}$$

By a change of variables, we have

$$\overline{\mathcal{U}}(t,x) = \left(P_t^\alpha(t,\cdot) \star \overline{v}\right)(x) = \int_{\mathbb{R}} \overline{v}(x-ty) H_\alpha(y) dy, \tag{25}$$

where

$$H_\alpha(y) = P_t^\alpha(1,y) = \frac{C_{1,\alpha} e^{-1/(4(-y))}}{(-y)^{1+\alpha}} \chi_{(-\infty,0)}(y). \tag{26}$$

Then, if we set $z_1 = (t_1, x_1), z_2 = (t_2, x_2) \in \overline{\mathbb{R}^2_+}$, we have the estimate

$$\begin{aligned} |\overline{\mathcal{U}}(z_1) - \overline{\mathcal{U}}(z_2)| &\le |z_1 - z_2|^{\alpha-\gamma-\frac{1}{p}} \parallel \overline{v} \parallel_{C^{\alpha-\gamma-\frac{1}{p}}(\mathbb{R},w)} \int_{\mathbb{R}} \max\{|y|^{\alpha-\gamma-\frac{1}{p}}, 1\} H_\alpha(y) dy, \\ &\le C|z_1 - z_2|^{\alpha-\gamma-\frac{1}{p}} \left(\|\overline{v}\|_{L^p(\mathbb{R},w)} + \left\|\overline{f}\right\|_{L^p(\mathbb{R},w)} \right). \end{aligned}$$

By direct computation from (25), and using Theorem 2, we have:

$$\|\overline{\mathcal{U}}\|_{L^p(\mathbb{R}^2_+,w)} \le \|\overline{v}\|_{L^p(\mathbb{R},w)} \le C \left\|\overline{f}\right\|_{L^p(\mathbb{R},w)}.$$

Therefore,

$$\|\overline{\mathcal{U}}\|_{C^{\alpha-\gamma-\frac{1}{p}}\left(\overline{\mathbb{R}^2_+}\right)} \le C \left(\|\overline{V}\|_{L^\infty(\mathbb{R}^2_+)} + \left\|\overline{f}\right\|_{L^\infty(\mathbb{R})} \right), \tag{27}$$

for a positive constant $C > 0$ depending only on α, p and γ.
Next we put $\widetilde{\mathcal{U}} = \mathcal{U} - \overline{\mathcal{U}}$, so that $\widetilde{\mathcal{U}}$ satisfies

$$\begin{cases} \mathcal{M}_\alpha \, \widetilde{\mathcal{U}} = 0 & \text{in } \mathcal{B}_2^+, \\ \lim_{t\to 0} \mathcal{N}_\alpha \, \widetilde{\mathcal{U}}(t,\cdot) = (1-\eta) f = 0 & \text{on } B_2. \end{cases}$$

Considering the even reflection $\widetilde{Z}$ of $\widetilde{\mathcal{U}}$ in the variable t, as described in ([37] (Lemma 4.1)), we have that

$$\mathcal{M}_\alpha \, \widetilde{Z} = 0 \text{ in } \mathcal{B}_2.$$

From the definition (7) of $\left(\mathscr{D}_{right}\right)$, and using ([5] (Corollary 1.13)) or ([39] (Corollary 1.5)), we have that for $x \in B_1$ and $t \in (-1,1)$ fixed,

$$\left|(\mathscr{D}_{right})\widetilde{Z}(t,x)\right| \le C \parallel \widetilde{Z}(t,\cdot) \parallel_{L^p(B_2,w)}. \tag{28}$$

Next, from the fact that

$$(\mathscr{D}_{right})\widetilde{Z} = \widetilde{Z}_{tt} + \frac{\mu}{|t|}\widetilde{Z}_t$$

and from the inequality (28), we obtain

$$\left|\widetilde{Z}_{tt} + \frac{\mu}{|t|}\widetilde{Z}_t\right| \le C \parallel \widetilde{Z} \parallel_{L^p(\mathcal{B}_2,w)}.$$

Therefore

$$\left|\left(|t|^\mu \widetilde{Z}_t\right)_t\right| \le C|t|^\mu \parallel \widetilde{Z} \parallel_{L^p(\mathcal{B}_2,w)}$$

Hence

$$\left|\widetilde{Z}_{tt}\right| \leq C \parallel \widetilde{Z} \parallel_{L^p(\mathcal{B}_2,w)}.$$

For any point $(t,x) \in \overline{\mathcal{B}}_1$, we have, by (28), that

$$\left|(\mathscr{D}_{right})\widetilde{Z}(t,x)\right| + \left|\widetilde{Z}_{tt}(t,x)\right| \leq C \parallel \widetilde{Z} \parallel_{L^p(\mathcal{B}_2,w)},$$

which implies that

$$\widetilde{Z} \in C^{1-\gamma-\frac{1}{p}}(\overline{\mathcal{B}}_1, w).$$

Thus, we have that $\widetilde{\mathcal{U}} \in C^{1-\gamma-\frac{1}{p}}(\overline{\mathcal{B}_1^+}, w)$ such that

$$\begin{aligned} \parallel \widetilde{\mathcal{U}} \parallel_{C^{1-\gamma-\frac{1}{p}}(\overline{\mathcal{B}_1^+},w)} &\leq C\left(\parallel \mathcal{U} \parallel_{L^p(\mathcal{B}_2^+,w)} + \parallel \overline{v} \parallel_{L^p(\mathbb{R})}\right) \\ &\leq C\left(\parallel \mathcal{U} \parallel_{L^p(\mathcal{B}_2^+,w)} + \parallel f \parallel_{L^p(B_2,w)}\right). \end{aligned}$$

We finally obtain

$$\begin{aligned} \parallel \mathcal{U} \parallel_{C^{\alpha-\gamma-\frac{1}{p}}(\overline{\mathcal{B}_1^+},w)} &\leq C\left(\parallel \widetilde{\mathcal{U}} \parallel_{C^{1-\gamma-\frac{1}{p}}(\overline{\mathcal{B}_1^+},w)} + \parallel \overline{\mathcal{U}} \parallel_{C^{\alpha-\gamma-\frac{1}{p}}(\overline{\mathbb{R}^2_+},w)}\right) \\ &\leq C\left(\parallel \mathcal{U} \parallel_{L^p(\mathcal{B}_2^+,w)} + \parallel f \parallel_{L^p(B_2),w}\right), \end{aligned}$$

since $\widetilde{\mathcal{U}} = \mathcal{U} - \overline{\mathcal{U}}$. This ends the proof of the first case.

Proof of (b): $f \in L^\infty(\mathbb{R})$. The proof here is similar to the first case above. By considering the same cut-off function $\eta \in C_c^\infty(B_2)$ with $\eta \equiv 1$ on B_2 and $0 \leq \eta \leq 1$ on $\mathbb{R}$, we let $\overline{v}$ be the unique solution to the equation

$$(\mathscr{D}_{right})^\alpha \overline{v} = \overline{f} \qquad \text{in} \quad \mathbb{R},$$

where $\overline{f} := \eta f$. Making use of the previous result Proposition 1, we have that $\overline{v} \in C^{\alpha-\gamma}(\mathbb{R})$ and

$$\|\overline{v}\|_{C^{\alpha-\gamma}(\mathbb{R})} \leq C\left(\|\overline{v}\|_{L^\infty(\mathbb{R})} + \left\|\overline{f}\right\|_{L^\infty(\mathbb{R})}\right),$$

where $C > 0$ is a constant that depends only on α and γ.

The next step is to consider the Bernardis–Reyes–Stinga–Torrea extension $\overline{\mathcal{U}}$ of $\overline{v}$, i.e.,

$$\overline{\mathcal{U}}(t,\cdot) = P_t^\alpha(t,\cdot) \star \overline{v},$$

which satisfies the equation

$$\begin{cases} \mathcal{M}_\alpha \overline{\mathcal{U}} = 0 & \text{in } \mathbb{R}^2_+, \\ \lim_{t\to 0} \mathcal{N}_\alpha \overline{\mathcal{U}}(t,x) = (\mathscr{D}_{right})^\alpha \overline{v}(x) = \overline{f}(x) & \text{on } \mathbb{R}. \end{cases}$$

Proceeding as in the previous case, it follows that

$$\|\overline{\mathcal{U}}\|_{C^{\alpha-\gamma}\left(\overline{\mathbb{R}}^2\right)} \leq C\left(\|\overline{\mathcal{U}}\|_{L^\infty(\mathbb{R}^2)} + \left\|\overline{f}\right\|_{L^\infty(\mathbb{R})}\right), \tag{29}$$

for a positive constant $C > 0$ depending only on α and γ.

Next we put $\widetilde{\mathcal{U}} = \mathcal{U} - \overline{\mathcal{U}}$, so that $\widetilde{\mathcal{U}}$ satisfies

$$\begin{cases} \mathcal{M}_\alpha \widetilde{\mathcal{U}} = 0 & \text{in } \mathcal{B}_2^+, \\ \lim_{t\to 0} \mathcal{N}_\alpha \widetilde{\mathcal{U}}(t,\cdot) = (1-\eta) f = 0 & \text{on } B_2. \end{cases}$$

We finally obtain

$$\| \mathcal{U} \|_{C^{\alpha-\gamma}(\overline{\mathcal{B}_1^+})} \leq C \left(\| \widetilde{\mathcal{U}} \|_{C^{1-\gamma}(\overline{\mathcal{B}_1^+})} + \| \overline{\mathcal{U}} \|_{C^{\alpha-\gamma}(\overline{\mathbb{R}^2_+})} \right) \leq C \left(\| \mathcal{U} \|_{L^\infty(\mathcal{B}_2^+)} + \| f \|_{L^\infty(B_2)} \right),$$

which ends the proof. □

4. Conclusions

Regularity theorems are an important result in the theory of PDEs, and their fractional counterparts also play a significance role in the study of problems involving nonlocal behaviour. As already observed in various papers [6,8,9,34,40–42] in the theory of fractional nonlocal PDEs, it is possible to find the qualitative behaviour of a solution. In this paper we have shown that the degenerate elliptic equation with mixed boundary conditions for a problem with fractional Marchaud derivative admits an interior regularity estimate. The current work fits in with some results obtained in the case of fractional Laplacians with Caffareli–Silvestre extensions. We stress that the types of regularity results proved herein form only a small subset of many possible versions of regularity theorems which can only be obtained using extension techniques.

Author Contributions: J.-D.D. and A.F. contributed equally to this work.

Funding: The second author's research was supported by a student grant from the Engineering and Physical Sciences Research Council, UK.

Conflicts of Interest: The authors declare no conflict of interest.

References

1. Dipierro, S.; Valdinoci, E. A Simple Mathematical Model Inspired by the Purkinje Cells: From Delayed Travelling Waves to Fractional Diffusion. *Bull. Math. Biol.* **2018**, *80*, 1849–1870. [CrossRef] [PubMed]
2. Marinov, T.; Santamaria, F. Modeling the effects of anomalous diffusion on synaptic plasticity. *BMC Neurosci.* **2013**, *14* (Suppl. 1), 343. [CrossRef]
3. Marinov, T.; Santamaria, F. Computational modeling of diffusion in the cerebellum. *Prog. Mol. Biol. Trans. Sci.* **2014**, *123*, 169–189.
4. Abatangelo, N.E.; Valdinoci, E. Getting acquainted with the fractional Laplacian. *arXiv* **2017**, arXiv:1710.11567.
5. Bernardis, A.; Martín-Reyes, F.J.; Stinga, P.R.; Torrea, J.L. Maximum principles, extension problem and inversion for nonlocal one-sided equations. *J. Differ. Equ.* **2016**, *260*, 6333–6362. [CrossRef]
6. Caffarelli, L.; Silvestre, L. An extension problem related to the fractional Laplacian. *Commun. Part. Differ. Equ.* **2007**, *32*, 1245–1260. [CrossRef]
7. Stinga, P.R.; Torrea, J.L. Problem and Harnack's Inequality for Some Fractional Operators. *Commun. Part. Differ. Equ.* **2010**, *35*, 2092–2122. [CrossRef]
8. Niang, A. Boundary regularity for a degenerate elliptic equation with mixed boundary conditions. *arXiv* **2018**, arXiv:1803.10641.
9. Cabré, X.; Sire, Y. Nonlinear equations for fractional Laplacians I: Regularity, maximum principles, and Hamiltonian estimates. *Ann. Inst. Henri Poincaré Anal. Non Linéaire* **2014**, *31*, 23–53. [CrossRef]
10. Björn, J. Regularity at infinity fot a mixed problem for degenerate elliptic operators in a half-cynlider. *Math. Scand.* **1997**, *81*, 101–126. [CrossRef]
11. Fabes, E.; Jerison, D.; Kenig, C. The Wiener test for degenerate elliptic equations. *Ann. Inst. Fourier (Grenoble)* **1982**, *32*, 151–182. [CrossRef]
12. Fabes, E.; Kenig, C.; Serapioni, R. The local regularity of solutions of degenerate elliptic equations. *Commun. Part. Differ. Equ.* **1982**, *7*, 77–116. [CrossRef]
13. Jin, T.; Li, Y.Y.; Xiong, J. On a fractional nirenberg problem Part I: Blow up analysis and compactness solutions. *J. Eur. Math. Soc.* **2014**, *16*, 1111–1171. [CrossRef]
14. Kassmann, M.; Madych, W.R. Difference quotients and elliptic mixed boundary value problems of second order. *Indiana Univ. Math. J.* **2007**, *56*, 1047–1082. [CrossRef]

15. Levendorskii, S. Degenerate Elliptic Equations. In *Mathematics and Its Applications*; Springer: Dordrecht, The Netherlands, 1993; Volume 258, 436p, ISBN 978-94-017-1215-6.
16. Savaré, G. Regularity and perturbation results for mixed second order elliptic problems. *Commun. Part. Differ. Equ.* **1997**, *22*, 869–899. [CrossRef]
17. Zaremba, S. Sur un problème mixte relatif à l'équation de Laplace. In *Bulletin International de L'Académie des Sciences de Cracovie. Classe des Sciences Mathématiques et Naturelles*; Serie A: Sciences Mathématiques; Impr. de l'Université: Cracovie, Poland, 1910; pp. 313–344.
18. Bucur, C.; Ferrari, F. An extension problem for the fractional derivative defined by Marchaud. *Fract. Calc. Appl. Anal.* **2016**, *19*, 867–887. [CrossRef]
19. Allen, M.; Caffarelli, L.; Vasseur, A. A parabolic problem with a fractional time derivative. *Arch. Ration. Mech. Anal.* **2016**, *221*, 603–630. [CrossRef]
20. Ferrari, F. Weyl and Marchaud derivatives: A forgotten history. *Mathematics* **2018**, *6*, 6. [CrossRef]
21. Samko, S.G.; Kilbas, A.A.; Marichev, O.I. *Fractional. Integrals and Derivatives, Theory and Applications*; Translated from the 1987 Russian Original; Gordon and Breach Science Publishers: Yverdon, Switzerland, 1993.
22. Kufner, A. *Weighted Sobolev Spaces*; Wiley: New York, NY, USA, 1985.
23. Maz'ya, V.G. *Sobolev Spaces with Applications to Elliptic Partial Differential Equations*; Volume 342 of Grundlehren der Mathematischen Wissenschaften [Fundamental Principles of Mathematical Sciences]; Springer: Heidelberg, Germany, 2011.
24. Triebel, H. *Interpolation Theory, Function Spaces, Differential Operators*, 2nd ed.; Johann Ambrosius Barth: Heidelberg, Germany, 1995.
25. Meyries, M.; Veraar, M. Sharp embedding results for spaces of smooth functions with power weights. *arXiv* **2011**, arXiv:1112.5388.
26. Köhne, M.; Prüss, J.; Wilke, M. On quasilinear parabolic evolution equations in weighted L_p-spaces. *J. Evol. Equ.* **2010**, *10*, 443–463.
27. Lunardi, A. *Analytic Semigroups and Optimal Regularity in Parabolic Problems*; Progress in Nonlinear Differential Equations and Their Applications; Birkhäuser Verlag: Basel, Switzerland, 1995; Volume 16.
28. Triebel, H. Theory of Function Spaces. In *Monographs in Mathematics*; Birkhäuser Verlag: Basel, Switzerland, 1983; Volume 78.
29. Coifman, R.R.; Fefferman, C. Weighted norm inequalities for maximal functions and singular integrals. *Stud. Math.* **1974**, *51*, 241–250. [CrossRef]
30. Muckenhoupt, B. Weighted norm inequalities for the Hardy maximal function. *Trans. Am. Math. Soc.* **1972**, *165*, 207–226. [CrossRef]
31. Opic, B.; Kufner, A. *Hardy-Type Inequalities*; Wiley: New York, NY, USA, 1990.
32. Stein, E.M. *Harmonic Analysis: Real-Variable Methods, Orthogonality, and Oscillatory Integrals*; Volume 43 of Princeton Mathematical Series, Monographs in Harmonic Analysis, III; Princeton University Press: Princeton, NJ, USA, 1993.
33. Lebedev, N.N. *Special Functions and Their Applications*; Prentice-Hall, Inc.: Englewood Cliffs, NJ, USA, 1965.
34. Fall, M.M. Regularity estimates for nonlocal Schrödinger equations. *arXiv* **2017**, arXiv:1711.02206.
35. Caffarelli, L.; Cabré, X. *Fully Nonlinear Elliptic Equations*; American Mathematical Society: Providence, RI, USA, 1995; Volume 43.
36. Silvestre, L. Regularity of the obstacle problem for a fractional power of the Laplace operator. *Commun. Pure Appl. Math.* **2007**, *60*, 67–112, . [CrossRef]
37. Bucur, C. Some nonlocal operators and effects due to nonlocality. *arXiv* **2017**, arXiv:1705.00953.
38. Fall, M.M.; Weth, T. Monotonicity and nonexistence results for some fractional elliptic problems in the half space. *arXiv* **2013**, arXiv:1309.7230.
39. Caffarelli, L.A.; Salsa, S.; Silvestre, L. Regularity estimates for the solution and the free boundary of the obstacle problem for the fractional Laplacian. *Invent. Math.* **2008**, *171*, 425–461. [CrossRef]
40. Silvestre, L. Regularity of the obstacle problem for a fractional power of the Laplace operator. *Comm. Pure Appl. Math.* **2007**, *60*, 67–112. [CrossRef]

41. Dipierro, S.; Soave, N.; Valdinoci, E. On fractional elliptic equations in Lipschitz sets and epigraphs: Regularity, monotonicity and rigidity results. *Math. Ann.* **2017**, *369*, 1283–1326. [CrossRef]
42. Fernandez, A. An elliptic regularity theorem for fractional partial differential operators. *Comput. Appl. Math.* **2018**, 1–12. [CrossRef]

axioms

MDPI

Article

Umbral Methods and Harmonic Numbers

Giuseppe Dattoli [1,†], Bruna Germano [2,†], Silvia Licciardi [1,*,†] and Maria Renata Martinelli [2,†]

1 ENEA—Frascati Research Center, Via Enrico Fermi 45, 00044 Rome, Italy; giuseppe.dattoli@enea.it
2 Department of Methods and Mathematic Models for Applied Sciences, University of Rome, La Sapienza, Via A. Scarpa, 14, 00161 Rome, Italy; bruna.germano@sbai.uniroma1.it (B.G.); martinelli@dmmm.uniroma1.it (M.R.M.)
* Correspondence: silviakant@gmail.com; Tel.: +39-392-509-6741
† These authors contributed equally to this work.

Received: 4 June 2018; Accepted: 24 August 2018; Published: 1 September 2018

Abstract: The theory of harmonic-based functions is discussed here within the framework of umbral operational methods. We derive a number of results based on elementary notions relying on the properties of Gaussian integrals.

Keywords: harmonic numbers 11K99; operators 44A99, 47B99; umbral methods 05A40; special functions 33C52, 33C65, 33C99, 33B10, 33B15; Hermite polynomials 33C45

---

1. Introduction

Methods employing the concepts and the formalism of umbral calculus have been exploited in [1] to guess the existence of generating functions involving harmonic numbers [2]. The conjectures put forward in [1] have been proven in [3,4] and further elaborated in [5], and these were extended to hyper-harmonic numbers in [6].

In this note, we use the same point of view as [1], by discussing the possibility of exploiting the formalism developed therein in a wider context.

The umbral methods we are going to describe have certain advantages with respect to the ordinary techniques. The key idea is that of exploiting the harmonic number index as a power exponent; such a "promotion" allows the possibility of reducing the associated computational technicalities to elementary algebraic manipulations. Series involving harmonic numbers can, e.g., be treated as formal series of known functions (exponential, Gaussian, rational, etc.), and the relevant properties can be exploited to carry out computations, which are significantly more cumbersome and involved when conventional methods are employed.

2. Harmonic Numbers and Generating Functions

The harmonic numbers are defined by means of the following partial sum [2]:

$$h_n := \sum_{r=1}^{n} \frac{1}{r}, \quad \forall n \in \mathbb{N}_0 \,. \tag{1}$$

The integral representation for this family of numbers can be derived using a standard procedure, tracing back to Euler, which is sketched below.

Proposition 1. *The use of elementary integral transform yields, for the finite sum in Equation (1), the identity:*

$$h_n = \sum_{r=1}^{n} \int_0^\infty e^{-s\,r} ds, \quad \forall n \in \mathbb{N}_0 \,, \tag{2}$$

thereby getting the n-th harmonic number through Euler's integral [7–9]:

$$h_n = \int_0^1 \frac{1-x^n}{1-x} dx. \tag{3}$$

Proof. $\forall n \in \mathbb{N}_0$, by applying the Laplace transform, the theorem of uniform convergence and the sum of a geometric series, we obtain:

$$\begin{aligned} h_n &= \sum_{r=1}^{n} \int_0^\infty e^{-sr} ds = \int_0^\infty \left[\left(\sum_{r=0}^{n} e^{-sr} \right) - 1 \right] ds \\ &= \int_0^\infty \frac{1-(e^{-s})^{n+1}}{1-e^{-s}} - 1\, ds = \int_{-\infty}^0 \frac{1-(e^{s})^{n+1}}{1-e^{s}} - 1\, ds \\ &= \int_{-\infty}^0 \frac{e^{(n+1)s} - e^s}{e^s - 1} ds \end{aligned}$$

and by applying the change of variables $e^s \to x$, we eventually end up with:

$$h_n = \int_0^1 \frac{1-x^n}{1-x} dx.$$

□

According to [8], from this point onwards, the definition in Equation (3) can be so extended to any real value of n, and therefore, it can be exploited as an alternative definition holding for n a positive real.

Definition 1. *The function:*

$$\varphi_h(z) := \int_0^1 \frac{1-x^z}{1-x} dx, \quad \forall z \in \mathbb{R}^+, \tag{4}$$

is called the harmonic number umbral vacuum, or simply the vacuum.

Definition 2. *The operator:*

$$\hat{h} := e^{\partial_z} \tag{5}$$

realizes the vacuum shift operator, zbeing the domain's variable of the function on which the operator acts For a deeper introduction to umbral calculus, see [10,11]

Theorem 1. *The umbral operator, $\hat{h}^n$, $\forall n \in \mathbb{R}^+$, defines the harmonic numbers, h_n, as the action of the shift operator (5) on the vacuum (4):*

$$\hat{h}^n \varphi_h(z)\Big|_{z=0} := \hat{h}^n \varphi_{h_z}\Big|_{z=0} = h_n \tag{6}$$

or simply:

$$\begin{aligned} \hat{h}^n &\equiv h_n, \\ h_0 &= 0. \end{aligned} \tag{7}$$

Proof. $\forall n \in \mathbb{R}^+$, by applying the shift operator (5) on the vacuum (4), we obtain:

$$\begin{aligned} \hat{h}^n \varphi_{h_0} &= \hat{h}^n \varphi_{h_z}\Big|_{z=0} = e^{n\partial_z} \varphi_{h_z}\Big|_{z=0} = \varphi_{h_{z+n}}\big|_{z=0} = \int_0^1 \frac{1-x^{z+n}}{1-x} dx \Big|_{z=0} \\ &= \int_0^1 \frac{1-x^n}{1-x} dx = h_n. \end{aligned}$$

□

Properties 1. *$\forall n, m \in \mathbb{R}^+$, we get:*

$$
\begin{aligned}
&(i)\ \ \hat{h}^n \hat{h}^m = \hat{h}^{n+m}\,, \\
&(ii)\ \left(\hat{h}^n\right)^m = \hat{h}^{nm}\,.
\end{aligned}
\tag{8}
$$

The proof is a fairly direct consequence of the realization given in Equation (5).

Definition 3. *We call the Harmonic-Based Exponential Function (HBEF) the series:*

$$
{}_he(x) := e^{\hat{h}\,x}\varphi_{h_0} = 1 + \sum_{n=1}^{\infty} \frac{h_n}{n!} x^n. \tag{9}
$$

This function, as already discussed in [1], has quite remarkable properties.

The relevant derivatives can accordingly be expressed as (see the concluding part of the paper for further comments):

$$
\begin{aligned}
&{}_he(x,m) := \left(\frac{d}{dx}\right)^m {}_he(x) = \hat{h}^m e^{\hat{h}\,x}\varphi_{h_0} = h_m + \sum_{n=1}^{\infty} \frac{h_{n+m}}{n!} x^n, \quad \forall x \in \mathbb{R}, \forall m \in \mathbb{N} \\
&{}_he(x,k+m) = \left(\frac{d}{dx}\right)^m {}_he(x,k), \quad \forall k \in \mathbb{N},
\end{aligned}
\tag{10}
$$

and according to Equation (9), we also find that:

$$
\int_0^{\infty} {}_he(-\alpha\,x)\,e^{-x}dx = \int_0^{\infty} e^{-(\alpha\,\hat{h}+1)\,x}dx = \frac{1}{\alpha\,\hat{h}+1}, \quad |\,\alpha\,| < 1. \tag{11}
$$

Corollary 1. *By expanding the umbral function on the r.h.s. of Equation (11), we obtain:*

$$
\frac{1}{\alpha\,\hat{h}+1} = 1 + \sum_{n=1}^{\infty}(-1)^n \alpha^n h_n, \quad |\,\alpha\,| < 1. \tag{12}
$$

Proof. By using the Taylor expansion and Equation (7), for $|\,\alpha\,| < 1$, we have:

$$
\frac{1}{\alpha\,\hat{h}+1} = \sum_{n=0}^{\infty}(-\alpha\hat{h})^n = 1 + \sum_{n=1}^{\infty}(-1)^n\alpha^n\hat{h}^n = 1 + \sum_{n=1}^{\infty}(-1)^n\alpha^n h_n,
$$

□

This is an expected conclusion, achievable by direct integration, underscored here to stress the consistency of the procedure.

A further interesting example comes from the following "Gaussian" integral.

$$
\int_{-\infty}^{\infty} {}_he(-\alpha\,x)\,e^{-x^2}dx = \int_{-\infty}^{\infty} e^{-(\alpha\,\hat{h}\,x + x^2)}\,dx = \sqrt{\pi}e^{\frac{\alpha^2\,\hat{h}^2}{4}}\ \forall \alpha \in \mathbb{R}. \tag{13}
$$

The last term in Equation (13) has been obtained by treating $\hat{h}$ as an ordinary algebraic quantity and then by applying the standard rules of the Gaussian integration.

We notice that, using Equation (9), we obtain:

$$
{}_{h^2}e\left(\frac{\alpha^2}{4}\right) := e^{\frac{\hat{h}^2\alpha^2}{4}}\varphi_{h_0} = 1 + \sum_{r=1}^{\infty}\frac{h_{2r}}{r!}\left(\frac{\alpha}{2}\right)^{2r}. \tag{14}
$$

Let us now consider the following slightly more elaborate example, involving the integration of two "Gaussians", namely the ordinary case and its analogous HBEF.

Example 1.

$$\int_{-\infty}^{\infty} {}_{h}e(-\alpha\,x^2)\,e^{-x^2}dx = \int_{-\infty}^{\infty} e^{-(\hat{h}\alpha+1)x^2}\,dx\;\varphi_{h_0} = \sqrt{\frac{\pi}{1+\alpha\,\hat{h}}}\varphi_{h_0}, \quad |\,\alpha\,| < 1. \tag{15}$$

This last result, obtained after applying elementary rules, can be worded as follows: the integral in Equation (15) depends on the operator function on its r.h.s., for which we should provide a computational meaning. The use of the Newton binomial yields:

$$\begin{aligned}\sqrt{\frac{\pi}{1+\alpha\hat{h}}}\varphi_{h_0} &= \sqrt{\pi}\sum_{r=0}^{\infty}\binom{-\frac{1}{2}}{r}\left(\alpha\,\hat{h}\right)^r\varphi_{h_0} = \sqrt{\pi}\left(1+\sqrt{\pi}\sum_{r=1}^{\infty}\frac{\alpha^r h_r}{\Gamma\left(\frac{1}{2}-r\right)\,r!}\right)\\ &= \sqrt{\pi}\left(1+\sum_{r=1}^{\infty}\binom{2r}{r}\frac{(-\alpha)^r\,h_r}{2^{2r}}\right),\\ &|\alpha| < 1\,.\end{aligned} \tag{16}$$

The correctness of this conclusion has been confirmed by the numerical check, as well.

It is evident that the examples we have provided show that the use of concepts borrowed from umbral theory offers a fairly powerful tool to deal with the "harmonic-based" functions.

3. Harmonic-Based Functions and Differential Equations

In the following, we will further push the formalism to stress the associated flexibility.

We note indeed that the function:

$${}_{\sqrt{\hat{h}}}e(x) := e^{\hat{h}^{\frac{1}{2}}\,x}\varphi_{h_0} = 1+\sum_{n=1}^{\infty}\frac{\left(\sqrt{\hat{h}}\,x\right)^n}{n!}\varphi_{h_0} = 1+\sum_{n=1}^{\infty}\frac{h_{n/2}}{n!}\,x^n, \forall x\in\mathbb{R}, \tag{17}$$

defines, $\forall\alpha\in\mathbb{R}$, an HBEF through the following Gauss transform:

$$\int_{-\infty}^{+\infty} {}_{\sqrt{\hat{h}}}e(\alpha\,x)\,e^{-x^2}dx = \int_{-\infty}^{+\infty} e^{\hat{h}^{\frac{1}{2}}\,\alpha\,x-x^2}dx\;\varphi_{h_0} = \sqrt{\pi}e^{\hat{h}\left(\frac{\alpha}{2}\right)^2}\varphi_{h_0} = \sqrt{\pi}\;{}_{h}e\left(\left(\frac{\alpha}{2}\right)^2\right). \tag{18}$$

On the other side, Equation (17) can be expressed in terms of the HBEF, ${}_{h}e(x)$, using appropriate integral transform methods [12].

Definition 4. *If:*

$$g_{\frac{1}{2}}(\eta) = \frac{1}{2\sqrt{\pi\eta^3}}e^{-\frac{1}{4\eta}}, \quad \forall\eta\in\mathbb{R}^+, \tag{19}$$

is the Levy distribution of order $\frac{1}{2}$, then [12]:

$$e^{-p^{\frac{1}{2}}\,x} = \int_0^{\infty} e^{-p\,\eta\,x^2}\,g_{\frac{1}{2}}(\eta)\,d\eta\,, \quad \forall p\in\mathbb{R}^+, \tag{20}$$

is the associated Levy integral transform.

The use of Equations (17) and (19) allows us to write the following identity.

Corollary 2.

$$\sqrt{\hat{h}}e(-x) = \int_0^\infty {}_he(-\eta\, x^2)\, g_{\frac{1}{2}}(\eta)\, d\eta. \tag{21}$$

Proof.

$$\sqrt{\hat{h}}e(-x) = e^{-\hat{h}^{\frac{1}{2}} x}\varphi_{h_0} = \int_0^\infty e^{-\hat{h}\eta x^2} g_{\frac{1}{2}}(\eta)\, d\eta\; \varphi_{h_0} = \int_0^\infty {}_he(-\eta\, x^2)\, g_{\frac{1}{2}}(\eta)\, d\eta\;.$$

□

The possibility of defining ${}_{\sqrt[k]{\hat{h}}}e(x)$ will be discussed elsewhere.

Theorem 2. *The function ${}_he(x)$ satisfies the first order non-homogeneous differential equation:*

$$\begin{cases} {}_he'(x) = \dfrac{d}{dx}{}_he(x) = {}_he(x) + \dfrac{e^x - x - 1}{x}, & \forall x \in \mathbb{R}_0, \\ {}_he(0) = 1. \end{cases} \tag{22}$$

Proof. Equation (10), for $m = 1$, yields:

$${}_he'(x) = {}_h e(x, 1) = 1 + \sum_{n=1}^\infty \frac{h_{n+1}}{n!}\, x^n\;. \tag{23}$$

Since $h_{n+1} = h_n + \frac{1}{n+1}$, we find:

$$1 + \sum_{n=1}^\infty \frac{h_{n+1}}{n!}\, x^n = {}_he(x) + \frac{1}{x}\,(e^x - x - 1) \tag{24}$$

and hence, Equation (22) follows. □

Corollary 3. *The solution of Equation (22) yields for the HBEF the explicit expression in terms of ordinary special functions $\forall x \in \mathbb{R}^+$:*

$$\begin{aligned} &{}_he(x) = 1 + e^x\,(\ln(x) + E_1(x) + \gamma)\,, \\ &E_1(x) = \int_x^\infty \frac{e^{-t}}{t} dt, \\ &(\ln(x) + E_1(x) + \gamma) = -\sum_{n=1}^\infty \frac{(-x)^n}{n\, n!}, \end{aligned} \tag{25}$$

where γ is the Euler–Mascheroni–constant .

The previous expression is the generating function of harmonic numbers originally derived by Gosper (see [2,13]).

By iterating the previous procedure, we find the following general recurrence.

Corollary 4.

$${}_he(x, m) = {}_he(x) + \sum_{r=0}^{m-1} \left(\frac{d}{dx}\right)^r \frac{e^x - 1 - x}{x}\;. \tag{26}$$

Definition 5. *The binomial expansion:*

$$h_n(x) := (x + \hat{h})^n \varphi_{h_0} = x^n + \sum_{s=1}^n \binom{n}{s} x^{n-s}\, h_s, \quad \forall x \in \mathbb{R}, \forall n \in \mathbb{N}_0, \tag{27}$$

specifies the harmonic polynomials.

They are easily shown to be linked to the HBEF by means of the generating function as follows.

Corollary 5.

$$\sum_{n=0}^{\infty} \frac{t^n}{n!} h_n(x) = e^{x\,t}{}_h e(t), \quad \forall x, t \in \mathbb{R}. \tag{28}$$

Proof. It is readily checked that:

$$\sum_{n=0}^{\infty} \frac{t^n}{n!} h_n(x) = \sum_{n=0}^{\infty} \frac{t^n}{n!} (x+\hat{h})^n \varphi_{h_0} = e^{t(x+\hat{h})} \varphi_{h_0} = e^{x\,t}{}_h e(t)\ .$$

□

According to Equation (28), $h_n(x)$ are recognized as Appél polynomials and satisfy the following recurrences.

Properties 2. *The properties below hold:*

$$(i)\ \ \frac{d}{dx} h_n(x) = n\, h_{n-1}(x), \quad \forall x \in \mathbb{R}, \tag{29}$$

$$\begin{aligned}(ii)\ \ & h_{n+1}(x) = (x+1)\, h_n(x) + f_n(x),\ , \\ & f_n(x) := \sum_{s=1}^{n} \frac{n!}{(n-s)!} \frac{x^{n-s}}{(s+1)!} = \int_0^1 (x+y)^n dy - x^n, \quad \forall x \in \mathbb{R}.\end{aligned} \tag{30}$$

Proof. The recurrence given in Equation (29) follows from the definition of the derivative itself since we treat h as an ordinary algebraic quantity. The proof of the identity (30) is slightly more elaborate; we note indeed that:

$$\begin{aligned}
h_{n+1}(x) &= (x+\hat{h})(x+\hat{h})^n \varphi_{h_0} = (x+\hat{h}) \left(x^n + \sum_{s=1}^{n} \binom{n}{s} x^{n-s}\, \hat{h}^s \right) \varphi_{h_0} \\
&= x\, h_n(x) + 1 \cdot x^n + \sum_{s=1}^{n} \binom{n}{s} x^{n-s}\, \hat{h}^{s+1} \varphi_{h_0} \\
&= x\, h_n(x) + \left(x^n + \sum_{s=1}^{n} \binom{n}{s} x^{n-s}\, \hat{h}^s \right) \varphi_{h_0} + \sum_{s=1}^{n} \frac{n!\, x^{n-s}}{(n-s)!(s+1)!} \\
&= (x+1)\, h_n(x) + \sum_{s=1}^{n} \frac{n!\, x^{n-s}}{(n-s)!(s+1)!}
\end{aligned}$$

and:

$$\begin{aligned}
\sum_{s=1}^{n} \frac{n!}{(n-s)!} \frac{x^{n-s}}{(s+1)!} &= \sum_{s=1}^{n} \frac{n!}{s!(n-s)!} \frac{x^{n-s}}{s+1} y^{s+1} \Bigg|_{y=1} \\
&= \sum_{s=1}^{n} \binom{n}{s} x^{n-s} \int_0^1 y^s dy = \int_0^1 \left(\sum_{s=0}^{n} \binom{n}{s} x^{n-s} y^s - x^n \right) dy \\
&= \int_0^1 (x+y)^n dy - x^n\ .
\end{aligned}$$

□

Corollary 6. *The identity:*

$$h_n(-1) = (-1)^n \left(1 - \frac{1}{n}\right), \quad \forall n \in \mathbb{N}, \tag{31}$$

follows from the Equation (30) after setting $x = -1$.

The further relationship:

$$h_n = 1 + \sum_{s=1}^{n} \binom{n}{s} h_s(-1), \quad \forall n \in \mathbb{N}_0, \tag{32}$$

is a consequence of the fact that $\hat{h}^n = ((\hat{h} - 1) + 1)^n$.

The harmonic Hermite polynomials (touched on in [1,3,14]) can also be written as follows.

Definition 6.

$$\begin{aligned} &\sum_{n=0}^{\infty} \frac{t^n}{n!} \, {}_hH_n(x) = e^{x\,t} {}_he(t^2), \quad \forall x, t \in \mathbb{R}, \\ &{}_hH_n(x) := H_n(x, \hat{h})\varphi_{h_0} = e^{\hat{h}\partial_x^2} x^n \varphi_{h_0} = x^n + n! \sum_{r=1}^{\lfloor \frac{n}{2} \rfloor} \frac{x^{n-2\,r} h_r}{(n-2\,r)!\,r!}. \end{aligned} \tag{33}$$

Properties 3. *The recurrences identity of the umbral Hermite polynomials:*

$$\begin{aligned} &(i)\ \frac{d}{dx} {}_hH_n(x) = n\, {}_hH_{n-1}(x)\ , \quad \forall x \in \mathbb{R}, \\ &(ii)\ {}_hH_{n+1}(x) = \left(x + 2\,\hat{h}\,\frac{d}{dx}\right) {}_hH_n(x)\varphi_{h_0} = \left(x + 2\frac{d}{dx}\right) {}_hH_n(x) + 2\,\alpha_n'(x)\ , \\ &\alpha_n(x) = n! \sum_{s=1}^{\lfloor \frac{n}{2} \rfloor} \frac{x^{n-2s}}{(s+1)!\,(n-2s)!}\ , \\ &\alpha_n'(x) = \frac{d}{dx}\alpha_n(x) = n\ \alpha_{n-1}(x)\ , \end{aligned} \tag{34}$$

are a by-product of the previous identities and a consequence of the monomiality principle discussed in [15].

Corollary 7. *The umbral Hermite satisfies the second order non-homogeneous ODE:*

$$\left(x\frac{d}{dx} + 2\left(\frac{d}{dx}\right)^2\right) {}_hH_n(x) = n\ {}_hH_n(x) - 2\,\alpha_n''(x). \tag{35}$$

4. Truncated Exponential Numbers and Final Comments

Before closing the paper, we want to stress the possibility of extending the present procedure to the truncated exponential numbers, namely:

$$e_n := \sum_{r=0}^{n} \frac{1}{r!}\ , \quad \forall n \in \mathbb{N}. \tag{36}$$

The relevant integral representation is written [16]:

$$e_\alpha := \frac{1}{\Gamma(\alpha+1)} \int_0^\infty e^{-s}(1+s)^\alpha ds\ , \tag{37}$$

which holds for $\alpha \in \mathbb{R}$, as well. For example, we find:

Example 2.

$$e_{-\frac{1}{2}} = \frac{e}{\sqrt{\pi}}\Gamma\left(\frac{1}{2},1\right) \tag{38}$$

with $\Gamma\left(\frac{1}{2}\,,1\right)$ *being the lower incomplete Gamma function.*

According to the previous discussion and to Equation (38), setting $\hat{e}^{\alpha} \leftrightarrow e_{\alpha}$, we also find that:

$$\begin{aligned} &\int_{-\infty}^{+\infty} e^{-\hat{e}\,x^2} dx = \sqrt{\pi} e_{-\frac{1}{2}}, \\ &e^{-\hat{e}\,x^2} = \sum_{r=0}^{\infty} (-1)^r \frac{e_r}{r!} x^{2r}\,. \end{aligned} \tag{39}$$

This last identity is a further proof that the implications offered by the topics treated in this paper are fairly interesting and deserve further and more detailed investigation, which will be more accurately treated elsewhere.

Author Contributions: G.D., B.G., S.L. and M.R.M. contributed to the realization of the article, as well as provided topics and obtained results.

Funding: This research received no external funding.

Conflicts of Interest: The authors declare no conflict of interest.

References

1. Dattoli, G.; Srivastava, H.M. A Note on Harmonic Numbers, Umbral Calculus and Generating Functions. *Appl. Math. Lett.* **2008**, *21*, 686–693. [CrossRef]
2. Weisstein, E.W. *CRC Concise Encyclopedia of Mathematics*; Chapman and Hall/CRC: Boca Raton, FL, USA, 2003; p. 3115, ISBN 1-58488-347-2.
3. Coffey, M.W. Expressions for Harmonic Number Generating Functions. In *Contemporary Mathematics, 517, Gems in Experimental Mathematics*; Amdeberhan, T., Medina, L.A, Moll, V.H., Eds.; AMS Special Session, Experimental Mathematics: Washington, DC, USA, 2009.
4. Cvijović, D. The Dattoli-Srivastava Conjectures Concerning Generating Functions Involving the Harmonic Numbers. *Appl. Math. Comput.* **2010**, *215*, 4040–4043. [CrossRef]
5. Mezo, I. Exponential Generating Function of Hyper-Harmonic Numbers Indexed by Arithmetic Progressions. *Cent. Eur. J. Math.* **2013**, *11*, 931–939.
6. Conway, J.H.; Guy, R.K. *The Book of Numbers*; Springer: New York, NY, USA, 1996.
7. Rochowicz, J.A., Jr. Harmonic Numbers: Insights, Approximations and Applications. *Spreadsheets Educ. eJSiE* **2015**, *8*, 4.
8. Lagarias, J.C. Euler's Constant: Euler's Work and Modern Developments. *Bulletin (New Series) of the American Mathematical Society*; S 0273-0979(2013)01423-X; pp. 527–628. Available online: http://citeseerx.ist.psu.edu/viewdoc/download?doi=10.1.1.363.9527&rep=rep1&type=pdf (accessed on 19 July 2013).
9. Cartier, P. Mathemagics (A Tribute to L. Euler and R. Feynman). In *Noise, Oscillators and Algebraic Randomness. Lecture Notes in Physics*; Planat, M. Ed.; Springer: Berlin, Heidelberg, 2000; Volume 550.
10. Licciardi, S. Umbral Calculus, A Different Mathematical Language. Ph.D. Thesis, Department of Mathematics and Computer Sciences, XXIX Cycle, University of Catania, Catania, Italy, 2018.
11. Roman, S.M.; Rota, G.-C. The umbral calculus. *Adv. Math* **1978**, *27*, 95–188. [CrossRef]
12. Doetsch, G. *Handbuch der Laplace Transformation*; Birkhnauser: Basel, Switzerland, 1950.
13. Mező I.; Dil, A. Euler-Seidel method for certain combinatorial numbers and a new characterization of Fibonacci sequence, *Cent. Eur. J. Math.* **2009**, *7* , 310–321. [CrossRef]
14. Zhukovsky, K.; Dattoli, G. Umbral Methods, Combinatorial Identities And Harmonic Numbers. *Appl. Math.* **2011**, *1*, 46. [CrossRef]

15. Dattoli, G. Generalized Polynomials, Operational Identities and their Applications. *J. Comput. Appl. Math* **2000**, *118*, 111–123. [CrossRef]
16. Dattoli, G.; Ricci, P.E.; Marinelli, L. Generalized Truncated Exponential Polynomials and Applications. In *An International Journal of Mathematics*; Rendiconti dell'Istituto di Matematica dell'Universitá di Trieste: Trieste, Italy, 2002; Volume 34, pp. 9–18.

axioms

MDPI

Article

Sub-Optimal Control in the Zika Virus Epidemic Model Using Differential Evolution

Nonthamon Chaikham [1,2] and Wannika Sawangtong [1,2,*]

1 Department of Mathematics, Faculty of Science, Mahidol University, Bangkok 10400, Thailand; nonthamon.chi@student.mahidol.ac.th
2 Centre of Excellence in Mathematics, Commission on Higher Education, Ministry of Education, 328 Sri Ayuthaya Road, Bangkok 10400, Thailand
* Correspondence: wannika.saw@mahidol.ac.th; Tel.: +66-2-201-5432

Received: 18 June 2018; Accepted: 18 August 2018; Published: 23 August 2018

Abstract: A dynamical model of Zika virus (ZIKV) epidemic with direct transmission, sexual transmission, and vertical transmission is developed. A sub-optimal control problem to counter against the disease is proposed including three controls: vector elimination, vector-to-human contact reduction, and sexual contact reduction. Each control variable is discretized into piece-wise constant intervals. The problem is solved by Differential Evolution (DE), which is one of the evolutionary algorithm developed for optimization. Two scenarios, namely four time horizons and eight time horizons, are compared and discussed. The simulations show that models with controls lead to decreasing the number of patients as well as epidemic period length. From the optimal solution, vector elimination is the prioritized strategy for disease control.

Keywords: Zika virus; sexual transmission; vertical transmission; sub-optimal control problem; differential evolution

---

1. Introduction

Despite being discovered in Africa and named after a forest in Uganda in 1947, there are only several reported cases of Zika virus (ZIKV) infection before 2007. The virus itself is a mosquito-borne virus from the genus flavivirus that is transmitted primarily through the bite of Aedes mosquitoes. At first, the ZIKV infection was not considered a threat due to its mild symptoms such as rashes, headache, malaise, and fever [1]. Besides, the mortality rate of ZIKV infection is considered negligible, unlike other Flavivirus infections such as dengue and yellow fever. However, the announcement [2] on the possible association between the ZIKV infection and the increasing number of newborn babies with microcephaly in 2015 raised the alarm and public awareness of this outbreak. There is some correlative evidence reported from Brazil, Colombia, and other Latin America countries [3,4].

Some previous studies suggest that ZIKV could pass from pregnant mothers to their baby through the intrauterine infection, and microcephaly also happened during this process. Beside vector-mediated transmission and intrauterine (or vertical) transmission, ZIKV could also transmit through sexual transmission [5]. Apart from being detected in blood and semen [6], researchers found that the virus persists in semen and urine after it disappeared from the serum [7]. Moreover, it could pass on even before the patient develops the symptoms [8]. To control the outbreak, it can be seen that only relying on reducing the number of mosquito population or preventing the contact between host and vectors might be insufficient. The control strategy should include reducing the sexual transmission and vertical transmission as well.

A dynamical model was a popular tool for describing and analyzing various diseases. It was used to describe the mechanism functions and causes abnormality in some non-infectious diseases such as diabetes [9,10]. In prior publications [11–13], models with systems of ordinary differential equations were utilized for describing the behavior of infectious virus such as Zika. Optimal control is one of the popular methods for finding the solution that will minimize the objective function value. According to some publications [14,15], optimal control can be used to find the best strategy associated with the implemented cost and the provided situations. The optimal solutions for some flavivirus control such as chikungunya [16] and dengue [17] have been studied.

In this paper, we develop the ZIKV virus infection model using some previous works [12,13,18]. The model consists of three populations: adult humans, newborn babies human, and mosquitoes. Three controls are selected from the suggestions of the World Health Organization(WHO) [19] and health authorities [20]: using larvicide and adulticide to reduce the mosquito population, using insect repellents to prevent mosquito biting, and abstaining from sexual activity or using protection to prevent pregnancy.

Regarding optimal control and sub-optimal control problem, there are direct and indirect methods for solving the problems. The indirect approaches, such as Pontryagin's Maximum Principle [21] and shooting method, are popular to find optimal solutions of some flavivirus control problems, for example, chikungunya [16] and dengue [17]. Previously, the authors also studied the optimal control problem of the Zika virus infection [18] by the indirect approach. However, the implementations for health authorities by these studies [16,17] are difficult to practically apply since the policies have to be varied continuously. Considering this drawback, we propose a sub-optimal control problem and find the solution that is both close to optimal and practical for real situation.

A sub-optimal control problem can be proposed through Model Predictive Control (MPC). MPC is the mathematical tool for controlling the optimization process in various fields such as industrial and medical technology [22]. It can be used from the simple to complex dynamic process, and the result control law is easy to implement. In our work, the control framework consists of a nonlinear system without constraints on spending cost or the final result of the system variables. This framework is related to the nonlinear unconstrained MPC schemes, which are those without terminal constraints and cost [23]. Furthermore, the study of Herty [24] estimated the suboptimality condition and suggested the guideline for the system containing a quadratic cost function.

In sub-optimal control problem, the previously continuous time horizon is discretized into N time horizons. Finding the solution of the controls in each time horizon helps archive the strategy which only applies the control at the constant level in each period. Caetano [25] and Yan [26] used the direct approach in finding the solutions to the sub-optimal disease control problem, and Yan [26] also used the Genetic Algorithm (GA).

Differential Evolution (DE) algorithm is an improved GA by Storn and Price [27]. DE algorithm is simple and fast, requires only a few control parameters and has high convergence attribute [28]. Thus, we find the solution of the sub-optimal control problem using the DE algorithm. The dynamical model is developed in Section 2 and the Differential Evolution is introduced in Section 3. The sub-optimal control problem is proposed through MPC and the numerical solutions are solved in Section 4. Lastly, the discussion and conclusion are presented in Section 5.

2. Dynamical Control Model for ZIKV Infection

To describe the behavior of ZIKV infection, the dynamical compartment model with control is developed from the models of Gao et al. [12] and Agusto et al. [13]. The developed model consists of two human populations, adults and babies, and one mosquito (vector) population. The total newborn baby population consists of six classes: susceptible $S_b(t)$, exposed $E_b(t)$, asymptomatically infected $A_b(t)$, symptomatically infected without microcephaly, referred to as infected $I_b(t)$, infected with microcephaly $I_{bm}(t)$, and recovered $R_b(t)$. The total adult population consists of six classes: susceptible $S_w(t)$, exposed $E_w(t)$, asymptomatically infected $A_w(t)$, symptomatically infected, referred to as infected $I_w(t)$, infected with microcephaly $I_{wm}(t)$, and recovered $R_w(t)$. The total vector population consists of three classes: susceptible $S_v(t)$, exposed $E_v(t)$, and infected $I_v(t)$.

Hence, the total newborn baby population is $N_b(t) = S_b(t) + E_b(t) + A_b(t) + I_b(t) + I_{bm}(t) + R_b(t)$, the total adult population is $N_w(t) = S_w(t) + E_w(t) + A_w(t) + I_w(t) + I_{wm}(t) + R_w(t)$, the total vector population is $N_v(t) = S_v(t) + E_v(t) + I_v(t)$, and the total human population is $N_h(t) = N_b(t) + N_w(t)$. The flow diagram of the compartment model is shown in Figure 1. Therefore, the ZIKV disease model with sexual transmission and vertical transmission is written by a system of differential equations under defined assumptions.

$$
\begin{aligned}
S_b'(t) =& \pi_b \left[(1-q_E)E_w(t) + (1-q_I)I_w(t) + (1-q_R)R_w(t)\right] \\
& - \lambda_b(I_v, N_b)S_b(t)(1-u_1(t)) - (\alpha+\mu_b)S_b(t) \\
E_b'(t) =& \pi_b q_E E_w(t) + p\lambda_b(I_v, N_b)S_b(t)(1-u_1(t)) \\
& - (\alpha+\sigma_b+\mu_b)E_b(t) \\
A_b'(t) =& (1-p)\lambda_b(I_v, N_b)S_b(t)(1-u_1(t)) - (\alpha+\gamma_b+\mu_b)A_b(t) \\
I_b'(t) =& \pi_b q_I I_w(t) + \sigma_b E_b(t) - (\alpha+\gamma_b+\mu_b)I_b(t) \\
I_{bm}'(t) =& (1-r)\pi_b q_R R_w(t) - (\alpha+\mu_b) \\
R_b'(t) =& r\pi_b q_R R_w(t) + \gamma_b(A_b(t)+I_b(t)) - (\alpha+\mu_b)R_b(t) \\
S_w'(t) =& \alpha S_b(t) - \lambda_w(I_v, N_w)S_w(t)(1-u_1(t)) \\
& - \lambda_s(E_w, I_w, N_w)S_w(t)(1-u_2(t)) - \mu_w S_w(t) \\
E_w'(t) =& pS_w(t)[\lambda_w(I_v, N_w)(1-u_1(t)) - \\
& \lambda_s(E_w, I_w, N_w)(1-u_2(t))] - (\sigma_w+\mu_w)E_w(t) \\
A_w'(t) =& (1-p)S_w(t)[\lambda_w(I_v, N_w)(1-u_1(t)) - \\
& \lambda_s(E_w, I_w, N_w)(1-u_2(t))] - (\gamma_w+\mu_w)A_w(t) \\
I_w'(t) =& \sigma_w E_w(t) - (\gamma_w+\mu_w)I_w(t) \\
I_{wm}'(t) =& \alpha I_{bm}(t) - \mu I_{wm}(t) \\
R_w'(t) =& \alpha R_b(t) + \gamma_w(A_w(t)+I_w(t)) - \mu_w R_w(t) \\
S_v'(t) =& \pi_v(1-u_3(t)) - \lambda_v(E_b, I_b, E_w, I_w, N_b, N_w)S_v(t)(1-u_1(t)) \\
& - (\mu_v + r_0 u_3(t))S_v(t) \\
E_v'(t) =& \lambda_v(E_b, I_b, E_w, I_w, N_b, N_w)S_v(t)(1-u_1(t)) - \\
& (\sigma_v + \mu_v + r_0 u_3(t))E_v(t) \\
I_v'(t) =& \sigma_v E_v(t) - (\mu_v + r_0 u_3(t))I_v(t)
\end{aligned}
\tag{1}
$$

where

$$
\begin{aligned}
&\lambda_b(I_v, N_b) = \frac{\eta\beta_b b I_v}{N_b}, \qquad \lambda_w(I_v, N_w) = \frac{\beta_w b I_v}{N_w}, \\
&\lambda_s(E_w, I_w, N_w) = \beta_s\left[\frac{I_w + \rho_s E_w}{N_w}\right], \\
&\lambda_v(E_b, I_b, E_w, I_w, N_b, N_w) = \beta_v b\left[\frac{I_w + \rho_w E_w + \eta(I_b + \rho_b Eb)}{N_w + N_b}\right]
\end{aligned}
$$

are the force of infection rates.

In the system, a susceptible adult can be infected either through the bite of an infected vector or sexual contact with an exposed or symptomatically infected adult. The infected class is more contagious due to the higher load of viremia [29], hence the infection modification parameters ρ_w, ρ_s are introduced for vector transmission and sexual transmission of an adult.

The newborn baby could be infected directly by mosquitoes or intrauterine via pregnant mother [30,31]. We assumed that the exposure rate of babies is different from adults, represented by the modification parameter $\eta > 0$. The infection modification ρ_b is introduced as well. In the vertical transmission in babies, it occurred in some cases that newborn baby is infected during pregnancy [31,32].

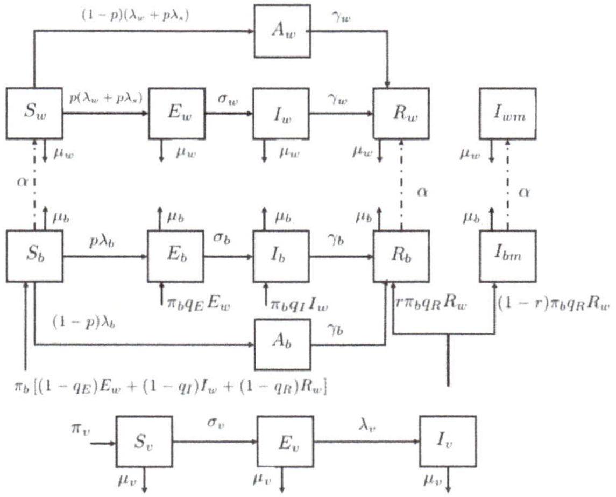

Figure 1. Flow diagram of the Zika virus model.

We assumed that babies are born with infection due to vertical transmission by the parameter $0 \leq q_E, q_I, q_R \leq 1$. Let the newborn birth rate be π_b, the fraction of healthy newborn babies are $\pi_b((1-q_E)E_w + (1-q_I)I_w + (1-q_R)R_w)$, and the infected babies are the remaining $\pi_b q_E E_w$, $\pi_b q_I Iw$, and $\pi_b q_R R_w$ fractions. A recovered mother might give birth to the baby with microcephaly [30], and the baby born with microcephaly is assumed by the fraction $(1-r)\pi_b q_R R_w$ in this paper.

Individuals with microcephaly experience a delay in development [33] and severe neurological disorders, so they are likely to have a short lifespan. Hence, we assumed that the adults with microcephaly do not reproduce or are involved in sexual transmission. The asymptomatically infected babies and adults are assumed to be noncontagious for both human and vector. Furthermore, the mortality rate from the disease is considered negligible. All parameter descriptions are summarized in Table 1.

Based on WHO [19] and Centers for Disease Control and Prevention (CDC) [20] suggestions, we considered vector, vertical, and sexual transmission the three control parameters selected for the model. The control variable $u_1(t)$ is the use of mosquito repellents to prevent or reduce the biting from mosquitoes. The control variable $u_2(t)$ is the procedure such as using protection for preventing pregnancy or unsafe sexual activity. The control variable $u_3(t)$ is the use of insecticide to control or eliminate the mosquito population. By applying control strategy, forces of infection in the adult and baby populations decrease by the fraction of $(1-u_1(t))$ and $(1-u_2(t))$. Correspondingly, the force of infection in vector population decreases by a fraction of $(1-u_1(t))$. Lastly, the recruitment rate of mosquito decreases by a fraction of $(1-u_3(t))$. It was also assumed that the mosquito death rate increases by $r_0 u_3(t)$, where $r_0 > 0$. Remark: If the value of each control parameters u_1, u_2, and u_3 is equal to zero for all given period, the model in the system in Equation (1) would be the Zika virus model without controls.

The controls u_1, u_2, and u_3 are determined to minimize the given objective function:

$$
\begin{aligned}
J(u_1, u_2, u_3) = \int_0^{t_f} \Big(& A_1 E_b + A_2 I_b + A_3 E_w + A_4 I_w + \\
& A_5 N_v + \frac{1}{2}(B_1 u_1^2 + B_2 u_2^2 + B_3 u_3^2) \Big) dt,
\end{aligned}
\tag{2}
$$

subject to the state of system in Equation (1).

Table 1. Definition of variables and parameters.

Variable	**Definition**
$S_b(t), S_w(t)$	Susceptible newborn babies and adults
$E_b(t), E_w(t)$	Exposed newborn babies and adults
$A_b(t), A_w(t)$	Asymptomatically infected newborn babies and adults
$I_b(t), I_w(t)$	Symptomatically infected newborn babies without microcephaly and symptomatically infected adults
$I_{bm}(t), I_{wm}(t)$	Newborn babies and adults with microcephaly
$R_b(t), R_w(t)$	Recovered newborn babies and adults
$S_v(t)$	Susceptible vectors
$E_v(t)$	Exposed vectors
$I_v(t)$	Infected vectors
Parameter	**Definition**
π_b	Birth rate of newborn babies
α	Maturity rate of babies
p	Fraction of symptomatic infection
$1-p$	Remaining fraction of symptomatic infection
q_E, q_I, q_R	Fraction of newborn babies who are infected from pregnant adult of each class
$1-r$	Fraction of newborn babies who are infected with microcephaly
β_b, β_w	Transmission probability per contact from infected vectors to susceptible newborn babies and adults
β_v	Transmission probability per contact from infected humans to susceptible vectors
β_s	Transmission probability per sexual contact from infected humans to susceptible humans
η	Exposure modification parameter in babies
ρ_b, ρ_w	Infectivity modification parameters in exposed babies and adults
ρ_s	Sexual infectivity modification parameters in exposed adults
σ_b, σ_w	Progression rate of exposed newborn babies and adults
γ_b, γ_w	Recovery rate of newborn babies and adults
μ_b, μ_w	Natural death rate of newborn babies and adults
π_v	Recruitment rate of mosquitoes
b	Biting rate of mosquitoes
σ_v	Progression rate of exposed mosquitoes
μ_v	Mortality rate of mosquitoes

In Section 4, we find a set of controls that minimize the number of exposed humans, symptomatically infected humans, all mosquitoes and the costs related to the controlling implementation. Firstly, we introduce constants A_1, A_2, A_3, A_4, and A_5 as the weighted constants related to the exposed newborn babies, symptomatically infected newborn babies, exposed adults, symptomatically infected adults, and all mosquitoes, respectively. The constants B_1, B_2, and B_3 are the weighted constants of the control variables u_1, u_2, and u_3, respectively. The terms $\frac{1}{2}B_1u_1^2$, $\frac{1}{2}B_2u_2^2$, and $\frac{1}{2}B_3u_3^2$ are the implementation costs of the three controls. The cost included in the first control might be the prices of using insect repellent, mosquito net, and herbal spray for instance. The cost included in the second control might be the prices of providing protections or warning leaflets about the safe sexual activity. The cost included in the last control could come from the expenses of using mosquito pesticides and the process implementation.

3. Differential Evolution

The differential evolution algorithm is a population-based optimization algorithm, introduced by Storn and Price [27,28]. It is also one of the Evolutionary algorithms developed to find the parameter values and optimize real parameters or functions. DE is suitable for various practical problems which are nonlinear, non-differentiable, and multi-dimensional. The process consists of four parts, namely initialization, mutation, crossover, and selection.

Suppose optimizing a problem consisting of D predicted parameters with N_p number of population. Then, a $D-$dimensional vector is solved by the following main steps until the best solution vector reaches the termination criteria.

Step 1. Initialization

The DE algorithm randomly selected a population of the parameter vectors $\{x_{1,i}^G, x_{2,i}^G, \ldots, x_{D,i}^G\}$, $i = 1, 2, \ldots, N_p\}$ where G is the generation number. Define the initial population as

$$x_{i,j}^G = x_j^L + \phi_i(x_j^U - x_j^L), \qquad j = 1, 2, \ldots, D$$

that is each parameter $x_{i,j}^G \in [x_j^L, x_j^U]$, where x_j^L is the lower bound and x_j^U is the upper bound of the $j-$th parameter and ϕ_i is a uniformly distributed random number between 0 and 1. Then, each of the N_p parameter vectors will undergo mutation, crossover, and selection.

Step 2. Mutation

This process helps expand the search space. For each target vector x_i^G, randomly select three vectors x_{r1}^G, x_{r2}^G, and x_{r3}^G such that each index i, $r1$, $r2$, and $r3$ is different. Then, a mutant vector v_i^G is generated by

$$v_i^G = x_{r1}^G + F(x_{r2}^G - x_{r3}^G)$$

where $F \in [0, 2]$ is the mutation vector and v_i^G is called the donor vector.

Step 3. Crossover

Crossover or recombination process involves successful solutions from the previous generation. First, a trial vectors u_i^{G+1} is produced from v_i^G and x_i^G by

$$u_{i,j}^G = \begin{cases} v_{i,j}^G & \text{if } rand_j \leq CR, \text{ or } j = I_{rand} \\ x_{i,j}^G & \text{if } rand_j > CR \text{ and } j \neq I_{rand} \end{cases}$$

where $rand_j \sim U[0, 1]$ for a comparison to the crossover rate $CR \in [0, 1]$, and I_{rand} is a random index in $\{1, \ldots, D\}$ that u_i^G gets at least one component.

Step 4. Selection

The next generation vectors are selected by comparing the target vector x_i^G and trial vector u_i^G and selecting the one with the lowest objective function value. The objective function is given by comparing the value of E_b, I_b, E_h, I_h, and N_v in the system in Equation (1) by applying vector u_i^G and x_i^G. Thus, the next generation vector x_i^{G+1} is replaced by u_i^G or x_i^G under the following condition,

$$x_i^{G+1} = \begin{cases} u_i^G & \text{if } J(u_i^G) < J(x_i^G), \\ x_i^G & \text{otherwise,} \end{cases}$$

where J is the objective function (see Section 4).

Step 5. Repeating

After obtaining a new target vector, the mutation, crossover, and selection steps are repeated until the termination criteria are met.

4. Numerical Simulations for Sub-Optimal Control Problem

4.1. Sub-Optimal Control Problem

Given the objective function as defined in Equation (2), we assume there exists a solution (u_1^*, u_2^*, u_3^*) of the optimal control problem characterized by the Pontryagins Maximum Principle [21].

Since the implemented result of the optimal control problem is complicated for the real situation, a sub-optimal control problem is proposed. The idea of sub-optimal control problem is based on the works of Rodrigues et. al. [16] and Moulay et al. [17], and the Model Predictive Control [22]. Regarding MPC, its strategy is based on the model of the system, a cost function which penalizes the undesirable behaviors, and constraints which represent the system limits. Furthermore, the stability and suboptimality of MPC can be estimated from the control horizon and the cost function [23,24]. For our nonlinear unconstrained model with a finite horizon, the continuous time horizon is discretized into N time horizon as

$$t_0 = 0 < t_1 < t_2 < \ldots < t_{N-1} = t_f.$$

Then, each of the controls $u_1(t)$, $u_2(t)$, and $u_3(t)$, in each time horizon, is approximated by a piece-wise constant control such that

$$\begin{aligned} u_1(t) &= u_{1i}, \text{ for } t_{i-1} \le t \le t_i,\ i = 1, \ldots, N \\ u_2(t) &= u_{2i}, \text{ for } t_{i-1} \le t \le t_i,\ i = 1, \ldots, N \\ u_3(t) &= u_{3i}, \text{ for } t_{i-1} \le t \le t_i,\ i = 1, \ldots, N \end{aligned} \tag{3}$$

where u_{1i}, u_{2i}, and u_{3i} are constants and are referred to as the control parameters. Thus, for our model with now N control parameters for each u_1, u_2, and u_3, we have total $3N$ parameters to be minimized. Let

$$z = [u_{11}, u_{21}, u_{31}, \ldots, u_{1N}, u_{2N}, u_{3N}] \in R^{3N}.$$

Then, the sub-optimal control problem is to

$$\min_z J(z) \tag{4}$$

subject to the system in Equation (1) and J is as defined in Equation (2).

4.2. Numerical Simulations

The simulations in this work are constructed using the parameter values in Table 2. Since the Zika virus is still under investigation and discovery, some parameter values are obtained from the previous study of other mosquito-borne viruses. The initial conditions are also selected for the theoretical sense and illustration purpose. The initial conditions for the system in Equation (1) are given by $S_b(0) = 10,000$, $E_b(0) = 100$, $A_b(0) = 100$, $I_b(0) = 40$, $I_{bm}(0) = 10$, $R_b(0) = 50$, $S_w(0) = 200,000$, $E_w(0) = 1000$, $A_w(0) = 1000$, $I_w(0) = 100$, $I_{wm}(0) = 50$, $R_w(0) = 1000$, $S_v(0) = 10,000$, $E_v(0) = 1000$, and $I_v(0) = 100$. The weight constants are given as $A_1 = 0.025$, $A_2 = 0.025$, $A_3 = 0.025$, $A_4 = 0.025$, $A_5 = 0.025$, $E_1 = 20$, $E_2 = 20$, and $E_3 = 20$. The total time span is 120 days, which is approximately four months.

From [24], the performance bound, which depends on the time horizon N and the cost function, could be computed to estimate the distance between MPC cost and the optimal cost. It also suggests that the time horizon N should not be too small. Hence, in our simulation with the total time of 120 days, we consider two scenarios with the different time horizon N as follows: (i) with $N = 4$; and (ii) with $N = 8$. All controls must be non-negative and let the maximum value be 0.9 since it is almost impossible to fully implement each control strategy. The numerical simulation from each scenario is generated with the DE algorithm with the maximum iteration of 150, $F = 0.850$, and $CR = 1.000$. Number of population N_p is 120 and 240 for Scenarios $N = 4$ and $N = 8$, respectively.

Table 2. Parameter values.

Parameter	Value	References
π_b	$\frac{1}{15\times365}$	[34]
α	$\frac{1}{16\times365}$	[13]
p	0.5	[13]
q_E, q_I, q_R	0.5	[13]
r	0.5	[13]
β_b, β_w	0.33	[13,35]
β_v	0.5	[12]
β_s	0.05	[12]
η	0.5	Assumed
ρ_b, ρ_w	0.5	[13]
ρ_s	0.6	[12]
σ_b, σ_w	$\frac{1}{7.5}$	[36]
γ_b, γ_w	$\frac{1}{8.5}$	[36]
μ_b	$\frac{1}{18.60\times365}$	[3]
μ_w	$\frac{1}{70\times365}$	[35]
π_v	500	[36]
b	0.5	[35,37,38]
σ_v	$\frac{1}{10}$	[37]
μ_v	$\frac{1}{21}$	[38]
r_0	0.1	[18]

(i) Scenario $N = 4$

In this scenario, the period is 30 days or one month for each time horizon. Figure 2 represents the simulation of three controls over 120 days. From the graph, the levels of u_1 and u_2 are at the highest around 0.36 and 0.37 in the first month, respectively. After that, the levels of both controls u_1 and u_2 discretely drop in the second month and become approximately zero in the third month. Differently, the level of control u_3 stays at the maximum value 0.9 during the entire time.

The objective function value is 2.293810×10^5, where the control values are:

$$u_1 = [0.379390 \quad 0.01558 \quad 0.00038 \quad 0.00007]$$
$$u_2 = [0.365622 \quad 0.08256 \quad 0.01328 \quad 0.00136]$$
$$u_3 = [0.900000 \quad 0.90000 \quad 0.90000 \quad 0.90000].$$

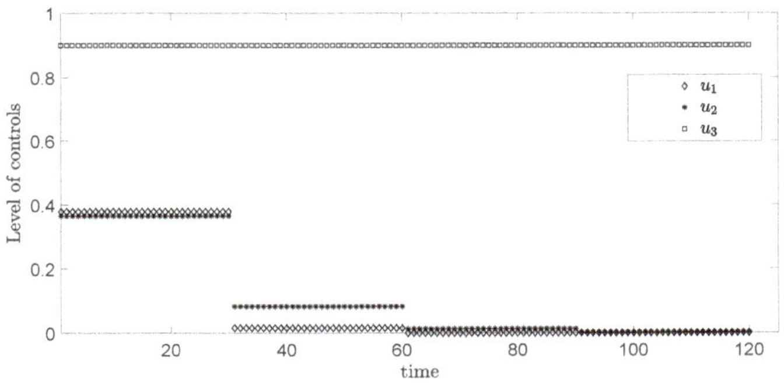

Figure 2. Level of controls from Scenario (i) $N = 4$.

(ii) Scenario $N = 8$

In this scenario, the period is 15 days or approximately two weeks for each time horizon. Figure 3 represents the simulation of three controls over 120 days. From the graph, u_1 is at the highest level around 0.6 in the first two weeks and discretely drops to approximately zero within one month. For u_2, it reaches around 0.45 for the first two weeks, then drops to nearly zero after the second month. Differently, the level of control u_3 stays at the maximum value 0.9 during the entire time.

The objective function value is 2.293640×10^5, where the control values are:

$$\begin{aligned} u_1 =&[0.59757 && 0.17349 && 0.02167 && 0.01069 \\ & 0.00873 && 0.00128 && 0.00201 && 0.00539] \\ u_2 =&[0.45277 && 0.25743 && 0.12570 && 0.04461 \\ & 0.01469 && 0.00008 && 0.01466 && 0.00918] \\ u_3 =&[0.89999 && 0.89998 && 0.89999 && 0.89999 \\ & 0.89997 && 0.89999 && 0.89999 && 0.89998]. \end{aligned}$$

The simulations in Figures 4–9 represent the number of exposed babies, infected babies, exposed adults, infected adults, exposed vectors and infected vectors, respectively. In each graph, the dotted line represents the model without control, the solid line represents the control model with Scenario $N = 4$, and the dashed line represents the control model with Scenario $N = 8$. In Figures 4–9, the numbers of exposed and infected individuals from the model with controls are less than those of the model without control. This implies that control strategy ceases the serve of the disease. The models from both scenarios yield very close results where $N = 8$ is slightly better. Notice in Figures 4–7 that the disease spreads heavily during the first month. Later, those patients decrease and become stable after two months, corresponding to the drop in control u_1 and u_2 after two months. Thus, both u_1 and u_2 contribute to the control of the disease. The control u_3 contributes the most and becomes the top priority in the strategy.

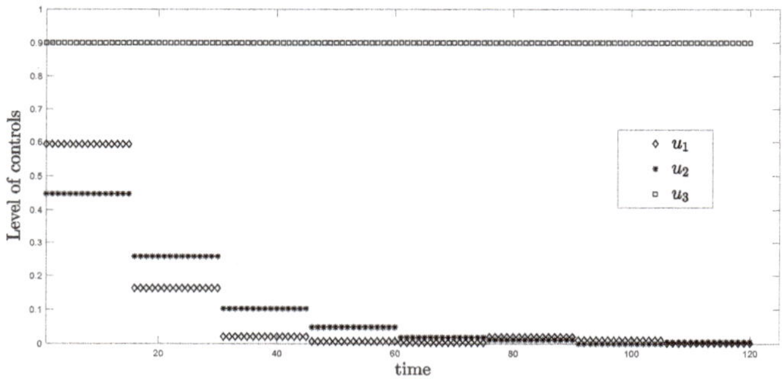

Figure 3. Level of controls from Scenario (ii) $N = 8$.

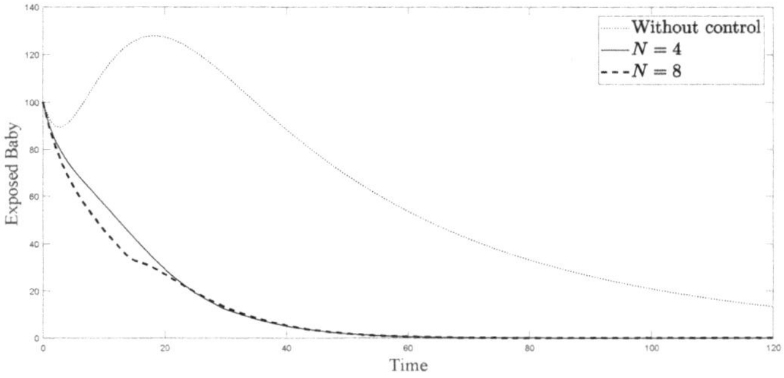

Figure 4. Simulation results of exposed newborn babies from: (i) the model without controls; (ii) the control model with $N = 4$; and (iii) the control model with $N = 8$.

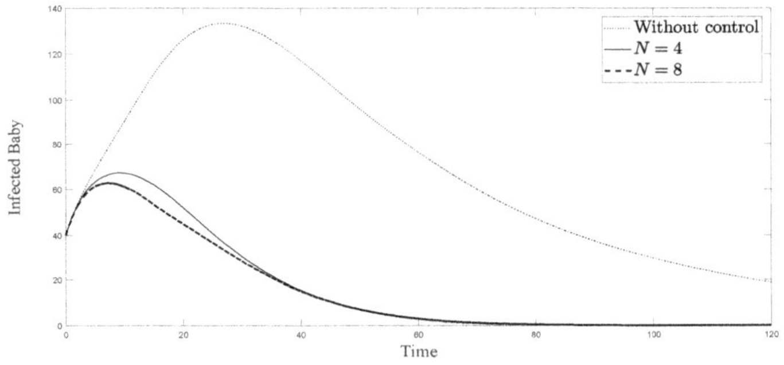

Figure 5. Simulation results of symptomatically infected newborn babies from: (i) the model without controls; (ii) the control model with $N = 4$; and (iii) the control model with $N = 8$.

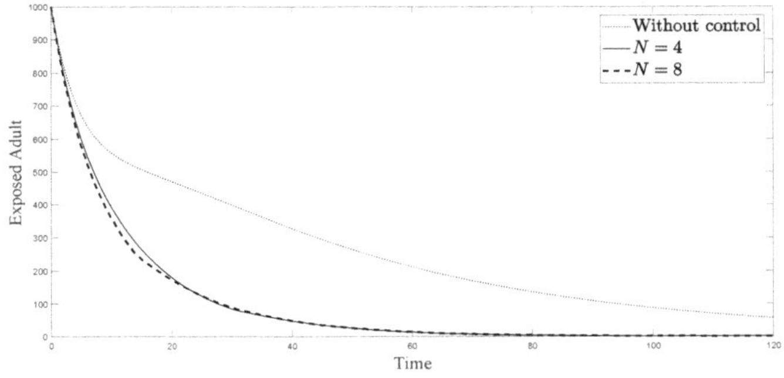

Figure 6. Simulation results of exposed adults from: (i) the model without controls; (ii) the control model with $N = 4$; and (iii) the control model with $N = 8$.

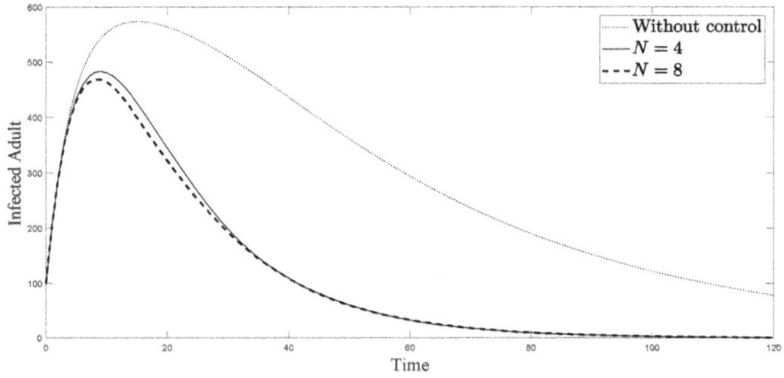

Figure 7. Simulation results of symptomatically infected adults from: (i) the model without controls; (ii) the control model with $N = 4$; and (iii) the control model with $N = 8$.

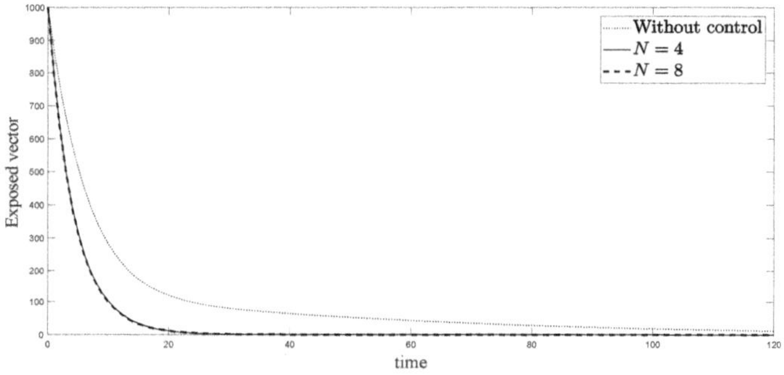

Figure 8. Simulation results of exposed vectors: (i) the model without controls; (ii) the control model with $N = 4$; and (iii) the control model with $N = 8$.

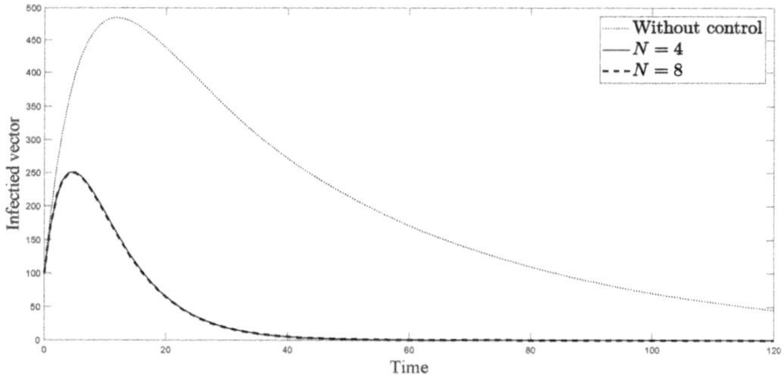

Figure 9. Simulation results of infected vectors from: (i) the model without controls; (ii) the control model with $N = 4$; and (iii) the control model with $N = 8$.

5. Conclusions

In this work, we developed the dynamical model of ZIKV disease with sexual transmission and vertical transmission. We proposed sub-optimal control problem with three control parameters: vector elimination, human–vector reduction, and human–human contact reduction. To find the control solution which is easy to implement, we partitioned the controls into discrete intervals. By using Differential Evolution, we solved and presented the numerical scenarios with four time horizons and eight time horizons. The simulations of both controls efficiently reduce the number of exposed and infected patients where the model with eight intervals yields a slightly better result. Controls u_1, using repellents to prevent mosquito biting, and u_2, avoiding pregnancy and unsafe sexual activity, should be applied mostly during the peak of the disease. After the disease dies out or becomes stable, the level of both controls will decrease. Mosquito elimination should be the first focus in controlling the disease, corresponding to the primary procedure announced by general health authorities. However, the cost associated with each control in this simulation is set to be equal. The consideration of adjusting the relative cost according to alternative scenarios might provide different results. Additionally, if more information is available in the future, the simulations should be conducted again using more precise data.

Author Contributions: Conceptualization, W.S.; Investigation, N.C. and W.S.; Methodology, W.S.; Visualization, N.C.; Writing—original draft, N.C.; and Writing—review and editing, W.S.

Funding: This research received no external funding.

Acknowledgments: This research was supported by Centre of Excellence in Mathematics, PERDO, CHE, Thailand and Golden Jubilee Scholarship from Faculty of science, Mahidol university, Thailand.

Conflicts of Interest: The authors declare no conflict of interest.

Abbreviations

The following abbreviations are used in this manuscript:

WHO	World Health Organization
GA	Genetic Algorithm
DE	Differential Evolution

References

1. Dasti, J.I. Zika virus infections: An overview of current scenario. *Asian Pac. J. Trop. Med.* **2016**, *7* , 621–625. [CrossRef] [PubMed]
2. European Centre for Disease Prevention and Control. Zika Virus Epidemic in the Americas: Potential Association with Microcephaly and Guillain-Barre Syndrome. Available online: https://ecdc.europa.eu/sites/portal/files/media/en/publications/Publications/zika-virus-americas-association-with-microcephaly-rapid-risk-assessment.pdf (accessed on 12 February 2017) .
3. World Health Organization. Microcephaly-Brazil: Disease Outbreak News. Available online: http://www.who.int/csr/don/archive/disease/microcephaly/en/ (accessed on 12 February 2017).
4. World Health Organization. Zika Situation Report. Available online: http://www.who.int/emergencies/zika-virus/situation-report/12-february-2016/en/ (accessed on 12 February 2017).
5. Coelho, F.C.; Durovni, B.; Saraceni, V.; Lemos, C.; Codeco, C.T.; Camargo, S.; de Carvalho, L.M.; Bastos, L.; Arduini, D.; Villela, D.A.; et al. Higher incidence of Zika in adult women than adult men in Rio de Janeiro suggests a significant contribution of sexual transmission from men to women. *Int. J. Infect. Dis.* **2016**, *51*, 128–132. [CrossRef] [PubMed]
6. Atkinson, B.; Hearn, P.; Afrough, B.; Lumley, S.; Carter, D.; Aarons, E.J.; Simpson, A.J.; Brooks, T.J.; Hewson, R. Detection of Zika virus in semen. *Emerg. Infect. Dis.* **2016**, *22*, 940. [CrossRef] [PubMed]
7. Musso, D.; Roche, C.; Robin, E.; Nhan, T.; Teissier, A.; Cao-Lormeau, V.-M. Potential sexual transmission of Zika virus. *Emerg. Infect. Dis.* **2015**, *21*, 359–361. [CrossRef] [PubMed]
8. Foy, B.D.; Kobylinski, K.C.; Foy, J.L.; Blitvich, B.J.; da Rosa, A.T.; Haddow, A.D.; Lanciotti, R.S.; Tesh, R.B. Probable non-vector-borne transmission of Zika virus. *Emerg. Infect. Dis.* **2011**, *17*, 880–882. [CrossRef] [PubMed]
9. Chalishajar, D. Mathematical Analysis of Insulin-glucose feedback system of Diabetes. *J. Eng. Appl. Sci.* **2014**, *5*, 36–58.
10. Chalishajar, D.; Geary, D.; Cox, G. Review Study of Detection of Diabetes Models through Delay Differential Equations. *Appl. Math.* **2016**, *7*, 1087–1102. [CrossRef]

11. Kucharski, A.J.; Funk, S.; Eggo, R.M.; Mallet, H.-P.; Edmunds, W.J.; Nilles, E.J. Transmission dynamics of Zika virus in island populations: A modelling analysis of the 2013-14 French Polynesia outbreak. *PLoS Negl. Trop. Dis.* **2016**, *10*, 0004726. [CrossRef] [PubMed]
12. Gao, D.; Lou, Y.; He, D.; Porco, T.C.; Kuang, Y.; Chowell, G.; Ruan, S. Prevention and control of Zika as a mosquito-borne and sexually transmitted disease: a mathematical modeling analysis. *Sci. Rep.* **2016**, *6*, 28070. [CrossRef] [PubMed]
13. Agusto, F.B.; Bewick, S.; Fagan, W.F. Mathematical model of Zika virus with vertical transmission. *Infect. Dis. Model.* **2017**, *2*, 244–267. [CrossRef] [PubMed]
14. Lashari, A.A.; Zaman, G. Optimal control of a vector borne disease with horizontal transmission. *Nonlinear Anal.* **2012**, *13*, 203–212. [CrossRef]
15. Yan, X.; Zou, Y.; Li, J. Optimal quarantine and isolation strategies in epidemics control. *World J. Model. Simul.* **2007**, *3*, 202–211.
16. Rodrigues, H.S.; Monteiro, M.T.T.; Torres, D.F. Vaccination models and optimal control strategies to dengue. *Math. Biosci.* **2014**, *247*, 1–12. [CrossRef] [PubMed]
17. Moulay, D.; Aziz-Alaoui, M.; Kwon, H.-D. Optimal control of chikungunya disease: Larvae reduction, treatment and prevention. *Math. Biosci. Eng.* **2012**, *9*, 369–392. [CrossRef] [PubMed]
18. Chaikham, N.; Sawangtong, W. Optimal control of Zika virus infection by vector elimination, vector-to-human and human-to-human contact reduction. *Adv. Differ. Equ.* **2017**, *2017*, 177. [CrossRef]
19. World Health Organization. Mosquito Control: Can It Stop Zika at Source? Available online: http://www.who.int/emergencies/zika-virus/articles/mosquito-control/en/ (accessed on 12 Frebruary 2017).
20. Centers for Disease Control and Prevention (CDC), Update: Interim Guidelines for Prevention of Sexual Transmission of Zika Virus-United States. 2016. Available online: https://emergency.cdc.gov/han/han00388.asp (accessed on 5 July 2017) .
21. Lenhart, S.; Workman, J.T. Basic Optimal Control Problems. In *Optimal Control Applied to Biological Models*; CRC Press: Boca Raton, FL, USA, 2007; pp. 1–18.
22. Camacho, E.F.; Alba, C.B. Introduction to Model Predictive Control. In *Model Predictive Control*; Springer Science & Business Media: Berlin/Heidelberg, Germany, 2007; pp. 1–30.
23. Grüne, L. Analysis and design of unconstrained nonlinear MPC schemes for finite and infinite dimensional systems. *SIAM J. Control Optim.* **2009**,*48*, 1206–1228. [CrossRef]
24. Herty, M.; Zanella, M. Performance bounds for the mean-field limit of constrained dynamics. *arXiv* **2015**, arXiv:1511.08364.
25. Caetano, M.A.L.; Yoneyama, T. Optimal and sub-optimal control in Dengue epidemics. *Optim. Control Appl. Methods* **2011**, *22*, 63–73. [CrossRef]
26. Yan, X.; Zou, Y. Optimal and sub-optimal quarantine and isolation control in SARS epidemics. *Math. Comput. Model.* **2008**, *47*, 235–245. [CrossRef]
27. Price, K.; Storn, R.; Lampinen, J. The Differential Evolution Algorithm. In *Differential Evolution : A Practical Approach to Global Optimization*; Springer: New York, NY, USA, 2005; pp. 37–134, ISBN 978-3-540-31306-9.
28. Storn, R.; Price, K. Differential evolution—A simple and efficient heuristic for global optimization over continuous spaces. *J. Glob. Optim.* **1997**, *11*, 341–359. [CrossRef]
29. Hills, S.L. Transmission of Zika virus through sexual contact with travelers to areas of ongoing transmission-continental United States, 2016. *Morb. Mortal. Wkly. Rep.* **2016**, *65*, 215–216. [CrossRef] [PubMed]
30. Moore, C.A.; Staples, J.E.; Dobyns, W.B.; Pessoa, A.; Ventura, C.V.; Fonseca, E.B.D.; Ribeiro, E.M.; Ventura, L.O.; Neto, N.N.; Arena, J.E.; et al. Characterizing the Pattern of Anomalies in Congenital Zika Syndrome for Pediatric Clinicians. *JAMA Pediatr.* **2017**, *171*, 288–295. [CrossRef] [PubMed]
31. Schuler-Faccini, L.; Ribeiro, E.M.; Feitosa, I.M.; Horovitz, D.D.; Cavalcanti, D.P.; Pessoa, A.; Doriqui, M.J.; Neri, J.I.; Neto, J.M.; Wanderley, H.Y.; et al. Possible Association Between Zika Virus Infection and Microcephaly—Brazil, 2015. *Morb. Mortal. Wkly. Rep.* **2016**, *65*, 59–62. [CrossRef] [PubMed]
32. Besnard, M.; Lastere, S.; Teissier, A.; Cao-Lormeau, V.M.; Musso, D. Evidence of perinatal transmission of Zika virus, French Polynesia, December 2013 and February 2014. *Euro Surveill.* **2014**, *19*, 20751. [CrossRef] [PubMed]

33. Carter, M.T.; Mirzaa, G.; McDonell, L.M.; Boycott, K.M. Microcephaly-Capillary Malformation Syndrome. 2013 December 12. *GeneReviews*®, Seattle (WA): University of Washington, Seattle, 1993–2018. Available online: https://www.ncbi.nlm.nih.gov/books/NBK174452/ (accessed on 5 July 2017).
34. Wikipedia. List of Sovereign States and Dependent Territories by Birth Rate. Available online: https://en.wikipedia.org/wiki/List_of_sovereign_states_and_dependencies_by_total_fertility_rate (accessed on 1 February 2017) .
35. Manore, C.; Hickmann, J.; Xu, S.; Wearing, H.; Hyman, J. Comparing dengue and chikungunya emergence and endemic transmission in a. aegypti and a. albopictus. *J. Theor. Biol.* **2014**, *356*, 174–191. [CrossRef] [PubMed]
36. Bewick, S.; Fagan, W.; Calabrese, J.; Agusto, F. Zika virus: Endemic versus epidemic dynamics and implications for disease spread in the Americas. *bioRxiv* **2016**, doi:10.1101/041897. [CrossRef]
37. Andraud, M.; Hens, N.; Marais, C.; Beutels, P. Dynamic epidemiological models for dengue transmission: A systematic review of structural approaches. *PLoS ONE* **2012**, *7*, 49085. [CrossRef] [PubMed]
38. Trpis, M.; Haussermann, W. Dispersal and other population parameters of aedes aegypti in an african village and their possible significance in epidemiology of vector-borne diseases. *Am. J. Trop. Med. Hyg.* **1986**, *35*, 1263–1279. [CrossRef] [PubMed]

axioms

MDPI

Article

Some Identities for Euler and Bernoulli Polynomials and Their Zeros

Taekyun Kim [1] and Cheon Seoung Ryoo [2,*]

[1] Department of Mathematics, Kwangwoon University, Seoul 139-701, Korea; tkkim@kw.ac.kr
[2] Department of Mathematics, Hannam University, Daejeon 306-791, Korea
* Correspondence: ryoocs@hnu.kr; Tel.: +82-42-629-7525

Received: 30 June 2018; Accepted: 11 August 2018; Published: 14 August 2018

Abstract: In this paper, we study some special polynomials which are related to Euler and Bernoulli polynomials. In addition, we give some identities for these polynomials. Finally, we investigate the zeros of these polynomials by using the computer.

Keywords: Appell sequence; Appell numbers and polynomials; Bernoulli and Euler polynomials; cosine–Bernoulli and cosine–Euler polynomials; sine–Bernoulli and sine–Euler polynomials

MSC: 11B68; 11S40; 11S80

---

1. Introduction

Many mathematicians have studied in the area of the Bernoulli numbers and polynomials, Euler numbers and polynomials, Genocchi numbers and polynomials, and tangent numbers and polynomials. The class of Appell polynomial sequences is one of the important classes of polynomial sequences. The Appell polynomial sequences arise in numerous problems of applied mathematics, mathematical physics and several other mathematical branches (see [1–14]). The Appell polynomials can be defined by considering the following generating function:

$$\begin{aligned} A(t)e^{xt} &= A_0(x) + A_1(x)\frac{t}{1!} + A_2(x)\frac{t^2}{2!} + \cdots + A_n(x)\frac{t^n}{n!} + \cdots \\ &= \sum_{n=0}^{\infty} A_n(x)\frac{t^n}{n!}, \text{ (see [5,7,8])}, \end{aligned} \tag{1}$$

where

$$A(t) = A_0 + A_1\frac{t}{1!} + A_2\frac{t^2}{2!} + \cdots + A_n\frac{t^n}{n!} + \cdots, A_0 \neq 0.$$

Alternatively, the sequence $A_n(x)$ is Appell sequence for $(g(t), t)$ if and only if

$$\frac{1}{g(t)}e^{xt} = \sum_{n=0}^{\infty} A_n(x)\frac{t^n}{n!}, \text{ (see [5,7,8])},$$

where

$$g(t) = \sum_{n=0}^{\infty} g_n\frac{t^n}{n!}, g_0 \neq 0.$$

Differentiating generating Equation (1) with respect to x and equating coefficients of $\frac{t^n}{n!}$, we have

$$\frac{d}{dx}A_n(x) = nA_{n-1}(x), n = 0, 1, 2, 3, \cdots.$$

The typical examples of Appell polynomials are the Bernoulli and Euler polynomials (see [1–14]). It is well known that the Bernoulli polynomials are defined by the generating function to be

$$\frac{t}{e^t-1}e^{xt}=\sum_{n=0}^{\infty}B_n(x)\frac{t^n}{n!}. \tag{2}$$

When $x=0, B_n=B_n(0)$ are called the Bernoulli numbers. The Euler polynomials are given by the generating function to be

$$\frac{2}{e^t+1}e^{xt}=\sum_{n=0}^{\infty}E_n(x)\frac{t^n}{n!}. \tag{3}$$

When $x=0, E_n=E_n(0)$ are called the Euler numbers.

The Bernoulli polynomials $\mathbf{B}_n^{(r)}(x)$ of order r are defined by the following generating function

$$\left(\frac{t}{e^t-1}\right)^r e^{xt}=\sum_{n=0}^{\infty}\mathbf{B}_n^{(r)}(x)\frac{t^n}{n!}, \quad (|t|<2\pi). \tag{4}$$

The Frobenius–Euler polynomials of order r, denoted by $\mathbf{H}_n^{(r)}(u,x)$, are defined as

$$\left(\frac{1-u}{e^t-u}\right)^r e^{xt}=\sum_{n=0}^{\infty}\mathbf{H}_n^{(r)}(u,x)\frac{t^n}{n!}. \tag{5}$$

The values at $x=0$ are called Frobenius–Euler numbers of order r; when $r=1$, the polynomials or numbers are called ordinary Frobenius–Euler polynomials or numbers.

In this paper, we study some special polynomials which are related to Euler and Bernoulli polynomials. In addition, we give some identities for these polynomials. Finally, we investigate the zeros of these polynomials by using the computer.

2. Cosine–Bernoulli, Sine–Bernoulli, Cosine–Euler and Sine–Euler Polynomials

In this section, we define the cosine–Bernoulli, sine–Bernoulli, cosine–Euler and sine–Euler polynomials. Now, we consider the Euler polynomials that are given by the generating function to be

$$\frac{2}{e^t+1}e^{(x+iy)t}=\sum_{n=0}^{\infty}E_n(x+iy)\frac{t^n}{n!}. \tag{6}$$

On the other hand, we observe that

$$e^{(x+iy)t}=e^{xt}e^{iyt}=e^{xt}(\cos yt+i\sin yt). \tag{7}$$

From Equations (6) and (7), we have

$$\sum_{n=0}^{\infty}E_n(x+iy)\frac{t^n}{n!}=\frac{2}{e^t+1}e^{(x+iy)t}=\frac{2}{e^t+1}e^{xt}(\cos yt+i\sin yt), \tag{8}$$

and

$$\sum_{n=0}^{\infty}E_n(x-iy)\frac{t^n}{n!}=\frac{2}{e^t+1}e^{(x-iy)t}=\frac{2}{e^t+1}e^{xt}(\cos yt-i\sin yt). \tag{9}$$

Thus, by (8) and (9), we can derive

$$\frac{2}{e^t+1}e^{xt}\cos yt=\sum_{n=0}^{\infty}\left(\frac{E_n(x+iy)+E_n(x-iy)}{2}\right)\frac{t^n}{n!}, \tag{10}$$

and

$$\frac{2}{e^t+1}e^{xt}\sin yt = \sum_{n=0}^{\infty}\left(\frac{E_n(x+iy)+E_n(x-iy)}{2i}\right)\frac{t^n}{n!}. \tag{11}$$

It follows that we define the following cosine–Euler polynomials and sine–Euler polynomials.

Definition 1. *The cosine–Euler polynomials $E_n^{(C)}(x,y)$ and sine–Euler polynomials $E_n^{(S)}(x,y)$ are defined by means of the generating functions*

$$\sum_{n=0}^{\infty} E_n^{(C)}(x,y)\frac{t^n}{n!} = \frac{2}{e^t+1}e^{xt}\cos yt, \tag{12}$$

and

$$\sum_{n=0}^{\infty} E_n^{(S)}(x,y)\frac{t^n}{n!} = \frac{2}{e^t+1}e^{xt}\sin yt, \tag{13}$$

respectively.

Note that $E_n^{(C)}(x,0) = E_n(x), E_n^{(S)}(x,0) = 0, (n \geq 0)$. The cosine–Euler and sine–Euler polynomials can be determined explicitly. A few of them are

$$\begin{aligned}
&E_0^{(C)}(x,y) = 1, \quad E_1^{(C)}(x,y) = -\frac{1}{2}+x,\\
&E_2^{(C)}(x,y) = -x+x^2-y^2,\\
&E_3^{(C)}(x,y) = \frac{1}{4}-\frac{3x^2}{2}+x^3+\frac{3y^2}{2}-3xy^2,\\
&E_4^{(C)}(x,y) = x-2x^3+x^4+6xy^2-6x^2y^2+y^4,
\end{aligned}$$

and

$$\begin{aligned}
&E_0^{(S)}(x,y) = 0, \quad E_1^{(S)}(x,y) = y,\\
&E_2^{(S)}(x,y) = -y+2xy,\\
&E_3^{(S)}(x,y) = -3xy+3x^2y-y^3,\\
&E_4^{(S)}(x,y) = y-6x^2y+4x^3y+2y^3-4xy^3.
\end{aligned}$$

By (10)–(13), we have

$$\begin{aligned}
E_n^{(C)}(x,y) &= \frac{E_n(x+iy)+E_n(x-iy)}{2},\\
E_n^{(S)}(x,y) &= \frac{E_n(x+iy)-E_n(x-iy)}{2i}.
\end{aligned}$$

Clearly, we can get the following explicit representations of $E_n(x+iy)$

$$\begin{aligned}
E_n(x+iy) &= \sum_{k=0}^{n}\binom{n}{k}(x+iy)^{n-k}E_k,\\
E_n(x+iy) &= \sum_{k=0}^{n}\binom{n}{k}(iy)^{n-k}E_k(x).
\end{aligned}$$

Let

$$e^{xt}\cos yt = \sum_{k=0}^{\infty} C_k(x,y)\frac{t^k}{k!}, \qquad e^{xt}\sin yt = \sum_{k=0}^{\infty} S_k(x,y)\frac{t^k}{k!}. \tag{14}$$

Then, by Taylor expansions of $e^{xt}\cos yt$ and $e^{xt}\sin yt$, we get

$$e^{xt}\cos yt = \sum_{k=0}^{\infty}\left(\sum_{m=0}^{[\frac{k}{2}]}\binom{k}{2m}(-1)^m x^{k-2m}y^{2m}\right)\frac{t^k}{k!} \tag{15}$$

and

$$e^{xt}\sin yt = \sum_{k=0}^{\infty}\left(\sum_{m=0}^{[\frac{k-1}{2}]}\binom{k}{2m+1}(-1)^m x^{k-2m-1}y^{2m+1}\right)\frac{t^k}{k!}, \tag{16}$$

where [ ] denotes taking the integer part. By (14)–(16), we get

$$C_k(x,y) = \sum_{m=0}^{[\frac{k}{2}]}\binom{k}{2m}(-1)^m x^{k-2m}y^{2m},$$

and

$$S_k(x,y) = \sum_{m=0}^{[\frac{k-1}{2}]}\binom{k}{2m+1}(-1)^m x^{k-2m-1}y^{2m+1}, (k \geq 0).$$

The two polynomials can be determined explicitly. A few of them are

$$\begin{aligned}
&C_0(x,y) = 1, \quad C_1(x,y) = x, \quad C_2(x,y) = x^2 - y^2,\\
&C_3(x,y) = x^3 - 3xy^2, \quad C_4(x,y) = x^4 - 6x^2y^2 + y^4,\\
&C_5(x,y) = x^5 - 10x^3y^2 + 5xy^4, \quad C_6(x,y) = x^6 - 15x^4y^2 + 15x^2y^4 - y^6,
\end{aligned}$$

and

$$\begin{aligned}
&S_0(x,y) = 0, \quad S_1(x,y) = y, \quad S_2(x,y) = 2xy,\\
&S_3(x,y) = 3x^2y - y^3, \quad S_4(x,y) = 4x^3y - 4xy^3,\\
&S_5(x,y) = 5x^4y - 10x^2y^3 + y^5, \quad S_6(x,y) = 6x^5y - 20x^3y^3 + 6xy^5.
\end{aligned}$$

Now, we observe that

$$\begin{aligned}
\frac{2}{e^t+1}e^{xt}\cos yt &= \left(\sum_{l=0}^{\infty}E_l\frac{t^l}{l!}\right)\left(\sum_{m=0}^{\infty}C_m(x,y)\frac{t^m}{m!}\right)\\
&= \sum_{n=0}^{\infty}\left(\sum_{l=0}^{n}\binom{n}{l}E_lC_{n-l}(x,y)\right)\frac{t^n}{n!}.
\end{aligned} \tag{17}$$

Therefore, we obtain the following theorem:

Theorem 1. *For $n \geq 0$, we have*

$$E_n^{(C)}(x,y) = \sum_{l=0}^{n}\binom{n}{l}E_lC_{n-l}(x,y)$$

and

$$E_n^{(S)}(x,y) = \sum_{l=0}^{n}\binom{n}{l}E_lS_{n-l}(x,y).$$

From (12), we have

$$
\begin{aligned}
2e^{xt}\cos yt &= \left(\sum_{l=0}^{\infty} E_l^{(C)}(x,y)\frac{t^l}{l!}\right)(e^t+1) \\
&= \sum_{n=0}^{\infty}\left(\sum_{l=0}^{n}\binom{n}{l}E_l^{(C)}(x,y)+E_n^{(C)}(x,y)\right)\frac{t^n}{n!}.
\end{aligned} \tag{18}
$$

By (14) and (18), we get

$$
C_n(x,y)=\frac{1}{2}\left(\sum_{l=0}^{n}\binom{n}{l}E_l^{(C)}(x,y)+E_n^{(C)}(x,y)\right). \tag{19}
$$

Therefore, we obtain the following theorem:

Theorem 2. *For $n \geq 0$, we have*

$$
C_n(x,y)=\frac{1}{2}\left(\sum_{l=0}^{n}\binom{n}{l}E_l^{(C)}(x,y)+E_n^{(C)}(x,y)\right),
$$

and

$$
S_n(x,y)=\frac{1}{2}\left(\sum_{l=0}^{n}\binom{n}{l}E_l^{(S)}(x,y)+E_n^{(S)}(x,y)\right).
$$

From (12), we note that

$$
\begin{aligned}
\sum_{n=0}^{\infty} E_n^{(C)}(1-x,y)\frac{t^n}{n!} &= \frac{2}{e^t+1}e^{(1-x)t}\cos yt \\
&= \frac{2}{e^{-t}+1}e^{-xt}\cos(-yt) \\
&= \left(\sum_{l=0}^{\infty}(-1)^l E_l\frac{t^l}{l!}\right)\left(\sum_{m=0}^{\infty}(-1)^m C_{m,}(x,y)\frac{t^m}{m!}\right) \\
&= \sum_{n=0}^{\infty}\left(\sum_{l=0}^{n}\binom{n}{l}E_l C_{n-l}(x,y)\right)\frac{(-1)^n}{n!}t^n.
\end{aligned} \tag{20}
$$

Therefore, we obtain the following theorem:

Theorem 3. *For $n \geq 0$, we have*

$$
\begin{aligned}
E_n^{(C)}(1-x,y) &= (-1)^n\sum_{l=0}^{n}\binom{n}{l}E_l C_{n-l}(x,y) \\
&= (-1)^n E_n^{(C)}(x,y),
\end{aligned}
$$

and

$$
\begin{aligned}
E_n^{(S)}(1-x,y) &= (-1)^{n+1}E_n^{(S)}(x,y) \\
&= (-1)^{n+1}\sum_{l=0}^{n}\binom{n}{l}E_l S_{n-l}(x,y).
\end{aligned}
$$

Now, we observe that

$$\begin{aligned}\sum_{n=0}^{\infty} E_n^{(C)}(x+1,y)\frac{t^n}{n!} &= \frac{2}{e^t+1}e^{(x+1)t}\cos yt \\ &= \frac{2}{e^t+1}e^{xt}(e^t-1+1)\cos yt \\ &= 2e^{xt}\cos yt - \frac{2}{e^t+1}e^{xt}\cos yt \\ &= \sum_{n=0}^{\infty}\left(2C_n(x,y) - E_n^{(C)}(x,y)\right)\frac{t^n}{n!}.\end{aligned} \tag{21}$$

By comparing the coefficients on the both sides, we get

$$E_n^{(C)}(x+1,q) + E_n^{(C)}(x,y) = 2C_n(x,y),\ (n \geq 0). \tag{22}$$

Therefore, we obtain the following theorem:

Theorem 4. *For $n \geq 0$, we have*

$$E_n^{(C)}(x+1,y) + E_n^{(C)}(x,y) = 2C_n(x,y),$$

and

$$E_n^{(S)}(x+1,y) + E_n^{(S)}(x,y) = 2S_n(x,y).$$

From (14) and (15), we have

$$\sum_{k=0}^{\infty} C_k(0,y)\frac{t^k}{k!} = \sum_{m=0}^{\infty}(-1)^m y^{2m}\frac{t^{2m}}{(2m)!}. \tag{23}$$

Therefore, by Theorem 4 and (23), we obtain the following corollary:

Corollary 1. *For $n \geq 0$, we have*

$$E_{2n}^{(C)}(1,y) + E_{2n}^{(C)}(0,y) = 2(-1)^n y^{2n},$$

and

$$E_{2n+1}^{(S)}(1,y) + E_{2n+1}^{(S)}(0,y) = 2(-1)^n y^{2n+1}.$$

By (12), we get

$$\begin{aligned}\sum_{n=0}^{\infty} E_n^{(C)}(x+r,y)\frac{t^n}{n!} &= \left(\frac{2e^{xt}}{e^t+1}\cos yt\right)e^{rt} \\ &= \left(\sum_{l=0}^{\infty} E_l^{(C)}(x,y)\frac{t^l}{l!}\right)\left(\sum_{k=0}^{\infty} r^k\frac{t^k}{k!}\right) \\ &= \sum_{n=0}^{\infty}\left(\sum_{k=0}^{n}\binom{n}{k}E_k^{(C)}(x,y)r^{n-k}\right)\frac{t^n}{n!}.\end{aligned} \tag{24}$$

Therefore, by comparing the coefficients on the both sides, we obtain the following theorem:

Theorem 5. *For $n \geq 0, r \in \mathbb{N}$, we have*

$$E_n^{(C)}(x+r,y) = \sum_{k=0}^{n} \binom{n}{k} E_k^{(C)}(x,y) r^{n-k},$$

and

$$E_n^{(S)}(x+r,y) = \sum_{k=0}^{n} \binom{n}{k} E_k^{(S)}(x,y) r^{n-k}.$$

Taking $r = 1$ in Theorem 5, we obtain the following corollary:

Corollary 2. *For $n \geq 0$, we have*

$$2C_n(x,y) = E_n^{(C)}(x,y) + \sum_{k=0}^{n} \binom{n}{k} E_k^{(C)}(x,y),$$

and

$$2S_n(x,y) = E_n^{(S)}(x,y) + \sum_{k=0}^{n} \binom{n}{k} E_k^{(S)}(x,y).$$

From Corollary 2, we note that

$$E_n^{(C)}(0,y) + \sum_{k=0}^{n} \binom{n}{k} E_k^{(C)}(0,y) = \begin{cases} 0, & \text{if } n = 2m+1, \\ 2(-1)^m y^{2m}, & \text{if } n = 2m, \end{cases} \tag{25}$$

and

$$E_n^{(S)}(0,y) + \sum_{k=0}^{n} \binom{n}{k} E_k^{(S)}(0,y) = \begin{cases} 2(-1)^m y^{2m+1}, & \text{if } n = 2m+1, \\ 0, & \text{if } n = 2m. \end{cases} \tag{26}$$

By (12), we get

$$\begin{aligned} \sum_{n=1}^{\infty} \frac{\partial}{\partial x} E_n^{(C)}(x,y) \frac{t^n}{n!} &= \frac{\partial}{\partial x} \left(\frac{2}{e^t+1} e^{xt} \cos yt \right) \\ &= \frac{2}{e^t+1} t e^{xt} \cos yt \\ &= \sum_{n=1}^{\infty} \left(n E_{n-1}^{(C)}(x,y) \right) \frac{t^n}{n!}. \end{aligned} \tag{27}$$

Comparing the coefficients on the both sides of (27), we have

$$\frac{\partial}{\partial x} E_n^{(C)}(x,y) = n E_{n-1}^{(C)}(x,y).$$

Similarly, for $n \geq 1$, we have

$$\begin{aligned} \frac{\partial}{\partial x} E_n^{(S)}(x,y) &= n E_{n-1}^{(S)}(x,y), \\ \frac{\partial}{\partial y} E_n^{(C)}(x,y) &= -n E_{n-1}^{(S)}(x,y), \\ \frac{\partial}{\partial y} E_n^{(S)}(x,y) &= n E_{n-1}^{(C)}(x,y). \end{aligned}$$

Now, we consider the Bernoulli polynomials that are given by the generating function to be

$$\frac{t}{e^t-1}e^{(x+iy)t}=\sum_{n=0}^{\infty}B_n(x+iy)\frac{t^n}{n!}.$$

We also have

$$\sum_{n=0}^{\infty}B_n(x+iy)\frac{t^n}{n!}=\frac{t}{e^t-1}e^{(x+iy)t}=\frac{t}{e^t-1}e^{xt}(\cos yt+i\sin yt), \tag{28}$$

and

$$\sum_{n=0}^{\infty}B_n(x-iy)\frac{t^n}{n!}=\frac{t}{e^t-1}e^{(x-iy)t}=\frac{t}{e^t-1}e^{xt}(\cos yt-i\sin yt). \tag{29}$$

Thus, by (28) and (29), we can derive

$$\frac{t}{e^t-1}e^{xt}\cos yt=\sum_{n=0}^{\infty}\left(\frac{B_n(x+iy)+B_n(x-iy)}{2}\right)\frac{t^n}{n!}, \tag{30}$$

and

$$\frac{t}{e^t-1}e^{xt}\sin yt=\sum_{n=0}^{\infty}\left(\frac{B_n(x+iy)+B_n(x-iy)}{2i}\right)\frac{t^n}{n!}. \tag{31}$$

It follows that we define the following cosine–Bernoulli and sine–Bernoulli polynomials.

Definition 2. *The cosine–Bernoulli polynomials $B_n^{(C)}(x,y)$ and sine–Bernoulli polynomials $B_n^{(S)}(x,y)$ are defined by means of the generating functions*

$$\sum_{n=0}^{\infty}B_n^{(C)}(x,y)\frac{t^n}{n!}=\frac{t}{e^t-1}e^{xt}\cos yt, \tag{32}$$

and

$$\sum_{n=0}^{\infty}B_n^{(S)}(x,y)\frac{t^n}{n!}=\frac{t}{e^t-1}e^{xt}\sin yt, \tag{33}$$

respectively.

By (30), (31), (32), and (33), we have

$$B_n^{(C)}(x,y)=\frac{B_n(x+iy)+B_n(x-iy)}{2},$$
$$B_n^{(S)}(x,y)=\frac{B_n(x+iy)-B_n(x-iy)}{2i}.$$

Note that $B_n^{(C)}(x,0)=B_n(x)$ are the Bernoulli polynomials. The cosine–Bernoulli and sine–Bernoulli polynomials can be determined explicitly. A few of them are

$$\begin{aligned}
&B_0^{(C)}(x,y)=1,\quad B_1^{(C)}(x,y)=-\frac{1}{2}+x,\\
&B_2^{(C)}(x,y)=\frac{1}{6}-x+x^2-y^2,\\
&B_3^{(C)}(x,y)=\frac{x}{2}-\frac{3x^2}{2}+x^3+\frac{3y^2}{2}-3xy^2,\\
&B_4^{(C)}(x,y)=-\frac{1}{30}+x^2-2x^3+x^4-y^2+6xy^2-6x^2y^2+y^4,
\end{aligned}$$

and

$$B_0^{(S)}(x,y) = 0, \quad B_1^{(S)}(x,y) = y, \quad B_2^{(S)}(x,y) = -y + 2xy,$$
$$B_3^{(S)}(x,y) = \frac{y}{2} - 3xy + 3x^2y - y^3,$$
$$B_4^{(S)}(x,y) = 2xy - 6x^2y + 4x^3y + 2y^3 - 4xy^3.$$

From (32), we have

$$\begin{aligned}\sum_{n=0}^{\infty} B_n^{(C)}(x,y)\frac{t^n}{n!} &= \frac{t}{e^t-1}e^{xt}\cos yt, \\ &= \left(\sum_{l=0}^{\infty} B_n \frac{t^l}{l!}\right)\left(\sum_{m=0}^{\infty} C_m(x,y)\frac{t^m}{m!}\right) \\ &= \sum_{n=0}^{\infty}\left(\sum_{l=0}^{n}\binom{n}{l}B_l C_{n-l}(x,y)\right)\frac{t^n}{n!}.\end{aligned} \tag{34}$$

Comparing the coefficients on the both sides of (34), we obtain the following theorem:

Theorem 6. *For $n \geq 0$, we have*

$$B_n^{(C)}(x,y) = \sum_{l=0}^{n}\binom{n}{l}B_l C_{n-l}(x,y),$$

and

$$B_n^{(S)}(x,y) = \sum_{l=0}^{n}\binom{n}{l}B_l S_{n-l}(x,y).$$

By replacing x by $1-x$ in (32), we get

$$\begin{aligned}\sum_{n=0}^{\infty} B_n^{(C)}(1-x,y)\frac{t^n}{n!} &= \frac{t}{e^t-1}e^{(1-x)t}\cos yt \\ &= \frac{t}{1-e^{-t}}e^{-xt}\cos yt \\ &= \sum_{n=0}^{\infty}(-1)^n B_n(x,y)\frac{t^n}{n!}.\end{aligned} \tag{35}$$

Therefore, we obtain the following theorem:

Theorem 7. *For $n \geq 0$, we have*

$$B_n^{(C)}(1-x,y) = (-1)^n B_n^{(C)}(x,y),$$

and

$$B_n^{(S)}(1-x,y) = (-1)^{n+1} B_n^{(S)}(x,y).$$

Now, we observe that

$$\begin{aligned}\sum_{n=0}^{\infty} B_n^{(C)}(x+1,q)\frac{t^n}{n!} &= \frac{t}{e^t-1}e^{(x+1)t}\cos yt \\ &= te^{xt}\cos yt + \frac{t}{e^t-1}e^{xt}\cos yt \\ &= \sum_{n=1}^{\infty} nC_{n-1}(x,y)\frac{t^n}{n!} + \sum_{n=0}^{\infty} B_n^{(C)}(x,y)\frac{t^n}{n!} \\ &= \sum_{n=0}^{\infty}\left(nC_{n-1}(x,y) + B_n^{(C)}(x,y)\right)\frac{t^n}{n!}.\end{aligned} \tag{36}$$

Thus, by (36), we get

$$B_n^{(C)}(x+1,y) = nC_{n-1}(x,y) + B_n^{(C)}(x,y), (n \geq 1). \tag{37}$$

Therefore, by (37), we obtain the following theorem:

Theorem 8. *For $n \geq 1$, we have*

$$B_n^{(C)}(x+1,y) - B_n^{(C)}(x,y) = nC_{n-1}(x,y),$$

and

$$B_n^{(S)}(x+1,y) - B_n^{(S)}(x,y) = nS_{n-1}(x,y).$$

Now, we define the new type polynomials that are given by the generating functions to be

$$\frac{2}{e^t+1}\cos yt = \sum_{n=0}^{\infty} E_n^{(C)}(y)\frac{t^n}{n!}, \tag{38}$$

and

$$\frac{2}{e^t+1}\sin yt = \sum_{n=0}^{\infty} E_n^{(S)}(y)\frac{t^n}{n!}, \tag{39}$$

respectively.

Note that $E_n^{(C)}(0) = E_n$, $E_n^{(S)}(0) = 0$, $E_n^{(C)}(0,y) = E_n^{(C)}(y)$, $E_n^{(S)}(0,y) = E_n^{(S)}(y)$, $(n \geq 0)$. The new type polynomials can be determined explicitly. A few of them are

$$\begin{aligned}&E_0^{(C)}(y) = 1, \quad E_1^{(C)}(x,y) = -\frac{1}{2}, \quad E_2^{(C)}(x,y) = -y^2, \quad E_3^{(C)}(y) = \frac{1}{4} + \frac{3y^2}{2}, \\ &E_4^{(C)}(y) = y^4, \quad E_5^{(C)}(y) = -\frac{1}{2} - \frac{5y^2}{2} - \frac{5y^4}{2}, \quad E_6^{(C)}(y) = -y^6,\end{aligned}$$

and

$$\begin{aligned}&E_0^{(S)}(x,y) = 0, \quad E_1^{(S)}(x,y) = y, \quad E_2^{(S)}(x,y) = -y, \quad E_3^{(S)}(x,y) = -y^3, \\ &E_4^{(S)}(x,y) = y + 2y^3 \quad E_5^{(S)}(x,y) = y^5, \quad E_6^{(S)}(x,y) = -3y - 5y^3 - 3y^5.\end{aligned}$$

From (38) and (39), we derive the following equations:

$$\frac{2}{e^t+1}\cos yt = \sum_{k=0}^{\infty}\left(\sum_{m=0}^{[\frac{k}{2}]}\binom{k}{2m}(-1)^m E_{k-2m}y^{2m}\right)\frac{t^k}{k!} \tag{40}$$

and

$$\frac{2}{e^t+1}\sin yt = \sum_{k=0}^{\infty}\left(\sum_{m=0}^{[\frac{k-1}{2}]}\binom{k}{2m+1}(-1)^m E_{k-2m-1}y^{2m+1}\right)\frac{t^k}{k!}. \quad (41)$$

By (38)–(41), we get

$$E_n^{(C)}(y) = \sum_{m=0}^{[\frac{n}{2}]}\binom{n}{2m}(-1)^m y^{2m}E_{n-2m}, \quad (42)$$

and

$$E_n^{(S)}(y) = \sum_{m=0}^{[\frac{n-1}{2}]}\binom{n}{2m+1}(-1)^m y^{2m+1}E_{n-2m-1}, (k \geq 0). \quad (43)$$

From (12), (13), (38) and (39), we derive the following theorem:

Theorem 9. *For $n \geq 0$, we have*

$$E_n^{(C)}(x,y) = \sum_{k=0}^{n}\binom{n}{k}x^{n-k}E_k^{(C)}(y),$$

and

$$E_n^{(S)}(x,y) = \sum_{k=0}^{n}\binom{n}{k}x^{n-k}E_k^{(S)}(y).$$

Now, we define the new type polynomials that are given by the generating functions to be

$$\frac{t}{e^t-1}\cos yt = \sum_{n=0}^{\infty}B_n^{(C)}(y)\frac{t^n}{n!}, \quad (44)$$

and

$$\frac{t}{e^t-1}\sin yt = \sum_{n=0}^{\infty}B_n^{(S)}(y)\frac{t^n}{n!}, \quad (45)$$

respectively.

Note that $B_n^{(C)}(0) = B_n$, $B_n^{(S)}(0) = 0$, $B_n^{(C)}(0,y) = B_n^{(C)}(y)$, $B_n^{(S)}(0,y) = B_n^{(S)}(y)$, $(n \geq 0)$. The new type polynomials can be determined explicitly. A few of them are

$$B_0^{(C)}(x,y) = 1, \quad B_1^{(C)}(x,y) = -\frac{1}{2}, \quad B_2^{(C)}(x,y) = \frac{1}{6}-y^2,$$
$$B_3^{(C)}(x,y) = \frac{3y^2}{2}, \quad B_4^{(C)}(x,y) = -\frac{1}{30}-y^2+y^4, \quad B_5^{(C)}(x,y) = \frac{5y^4}{2},$$

and

$$B_0^{(S)}(x,y) = 0, \quad B_1^{(S)}(x,y) = y, \quad B_2^{(S)}(x,y) = -y,$$
$$B_3^{(S)}(x,y) = \frac{y}{2}-y^3, \quad B_4^{(S)}(x,y) = 2y^3, \quad B_5^{(S)}(x,y) = -\frac{y}{6}-\frac{5y^3}{3}+y^5.$$

From (44) and (45), we derive the following equations:

$$\frac{t}{e^t-1}\cos yt = \sum_{k=0}^{\infty}\left(\sum_{m=0}^{[\frac{k}{2}]}\binom{k}{2m}(-1)^m B_{k-2m}y^{2m}\right)\frac{t^k}{k!} \quad (46)$$

and

$$\frac{t}{e^t-1}\sin yt = \sum_{k=0}^{\infty}\left(\sum_{m=0}^{[\frac{k-1}{2}]}\binom{k}{2m+1}(-1)^m B_{k-2m-1}y^{2m+1}\right)\frac{t^k}{k!}. \quad (47)$$

By (44)–(47), we get

$$B_n^{(C)}(y) = \sum_{m=0}^{[\frac{n}{2}]} \binom{n}{2m} (-1)^m y^{2m} B_{n-2m}, \tag{48}$$

and

$$B_n^{(S)}(y) = \sum_{m=0}^{[\frac{n-1}{2}]} \binom{n}{2m+1} (-1)^m y^{2m+1} B_{n-2m-1}, (k \geq 0). \tag{49}$$

From (32), (33), (44) and (45), we derive the following theorem:

Theorem 10. *For $n \geq 0$, we have*

$$B_n^{(C)}(x,y) = \sum_{k=0}^{n} \binom{n}{k} x^{n-k} B_k^{(C)}(y),$$

and

$$B_n^{(S)}(x,y) = \sum_{k=0}^{n} \binom{n}{k} x^{n-k} B_k^{(S)}(y).$$

We remember that the classical Stirling numbers of the first kind $S_1(n,k)$ and $S_2(n,k)$ are defined by the relations (see [12])

$$(x)_n = \sum_{k=0}^{n} S_1(n,k) x^k \text{ and } x^n = \sum_{k=0}^{n} S_2(n,k)(x)_k, \tag{50}$$

respectively. Here, $(x)_n = x(x-1)\cdots(x-n+1)$ denotes the falling factorial polynomial of order n. The numbers $S_2(n,m)$ also admit a representation in terms of a generating function

$$(e^t-1)^m = m! \sum_{n=m}^{\infty} S_2(n,m) \frac{t^n}{n!}. \tag{51}$$

By (12), (51) and by using Cauchy product, we get

$$\begin{aligned}
\sum_{n=0}^{\infty} E_n^{(C)}(x,y) \frac{t^n}{n!} &= \left(\frac{2}{e^t+1}\right) (1-(1-e^{-t}))^{-x} \cos yt \\
&= \left(\frac{2}{e^t+1}\right) \cos yt \sum_{l=0}^{\infty} \binom{x+l-1}{l} (1-e^{-t})^l \\
&= \sum_{l=0}^{\infty} < x >_l \frac{(e^t-1)^l}{l!} \left(\frac{2}{e^t+1}\right) e^{-lt} \cos yt \\
&= \sum_{l=0}^{\infty} < x >_l \sum_{n=0}^{\infty} S_2(n,l) \frac{t^n}{n!} \sum_{n=0}^{\infty} E_n^{(C)}(-l,y) \frac{t^n}{n!} \\
&= \sum_{n=0}^{\infty} \left(\sum_{l=0}^{\infty} \sum_{i=l}^{n} \binom{n}{i} S_2(i,l) E_{n-i}^{(C)}(-l,y) < x >_l \right) \frac{t^n}{n!},
\end{aligned} \tag{52}$$

where $< x >_l = x(x+1)\cdots(x+l-1) (l \geq 1)$ with $< x >_0 = 1$.

By comparing the coefficients on both sides of (52), we have the following theorem:

Theorem 11. *For $n \in \mathbb{Z}_+$, we have*

$$E_n^{(C)}(x,y) = \sum_{l=0}^{\infty}\sum_{i=l}^{n}\binom{n}{i} S_2(i,l) E_{n-i}^{(C)}(-l,y) < x >_l,$$
$$E_n^{(S)}(x,y) = \sum_{l=0}^{\infty}\sum_{i=l}^{n}\binom{n}{i} S_2(i,l) E_{n-i}^{(S)}(-l,y) < x >_l .$$

By (12), (38), (50), (51) and by using Cauchy product, we have

$$\begin{aligned}
\sum_{n=0}^{\infty} E_n^{(C)}(x,y)\frac{t^n}{n!} &= \left(\frac{2}{e^t+1}\right)((e^t-1)+1)^x \cos(yt) \\
&= \frac{2}{e^t+1}\cos(yt)\sum_{l=0}^{\infty}\binom{x}{l}(e^t-1)^l \\
&= \sum_{l=0}^{\infty}(x)_l\frac{(e^t-1)^l}{l!}\left(\frac{2}{e^t+1}\cos(yt)\right) \\
&= \sum_{l=0}^{\infty}(x)_l\sum_{n=0}^{\infty}S_2(n,l)\frac{t^n}{n!}\sum_{n=0}^{\infty}E_n^{(C)}(y)\frac{t^n}{n!} \\
&= \sum_{n=0}^{\infty}\left(\sum_{l=0}^{\infty}\sum_{i=l}^{n}\binom{n}{i}(x)_l S_2(i,l)E_{n-i}^{(C)}(y)\right)\frac{t^n}{n!}.
\end{aligned} \tag{53}$$

By comparing the coefficients on both sides of (53), we have the following theorem:

Theorem 12. *For $n \in \mathbb{Z}_+$, we have*

$$E_n^{(C)}(x,y) = \sum_{l=0}^{\infty}\sum_{i=l}^{n}\binom{n}{i}(x)_l S_2(i,l) E_{n-i}^{(C)}(y),$$
$$E_n^{(S)}(x,y) = \sum_{l=0}^{\infty}\sum_{i=l}^{n}\binom{n}{i}(x)_l S_2(i,l) E_{n-i}^{(S)}(y).$$

By (4), (12), (38), (50), (51) and by using Cauchy product, we have

$$\begin{aligned}
&\sum_{n=0}^{\infty} E_n^{(C)}(x,y)\frac{t^n}{n!} \\
&= \left(\frac{2}{e^t+1}\right)e^{xt}\cos(yt) \\
&= \frac{(e^t-1)^r}{r!}\frac{r!}{t^r}\left(\frac{t}{e^t-1}\right)^r e^{xt}\sum_{n=0}^{\infty}E_n^{(C)}(y)\frac{t^n}{n!} \\
&= \frac{(e^t-1)^r}{r!}\left(\sum_{n=0}^{\infty}\mathbf{B}_n^{(r)}(x)\frac{t^n}{n!}\right)\left(\sum_{n=0}^{\infty}E_n^{(C)}(y)\frac{t^n}{n!}\right)\frac{r!}{t^r} \\
&= \sum_{n=0}^{\infty}\left(\sum_{l=0}^{n}\frac{\binom{n}{l}}{\binom{l+r}{r}}S_2(l+r,r)\sum_{i=0}^{n-l}\binom{n-l}{i}\mathbf{B}_i^{(r)}(x)E_{n-l-i}^{(C)}(y)\right)\frac{t^n}{n!}.
\end{aligned}$$

By comparing the coefficients on both sides, we have the following theorem:

Theorem 13. *For $n \in \mathbb{Z}_+$ and $r \in \mathbb{N}$, we have*

$$E_n^{(C)}(x,y) = \sum_{l=0}^{n} \frac{\binom{n}{l}}{\binom{l+r}{r}} S_2(l+r,r) \sum_{i=0}^{n-l} \binom{n-l}{i} E_{n-l-i}^{(C)}(y)\mathbf{B}_i^{(r)}(x),$$

$$E_n^{(S)}(x,y) = \sum_{l=0}^{n} \frac{\binom{n}{l}}{\binom{l+r}{r}} S_2(l+r,r) \sum_{i=0}^{n-l} \binom{n-l}{i} E_{n-l-i}^{(S)}(y)\mathbf{B}_i^{(r)}(x).$$

By (5), (12), (38), (50), (51) and by using the Cauchy product, we get

$$\begin{aligned}
&\sum_{n=0}^{\infty} E_n^{(C)}(x,y)\frac{t^n}{n!} = \left(\frac{2}{e^t+1}\right) e^{xt}\cos(yt) \\
&= \frac{(e^t-u)^r}{(1-u)^r}\left(\frac{1-u}{e^t-u}\right)^r e^{xt}\left(\frac{2}{e^t+1}\right)\cos(yt) \\
&= \sum_{n=0}^{\infty} \mathbf{H}_n^{(r)}(u,x)\frac{t^n}{n!}\sum_{i=0}^{r}\binom{r}{i} e^{it}(-u)^{r-i}\frac{1}{(1-u)^r}\left(\frac{2}{e^t+1}\right)\cos(yt) \\
&= \frac{1}{(1-u)^r}\sum_{i=0}^{r}\binom{r}{i}(-u)^{r-i}\sum_{n=0}^{\infty}\mathbf{H}_n^{(r)}(u,x)\frac{t^n}{n!}\sum_{n=0}^{\infty}E_n^{(C)}(i,y)\frac{t^n}{n!} \\
&= \sum_{n=0}^{\infty}\left(\frac{1}{(1-u)^r}\sum_{i=0}^{r}\binom{r}{i}(-u)^{r-i}\sum_{l=0}^{n}\binom{n}{l}\mathbf{H}_l^{(r)}(u,x)E_{n-l}^{(C)}(i,y)\right)\frac{t^n}{n!}.
\end{aligned}$$

By comparing the coefficients on both sides, we have the following theorem:

Theorem 14. *For $n \in \mathbb{Z}_+$ and $r \in \mathbb{N}$, we have*

$$E_n^{(C)}(x,y) = \frac{1}{(1-u)^r}\sum_{i=0}^{r}\sum_{l=0}^{n}\binom{r}{i}\binom{n}{l}(-u)^{r-i}\mathbf{H}_l^{(r)}(u,x)E_{n-l}^{(C)}(i,y),$$

$$E_n^{(S)}(x,y) = \frac{1}{(1-u)^r}\sum_{i=0}^{r}\sum_{l=0}^{n}\binom{r}{i}\binom{n}{l}(-u)^{r-i}\mathbf{H}_l^{(r)}(u,x)E_{n-l}^{(S)}(i,y).$$

By Theorems 12–14, we have the following corollary.

Corollary 3. *For $n \in \mathbb{Z}_+$ and $r \in \mathbb{N}$, we have*

$$\begin{aligned}
&\sum_{l=0}^{\infty}\sum_{i=l}^{n}\binom{n}{i}(x)_l S_2(i,l)E_{n-i}^{(C)}(y) \\
&= \frac{1}{(1-u)^r}\sum_{i=0}^{r}\sum_{l=0}^{n}\binom{r}{i}\binom{n}{l}(-u)^{r-i}\mathbf{H}_l^{(r)}(u,x)E_{n-l}^{(C)}(i,y) \\
&= \sum_{l=0}^{n}\frac{\binom{n}{l}}{\binom{l+r}{r}}S_2(l+r,r)\sum_{i=0}^{n-l}\binom{n-l}{i}E_{n-l-i}^{(C)}(y)\mathbf{B}_i^{(r)}(x).
\end{aligned}$$

3. Distribution of Zeros of the Cosine–Euler and Sine–Euler Polynomials

This section aims to demonstrate the benefit of using numerical investigation to support theoretical prediction and to discover a new interesting pattern of the zeros of the cosine–Euler and sine–Euler polynomials. Using a computer, a realistic study for the cosine–Euler polynomials $E_n^{(C)}(x,y)$ and sine–Euler polynomials $E_n^{(S)}(x,y)$ is very interesting. It is the aim of this paper to observe an interesting phenomenon of "scattering" of the zeros of the the cosine–Euler polynomials $E_n^{(C)}(x,y)$ and sine–Euler polynomials $E_n^{(S)}(x,y)$ in a complex plane. We investigate the beautiful zeros of the cosine–Euler and sine–Euler polynomials by using a computer. We plot the zeros of the cosine–Euler polynomials $E_n^{(C)}(x,y)$ (Figure 1).

In Figure 1 (top-left), we choose $n = 30$ and $y = -3$. In Figure 1 (top-right), we choose $n = 30$ and $y = 0$. In Figure 1 (bottom-left), we choose $n = 30$ and $y = 1/2$. In Figure 1 (bottom-right), we choose $n = 30$ and $y = 3$.

We plot the zeros of the sine–Euler polynomials $E_n^{(S)}(x,y)$ (Figure 2).

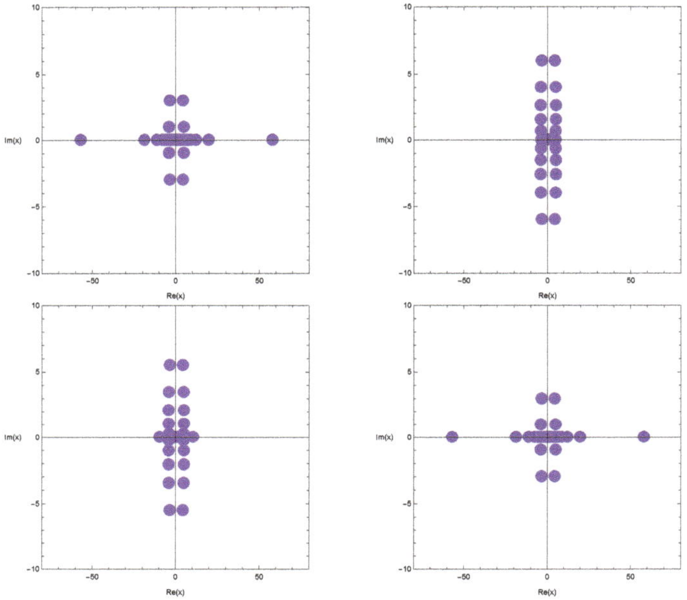

Figure 1. Zeros of $E_n^{(C)}(x,y)$.

In Figure 2 (top-left), we choose $n = 30$ and $x = -3$. In Figure 2 (top-right), we choose $n = 30$ and $x = -1$. In Figure 2 (bottom-left), we choose $n = 30$ and $x = 1$. In Figure 2 (bottom-right), we choose $n = 30$ and $x = 3$.

We observe that $E_n^{(C)}(x,a), x \in \mathbb{C}$ has $Re(x) = \frac{1}{2}$ reflection symmetry in addition to the usual $Im(x) = 0$ reflection symmetry analytic complex functions, where $a \in \mathbb{R}$(Figures 1 and 2).

Since

$$\begin{aligned}\sum_{n=0}^{\infty} E_n^{(C)}(1-x,-y)\frac{(-1)^n t^n}{n!} &= \frac{2}{e^{-t}+1}e^{(1-x)(-t)}\cos yt \\ &= \frac{2}{e^t+1}e^{xt}\cos yt = \sum_{n=0}^{\infty} E_n^{(C)}(x,y)\frac{t^n}{n!},\end{aligned}$$

we obtain

$$\begin{aligned}E_n^{(C)}(x,y) &= (-1)^n E_n^{(C)}(1-x,-y), \quad E_n^{(C)}(x,y) = (-1)^n E_n^{(C)}(1-x,y), \\ E_n^{(S)}(x,y) &= (-1)^n E_n^{(S)}(1-x,-y), \quad E_n^{(S)}(x,y) = (-1)^{n+1} E_n^{(S)}(1-x,y).\end{aligned}$$

Hence, we have the following theorem:

Theorem 15. *If* $n \equiv 1 \pmod 2$, *then*

$$E_n^{(C)}(1/2, y) = 0, \quad B_n^{(C)}(1/2, y) = 0, \text{ for } n \in \mathbb{N}.$$

If $n \equiv 0 \pmod 2$, *then*

$$E_n^{(S)}(1/2, y) = 0, \quad B_n^{(S)}(1/2, y) = 0, \text{ for } n \in \mathbb{N}.$$

Our numerical results for numbers of real and complex zeros of the cosine–Euler polynomials $E_n^{(C)}(x, y) = 0$ are displayed (Table 1).

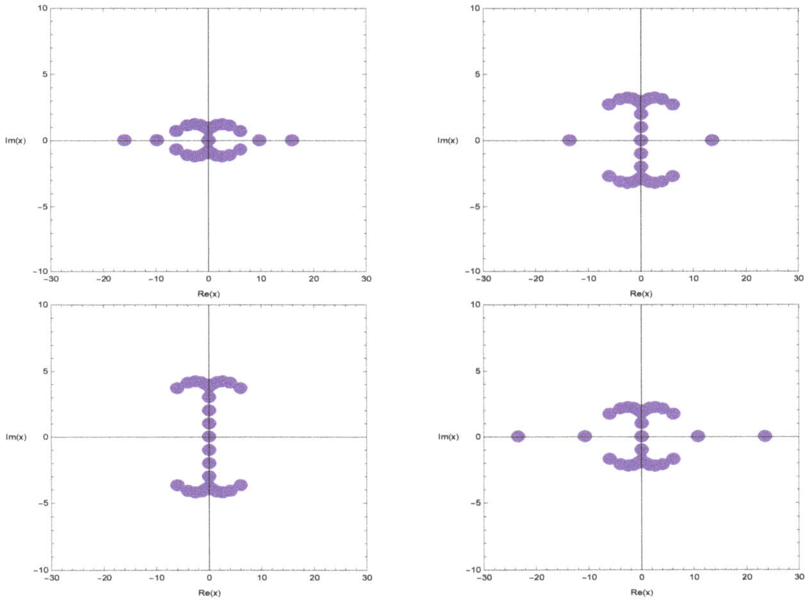

Figure 2. Zeros of $E_n^{(S)}(x, y)$.

Table 1. Numbers of real and complex zeros of $E_n^{(C)}(x, y)$.

Degree n	$y = -3$		$y = 2$	
	Real Zeros	**Complex Zeros**	**Real Zeros**	**Complex Zeros**
1	1	0	1	0
2	2	0	2	0
3	3	0	3	0
4	4	0	4	0
5	5	0	5	0
6	6	0	6	0
7	7	0	7	0
8	8	0	8	0
9	9	0	9	0
10	10	0	10	0

Our numerical results for numbers of real and complex zeros of the sine–Euler polynomials $E_n^{(S)}(x, y) = 0$ are displayed (Table 2).

Stacks of zeros of the cosine–Euler polynomials $E_n^{(C)}(x,y)$ for $1 \le n \le 40$ from a 3D structure are presented (Figure 3).

In Figure 3 (left), we choose $y = -3$. In Figure 3 (right), we choose $y = 1/2$. The plot of real zeros of the cosine–Euler polynomials $E_n^{(C)}(x,y)$ for $1 \le n \le 40$ structure are presented (Figure 4).

In Figure 4 (left), we choose $y = -3$. In Figure 4 (right), we choose $y = 1/2$. Stacks of zeros of the sine–Euler polynomials $E_n^{(S)}(x,y)$ for $1 \le n \le 40$ from a 3D structure are presented (Figure 5).

Table 2. Numbers of real and complex zeros of $E_n^{(S)}(x,y)$.

Degree n	$x = -3$		$x = 1$	
	Real Zeros	Complex Zeros	Real Zeros	Complex Zeros
1	1	0	1	0
2	1	0	1	0
3	3	0	3	0
4	3	0	3	0
5	5	0	5	0
6	5	0	1	4
7	7	0	7	0
8	7	0	1	6
9	9	0	9	0
10	9	0	1	8

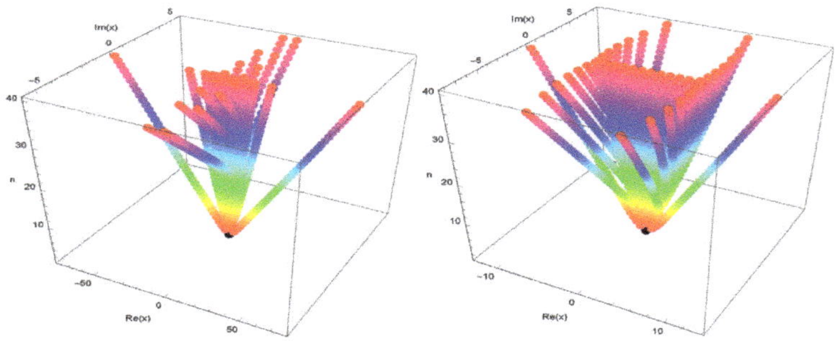

Figure 3. Stacks of zeros of $E_n^{(C)}(x,y), 1 \le n \le 40$.

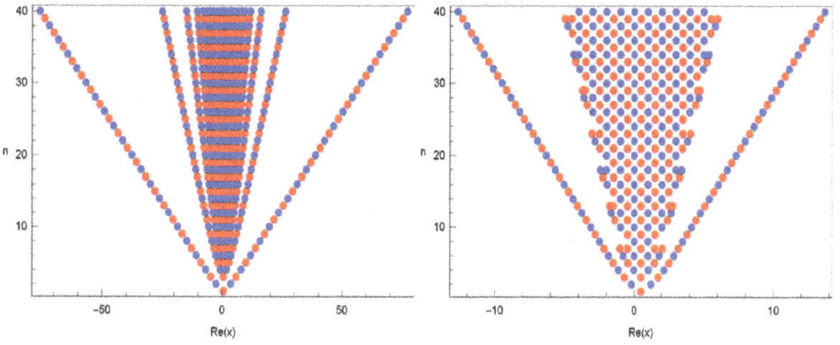

Figure 4. Real zeros of $E_n^{(C)}(x,y)$.

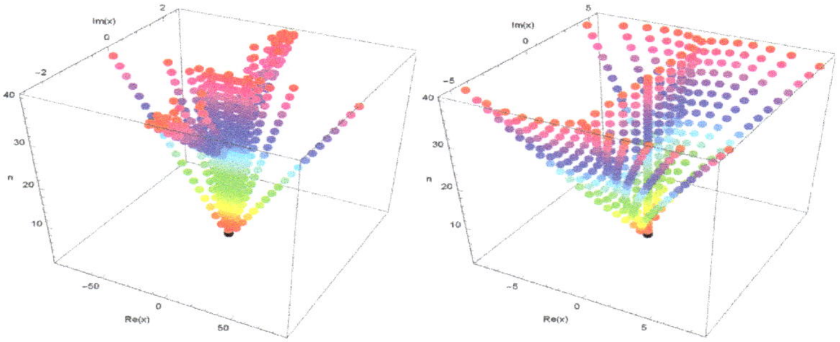

Figure 5. Stacks of zeros of $E_n^{(S)}(x,y), 1 \leq n \leq 40$.

In Figure 5 (left), we choose $x = -3$. In Figure 3 (right), we choose $x = 1$. The plot of real zeros of the sine–Euler polynomials $E_n^{(S)}(x,y)$ for $1 \leq n \leq 40$ structure are presented (Figure 6).

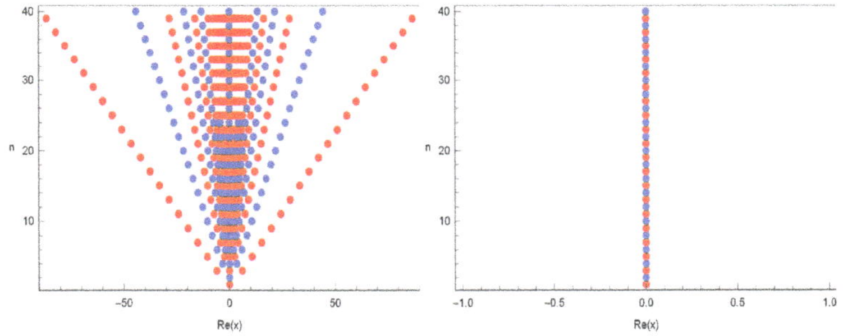

Figure 6. Real zeros of $E_n^{(S)}(x,y)$.

In Figure 6 (left), we choose $x = -3$. In Figure 6 (right), we choose $x = 1$.

We observe a remarkable regular structure of the complex roots of the cosine–Euler polynomials $E_n^{(C)}(x,y)$. We also hope to verify a remarkable regular structure of the complex roots of the cosine–Euler polynomials $E_n^{(C)}(x,y)$. Next, we calculated an approximate solution satisfying $E_n^{(C)}(x,y) = 0, x \in \mathbb{R}$. The results are given in Table 3.

Table 3. Approximate solutions of $E_n^{(C)}(x,-3) = 0, x \in \mathbb{R}$.

Degree n				x			
1				0.50000			
2			−2.5414,	3.5414			
3			−4.7678,	0.50000,	5.7678		
4		−6.8305,	−0.82832,	1.8283,	7.8305		
5		−8.8303,	−1.8336,	0.50000,	2.8336,	9.8303	
6		−10.799,	−2.7017,	−0.40666,	1.4067,	3.7017,	11.799
7	−12.751,	−3.4960,	−1.1389,	0.50000,	2.1389,	4.4960,	13.751

Next, we calculated an approximate solution satisfying $E_n^{(S)}(x,y) = 0, y \in \mathbb{R}$. The results are given in Table 4.

Table 4. Approximate solutionsof $E_n^{(S)}(-3, y) = 0, y \in \mathbb{R}$.

Degree n				y			
1				0.00000			
2				0.00000			
3			−6.0000,	0,	6.0000		
4			−3.3912,	0,	3.3912		
5		−10.687,	−2.4038,	0,	2.4038,	10.687	
6		−5.9045,	−1.8630,	0,	1.8630,	5.9045	
7	−15.241,	−4.1727,	−1.5184,	0,	1.5184,	4.1727,	15.241

Author Contributions: T.K. and C.S.R. wrote and checked the results of the paper; C.S.R. conducted numerical experiments of this paper; T.K. completed the revision of the article.

Funding: This work was supported by the National Research Foundation of Korea (NRF) grant funded by the Korea government (MEST) (No. 2017R1A2B4006092).

Acknowledgments: The authors would like to thank the referees for their valuable comments.

Conflicts of Interest: The authors declare no conflict of interest.

References

1. Ağyüz, E.; Acikgoz, M.; Araci, S. A symmetric identity on the q-Genocchi polynomials of higher-order under third dihedral group D3. *Proc. Jangjeon Math. Soc.* **2015**, *18*, 177–187
2. Bayad, A.; Chikhi, J. Non linear recurrences for Apostol-Bernoulli-Euler numbers of higher order. *Adv. Stud. Contemp. Math. (Kyungshang)* **2012** , *22*, 1–6.
3. Carlitz, L. Recurrences for the Bernoulli and Euler numbers. II. *Math. Nachr.* **1965**, *29*, 151–160. [CrossRef]
4. He, Y.; Kim, T. A higher-order convolution for Bernoulli polynomials of the second kind. *Appl. Math. Comput.* **2018**, *324*, 51–58. [CrossRef]
5. Kim, D.S.; Kim, T. Some identities of Bernoulli and Euler polynomials arising from umbral calculus. *Adv. Stud. Contemp. Math. (Kyungshang)* **2013**, *23*, 159–171. [CrossRef]
6. Kim, D.S.; Kim, T.; Kim, Y.H.; Lee, S.H. Some arithmetic properties of Bernoulli and Euler numbers. *Adv. Stud. Contemp. Math. (Kyungshang)* **2012**, *22*, 467–480.
7. Kim, D.S.; Kim, T.; Kim, Y.-H.; Dolgy, D.V. A note on Eulerian polynomials associated with Bernoulli and Euler numbers and polynomials. *Adv. Stud. Contemp. Math. (Kyungshang)* **2012**, *22*, 379–389.
8. Roman, S. *The Umbral Calculu, Pure and Applied Mathematics, 111*; Academic Press, Inc.: New York, NY, USA; Harcourt Brace Jovanovich Publishers: San Diego, CA, USA, 1984; ISBN 0-12-594380-6.
9. Ryoo, C.S.; Kim, Y.H. A numerical investigation on the structure of the roots of the twisted q-Euler polynomials. *Adv. Stud. Contemp. Math. (Kyungshang)* **2009**, *19*, 131–141.
10. Ryoo, C.S.; Agarwal, R.P. Some identities involving q-poly-tangent numbers and polynomials and distribution of their zeros. *Adv. Differ. Equ.* **2017**, *213*. [CrossRef]
11. Simsek, Y. Identities on the Changhee numbers and Apostol-type Daehee polynomials. *Adv. Stud. Contemp. Math. (Kyungshang)* **2017**, *27*, 199–212.
12. Srivastava, H.M.; Pintér, Á. Remarks on some relationships between the Bernoulli and Euler polynomials. *Appl. Math. Lett.* **2004**, *17*, 375–380. [CrossRef]
13. Srivastava, H.M.; Pintér, Á. Addition theorems for the Appell polynomials and the associated classes of polynomial expansions. *Aequ. Math.* **2013**, *85*, 483–495.
14. Young, P.T. Degenerate Bernoulli polynomials, generalized factorial sums, and their applications. *J. Number Theor.* **2008**, *128*, 738–758 [CrossRef]

axioms MDPI

Article

A New Type of Generalization on W—Asymptotically $\mathcal{J}_\lambda$—Statistical Equivalence with the Number of α

Hafize Gümüş [1,*] and Nihal Demir [2]

[1] Department of Math Education, Faculty of Eregli Education, Necmettin Erbakan University, Konya 42060, Turkey
[2] İkizli Middle School, Konya 42060, Turkey; nihaldemircise@gmail.com
* Correspondence: hgumus@konya.edu.tr

Received: 19 June 2018; Accepted: 31 July 2018; Published: 2 August 2018

Abstract: In our paper, by using the concept of $W-$asymptotically $\mathcal{J}-$ statistical equivalence of order α which has been previously defined, we present the definitions of $W-$asymptotically $\mathcal{J}_\lambda-$statistical equivalence of order α, $W-$strongly asymptotically $\mathcal{J}_\lambda-$statistical equivalence of order α, and $W-$strongly Cesáro asymptotically $\mathcal{J}-$statistical equivalence of order α where $0 < \alpha \leq 1$. We also extend these notions with a sequence of positive real numbers, $p = (p_k)$, and we investigate how our results change if p is constant.

Keywords: $\mathcal{J}-$statistical convergence; asymptotic equivalence; set sequences; $\lambda = (\lambda_n)$ sequence

MSC: 40G15; 40A35

---

1. Introduction and Background

To make it easier to understand, we prefer to give introduction section in five parts. In first part, we give the main definitions related to statistical convergence: $\lambda-$statistical convergence, $\mathcal{J}$—convergence, $\mathcal{J}$—statistical convergence, and $\mathcal{J}_\lambda-$ statistical convergence. In the second part, we mention asymptotic equivalence and S_λ^L-asymptotic equivalence. In the third part, we explain set sequences, and we give some important definitions for set sequences in the Wijsman sense. In the fourth part, we explain how statistical convergence and $\mathcal{J}-$convergence were expanded using the number of α $(0 < \alpha < 1)$. Finally, in last part, we explain the purpose and innovations of our study.

1.1. $\mathcal{J}_\lambda-$Statistical Convergence

Statistical convergence is a concept which was formally introduced by Fast [1] and Steinhaus [2], independently. Later on, Schoenberg reintroduced this concept in his own study [3]. This new type of convergence has been used in different areas by several authors in references [4–8]. Statistical convergence is based on the definition of the natural density of the set $K \subseteq \mathbb{N}$ and we define the natural density of K by $d(K) = \lim_{n\to\infty} \frac{|K_n|}{n}$. In this definition, $K_n = \{k \in K : k \leq n\}$ and $|K_n|$ gives the number of elements in K_n.

Using this information, we say that a sequence (x) of real numbers is statistically convergent to the number L if $d(K(\varepsilon)) = d(\{k \leq n : |x_k - L| \geq \varepsilon\}) = 0$. In this case we write $st - \lim x_k = L$, and usually, S denotes the set of all statistical convergent sequences.

Let $\lambda = (\lambda_n)$ be a positive number sequence which is non-decreasing and tending to ∞. Also, for this sequence $\lambda_{n+1} \leq \lambda_n + 1,\ \lambda_1 = 1$. We denote the set of this kind of sequence by Λ, and we have the interval $I_n = [n - \lambda_n + 1, n]$. Mursaleen [9] defined $\lambda-$statistical convergence such that

$$\lim_{n\to\infty} \frac{1}{\lambda_n} |\{k \in I_n : |x_k - L| \geq \varepsilon\}| = 0$$

for any $\varepsilon > 0$, and he denoted this new method by S_λ. On the other hand, Kostyrko, Šalát and Wilezyński [10] introduced a new type of convergence which is defined in a metric space and is called $\mathcal{J}$—convergence. This type of convergence is based on the definition of an ideal $\mathcal{J}$ in $\mathbb{N}$.

A family of sets, $\mathcal{J} \subseteq 2^{\mathbb{N}}$, is an ideal if the following properties are provided:

(i) $\emptyset \in \mathcal{J}$; (ii) $A, B \in \mathcal{J}$ implies $A \cup B \in \mathcal{J}$; and (iii) for each $A \in \mathcal{J}$ and each $B \subseteq A$ implies $B \in \mathcal{J}$.

We say that $\mathcal{J}$ is non-trivial if $\mathbb{N} \notin \mathcal{J}$ and $\mathcal{J}$ is admissible if $\{n\} \in \mathcal{J}$ for every $n \in \mathbb{N}$.

A family of sets, $\mathcal{F} \subseteq 2^{\mathbb{N}}$, is a filter if the following properties are provided:

(i) $\emptyset \notin \mathcal{F}$; (ii) if $A, B \in \mathcal{F}$ then we have $A \cap B \in \mathcal{F}$; and (iii) for each $A \in \mathcal{F}$ and each $A \subseteq B$, we have $B \in \mathcal{F}$.

If $\mathcal{J}$ is an ideal in $\mathbb{N}$, then we have,

$$F(\mathcal{J}) = \{A \subset \mathbb{N} : \mathbb{N} \backslash A \in \mathcal{J}\}$$

is a filter in $\mathbb{N}$.

Definition 1. *([10]) A sequence of reals $x = (x_k)$ is $\mathcal{J}-$ convergent to $L \in \mathbb{R}$ if and only if the set*

$$A_\varepsilon = \{k \in \mathbb{N} : |x_k - L| \geq \varepsilon\} \in \mathcal{J}$$

for each $\varepsilon > 0$. In this case, we say that L is the $\mathcal{J}-$limit of the sequence (x).

$\mathcal{J}$—convergence generalizes many types of convergence such as usual convergence and statistical convergence. If we choose the ideals $\mathcal{J}_f = \{A \subset \mathbb{N} : A \text{ is finite}\}$ and $\mathcal{J}_d = \{A \subset \mathbb{N} : d(A) = 0\}$, then we obtain usual convergence and statistical convergence, respectively.

Based on the statistical convergence and $\mathcal{J}-$convergence, an important role was located in this area, $\mathcal{J}$—statistical convergence, which was introduced by Das, Savaş and Ghosal [11] as follows:

Definition 2. *([11]) A sequence $x = (x_k)$ is $\mathcal{J}-$statistically convergent to L if*

$$\left\{n \in \mathbb{N} : \frac{1}{n} |\{k \leq n : |x_k - L| \geq \varepsilon\}| \geq \delta\right\} \in \mathcal{J}$$

for every $\varepsilon > 0$ and $\delta > 0$.

1.2. Asymptotic Equivalence

Asymptotic equivalence was first introduced by Pobyvanets [12] and some main definitions and asymptotic reguler matrices were given by Marouf [13]. Bilgin [14] defined $f-$Asymptotically equivalent sequences, and on the other hand, asymptotically statistically equivalent sequences were presented by Patterson [15]. Gümüş and Savaş [16] gave the definition of $\mathcal{J}$—asymptotically $\lambda-$statistically equivalent sequences by using the $\lambda = (\lambda_n)$ sequence, and they were also interested in some inclusion relations between other related spaces.

According to Marouf, if $x = (x_k)$ and $y = (y_k)$ are two non-negative sequences, we say that they are asymptotically equivalent if

$$\lim_k \frac{x_k}{y_k} = 1.$$

This is denoted by $x \sim y$.

Definition 3. *([16]) Let $\mathcal{J}$ be an admissible ideal and $\lambda \in \Lambda$. Two number sequences $x = (x_k)$ and $y = (y_k)$ are $S^L(\mathcal{J})$−asymptotically equivalent of multiple L (or $\mathcal{J}$— asymptotically λ−statistically equivalent) if every $\delta, \varepsilon > 0$,*

$$\left\{n \in \mathbb{N} : \frac{1}{\lambda_n}\left|\left\{k \in I_n : \left|\frac{x_k}{y_k} - L\right| \geq \varepsilon\right\}\right| \geq \delta\right\} \in \mathcal{J}.$$

1.3. Set Sequences

In recent years, studies on set sequences has become popular. Firstly, usual convergence has been extended to convergence of sequences of sets. The first definitions of this subject were based on Baronti and Papini's [17] work in 1986. Now, we revisit the definitions of convergence, boundedness, and the Cesáro summability of set sequences. Throughout the paper, (X, ρ) is a metric space, and $A, A_k, B_k \subseteq X$ represents non-empty closed subsets of X for all $k \in \mathbb{N}$.

(X, ρ) is a metric, $x \in X$ is a point in X, and A is any non-empty subset of X. The distance from x to A is defined by

$$\rho(x, A) = \inf_{a \in A} \rho(x, A).$$

Definition 4. *([17]) In any metric space, the set sequence $\{A_k\}$ is Wijsman convergent to A if*

$$\lim_{k \to \infty} \rho(x, A_k) = \rho(x, A)$$

for each $x \in X$. We write for this case $W - \lim_{k \to \infty} A_k = A$.

We would like to give a well known example of this subject.

Example 1. *In the (x, y)-plane, consider the $A_k = \{(x, y) : x^2 + y^2 + 2kx = 0\}$ sequence of circles. We can easily see that for $k \to \infty$, this sequence is Wijsman convergent to the y-axis, i.e., $A = \{(x, y) : x = 0\}$.*

Definition 5. *([17]) In any metric space, the set sequence $\{A_k\}$ is bounded if*

$$\sup_k \rho(x, A_k) < \infty$$

for each $x \in X$. This is shown as $\{A_k\} \in L_\infty$.

Definition 6. *([17]) In any metric space, the set sequence $\{A_k\}$ is Wijsman Cesáro summable to A if*

$$\lim_{n \to \infty} \frac{1}{n} \sum_{k=1}^{n} \rho(x, A_k) = d(x, A)$$

for each $x \in X$, and $\{A_k\}$ is Wijsman strongly Cesáto summable to A if

$$\lim_{n \to \infty} \frac{1}{n} \sum_{k=1}^{n} |\rho(x, A_k) - \rho(x, A)| = 0$$

for each $x \in X$.

Nuray and Rhodes [18] introduced Wijsman statistical convergence for set sequences by combining statistical convergence with this new concept. Similarly, Kisi and Nuray [19] defined Wijsman $\mathcal{J}$−convergence for set sequences with an ideal $\mathcal{J}$.

Definition 7. *([18]) Let (X, ρ) be a metric space. For any non-empty closed subsets, $A, A_k \subseteq X$, we say that the sequence $\{A_k\}$ is Wijsman statistically convergent to A if $\{d(x, A_k)\}$ is statistically convergent to $d(x, A)$, i.e., $\varepsilon > 0$ and $x \in X$,*

$$\lim_{n\to\infty} \frac{1}{n} |\{k \leq n : |\rho(x, A_k) - \rho(x, A)| \geq \varepsilon\}| = 0.$$

In this case, we write $st - \lim_W A_k = A$ or $A_k \to A(WS)$. We denote the set of all Wijsman statistically convergent sequences by WS.

Definition 8. *([19]) Let (X, ρ) be a metric space and $\mathcal{J} \subseteq 2^{\mathbb{N}}$ be a proper ideal in $\mathbb{N}$. For any non-empty closed subsets, $A, A_k \subset X$, we say that the sequence $\{A_k\}$ is Wijsman $\mathcal{J}-$convergent to A, if for each $\varepsilon > 0$, and $x \in X$,, the set is*

$$A(x, \varepsilon) = \{k \in \mathbb{N} : |\rho(x, A_k) - \rho(x, A)| \geq \varepsilon\} \in \mathcal{J}.$$

In this case, we write $\mathcal{I}_W - \lim A_k = A$ or $A_k \to A(\mathcal{I}_W)$, and we denote the set of all Wijsman $\mathcal{J}-$convergent sequences by $\mathcal{J}_W$.

Example 2. *Let $X = \mathbb{R}^2$ and $\{A_k\}$ be a sequence as follows:*

$$A_k = \begin{cases} \{(x, y) \in \mathbb{R}^2 : x^2 + y^2 - 2ky = 0\} & \text{if, } k \neq n^2 \\ \{(x, y) \in \mathbb{R}^2 : y = -1\} & \text{if, } k = n^2 \end{cases}$$

and

$$A = \left\{(x, y) \in \mathbb{R}^2 : y = 0\right\}.$$

The sequence $\{A_k\}$ is not Wijsman convergent to the set A. Howeverm if we choose the ideal $\mathcal{J} = \mathcal{J}_d$, then $\{A_k\}$ is Wijsman $\mathcal{J}-$convergent to set A,, where $\mathcal{J}_d = \{T \subseteq \mathbb{N} : d(T) = 0\}$, and where d is the natural density.

Definition 9. *([19]) In any metric space, let $\mathcal{J} \subseteq 2^{\mathbb{N}}$ be a non-trivial ideal and $A, A_k \subset X$. The sequence $\{A_k\}$ is said to be Wijsman $\mathcal{J}-$statistically convergent to A or $S(\mathcal{J}_W)$-convergent to A if*

$$\left\{n \in \mathbb{N} : \frac{1}{n} |\{k \leq n : |\rho(x, A_k) - \rho(x, A)| \geq \varepsilon\}| \geq \delta\right\} \in \mathcal{J}$$

for each $\varepsilon > 0$ and each $x \in X$ and $\delta > 0$, and we write $A_k \to A(S(\mathcal{J}_W))$. The class of all Wijsman $\mathcal{J}-$ statistically convergent sequences is denoted by $S(\mathcal{J}_W)$.

Recently, Hazarika and Esi [20] and Savas [21] obtained some results about asymptotically $\mathcal{J}-$statistically equivalent set sequences.

1.4. The Number α

In recent years, many concepts that are considered essential in this area has been reworked using the alpha number. In references [22,23], by using the natural density of order α, the statistical convergence of order α $(0 < \alpha < 1)$ was introduced. The new definition is not exactly parallel to that of statistical convergence. Some other applications of this concept are the $\lambda-$statistical convergence of order α by Çolak and Bektaş [24], the lacunary statistical convergence of order α by Şengül and Et [25], the weighted statistical convergence of order α and its applications by Ghosal [26], and the almost statistical convergence of order α by Et, Altın and Çolak [27]. $\mathcal{J}-$statistical convergence and $\mathcal{J}-$lacunary statistical convergence of order α were introduced by Das and Savaş in 2014 [28]. In all of these studies, n was replaced by n^{α} in the denominator in the definition of natural density, and a different direction was given.

In 2017, Savas [21] gave a new definition about Wijsman asymptotically $\mathcal{J}-$statistical equivalence of order α $(0 < \alpha \leq 1)$ as follows:

Definition 10. *([21]) In any metric space, let $A_k, B_k \subseteq X$ be any non-empty closed subsets such that $d(x, A_k) > 0$ and $d(x, B_k) > 0$ for all $x \in X$. We say that the sequences $\{A_k\}$ and $\{B_k\}$ are Wijsman asymptotically $\mathcal{J}-$ statistically equivalent of order α $(0 < \alpha \leq 1)$ to multiple L if for each $\varepsilon > 0$, $\delta > 0$, and $x \in X$,*

$$\left\{n \in \mathbb{N} : \frac{1}{n^\alpha}\left|\left\{k \leq n : \left|\frac{\rho(x, A_k)}{\rho(x, B_k)} - L\right| \geq \varepsilon\right\}\right| \geq \delta\right\} \in \mathcal{J}.$$

In this case, we write $\{A_k\} \overset{[S^L(\mathcal{J}_W)]^\alpha}{\sim} \{B_k\}$. It is obvious that $[S^L(\mathcal{J}_W)]^\alpha$ denotes the the set of all sequences such that $\{A_k\} \overset{[S^L(\mathcal{J}_W)]^\alpha}{\sim} \{B_k\}$.

1.5. Present Study

It should be mentioned that the generalization of the concept of $W-$asymptotically $\mathcal{J}-$statistical equivalence of order α for $\lambda = (\lambda_n)$ sequences has not been studied until now. So, this brings to mind the question of how our new results will be if we use λ and p sequences. This makes the study interesting. In this study, we searched for the answer to this question. We generalized $W-$asymptotically $\mathcal{J}_\lambda-$statistical equivalence of order α and compared the properties of this new concept with the other type of convergences without λ.

2. Main Results

Following this information, we now consider our main definitions and results. Throughout the paper, (X, ρ) is a metric space, $\mathcal{J} \subseteq 2^{\mathbb{N}}$ is an admissible ideal, $(\lambda_n) \in \Lambda$ and λ_n^α is the αth power of $(\lambda_n)^\alpha$ of λ_n, that is $\lambda^\alpha = (\lambda_n)^\alpha = (\lambda_1^\alpha, \lambda_2^\alpha, ..., \lambda_n^\alpha, ...)$, and $p = (p_k)$ is a positive real number sequence. We use the W $(Wijsman)$ symbol since our expressions are defined for set sequences.

Definition 11. *Let $A_k, B_k \subseteq X$ be non-empty closed subsets such that $d(x, A_k) > 0$ and $d(x, B_k) > 0$ for all $x \in X$. Then, the sequences $\{A_k\}$ and $\{B_k\}$ are $W-$strongly Cesáro asymptotically $\mathcal{J}-$statistically equivalent of order α $(0 < \alpha \leq 1)$ to multiple L if for each $\delta > 0$ and $x \in X$,*

$$\left\{n \in \mathbb{N} : \frac{1}{n^\alpha}\sum_{k=1}^{n}\left|\frac{\rho(x, A_k)}{\rho(x, B_k)} - L\right|^{p_k} \geq \delta\right\} \in \mathcal{J}.$$

For this situation, we write $\{A_k\} \overset{[\sigma^{L(p)}(\mathcal{J}_W)]^\alpha}{\sim} \{B_k\}$, and $\left[\sigma^{L(p)}(\mathcal{J}_W)\right]^\alpha$ denotes the set of all sequences $\{A_k\}$ and $\{B_k\}$ such that $\{A_k\} \overset{[\sigma^{L(p)}(\mathcal{J}_W)]^\alpha}{\sim} \{B_k\}$.

Now let us give our definitions with the λ sequence.

Definition 12. *Let $A_k, B_k \subseteq X$ be non-empty closed subsets such that $d(x, A_k) > 0$ and $d(x, B_k) > 0$ for all $x \in X$. Then, the sequences $\{A_k\}$ and $\{B_k\}$ are $W-$ asymptotically $\mathcal{J}_\lambda-$statistical equivalent of order α $(0 < \alpha \leq 1)$ to multiple L and denoted by $\{A_k\} \overset{[S_\lambda^L(\mathcal{J}_W)]^\alpha}{\sim} \{B_k\}$ if for each $\varepsilon > 0$, $\delta > 0$, and $x \in X$,*

$$\left\{n \in \mathbb{N} : \frac{1}{\lambda_n^\alpha}\left|\left\{k \in I_n : \left|\frac{\rho(x, A_k)}{\rho(x, B_k)} - L\right| \geq \varepsilon\right\}\right| \geq \delta\right\} \in \mathcal{J}.$$

We denote the set of all sequences of $\{A_k\}$ and $\{B_k\}$ such that $\{A_k\} \overset{[S_\lambda^L(\mathcal{J}_W)]^\alpha}{\sim} \{B_k\}$ by $[S_\lambda^L(\mathcal{J}_W)]^\alpha$.

Definition 13. *Let $A_k, B_k \subseteq X$ be non-empty closed subsets such that $d(x, A_k) > 0$ and $d(x, B_k) > 0$ for all $x \in X$. Then, the sequences $\{A_k\}$ and $\{B_k\}$ are $W-$ strongly asymptotically $\mathcal{J}_\lambda-$statistically equivalent of order α $(0 < \alpha \leq 1)$ to multiple L if for each $\delta > 0$ and each $x \in X$,*

$$\left\{ n \in \mathbb{N} : \frac{1}{\lambda_n^\alpha} \sum_{k \in I_n} \left| \frac{\rho(x, A_k)}{\rho(x, B_k)} - L \right|^{p_k} \geq \delta \right\} \in \mathcal{J}.$$

We denote the set of this kind of sequence by $\left[V_\lambda^{L(p)}(\mathcal{J}_W) \right]^\alpha$.

The next theorem examines the relation between Savas' definition and our second definition.

Theorem 1. *(i) If $\liminf_{n\to\infty} \frac{\lambda_n^\alpha}{n^\alpha} > 0$ then $\left[S^L(\mathcal{J}_W)\right]^\alpha \subset \left[S_\lambda^L(\mathcal{J}_W)\right]^\alpha$.*
(ii) If $\liminf_{n\to\infty} \frac{\lambda_n^\alpha}{n^\alpha} = 1$ then $\left[S_\lambda^L(\mathcal{J}_W)\right]^\alpha \subset \left[S^L(\mathcal{J}_W)\right]^\alpha$.

Proof. (i) Assume that $\frac{\lambda_n^\alpha}{n^\alpha} > 0$ and $\{A_k\} \overset{S^L(\mathcal{J}_W)}{\sim} \{B_k\}$. Then, there exists a $\eta > 0$ such that $\frac{\lambda_n^\alpha}{n^\alpha} \geq \eta$ for sufficiently large n. For every $\varepsilon > 0$, we have,

$$\frac{1}{n^\alpha} \left\{ k \leq n : \left| \frac{\rho(x,A_k)}{\rho(x,B_k)} - L \right| \geq \varepsilon \right\} \supseteq \frac{1}{n^\alpha} \left\{ k \in I_n : \left| \frac{\rho(x,A_k)}{\rho(x,B_k)} - L \right| \geq \varepsilon \right\}.$$

If we think about the number of elements of the sets that provide this relation,

$$\begin{aligned}
\frac{1}{n^\alpha} \left| \left\{ k \leq n : \left| \frac{\rho(x,A_k)}{\rho(x,B_k)} - L \right| \geq \varepsilon \right\} \right| &\geq \frac{1}{n^\alpha} \left| \left\{ k \in I_n : \left| \frac{\rho(x,A_k)}{\rho(x,B_k)} - L \right| \geq \varepsilon \right\} \right| \\
&\geq \frac{\lambda_n^\alpha}{n^\alpha} \frac{1}{\lambda_n^\alpha} \left| \left\{ k \in I_n : \left| \frac{\rho(x,A_k)}{\rho(x,B_k)} - L \right| \geq \varepsilon \right\} \right| \\
&\geq \eta \frac{1}{\lambda_n^\alpha} \left| \left\{ k \in I_n : \left| \frac{\rho(x,A_k)}{\rho(x,B_k)} - L \right| \geq \varepsilon \right\} \right|,
\end{aligned}$$

we get, for any $\delta > 0$,

$$\begin{aligned}
&\left\{ n \in \mathbb{N} : \frac{1}{\lambda_n^\alpha} \left| \left\{ k \in I_n : \left| \frac{\rho(x,A_k)}{\rho(x,B_k)} - L \right| \geq \varepsilon \right\} \right| \geq \delta \right\} \\
&\subseteq \left\{ n \in \mathbb{N} : \frac{1}{n^\alpha} \left| \left\{ k \leq n : \left| \frac{\rho(x,A_k)}{\rho(x,B_k)} - L \right| \geq \varepsilon \right\} \right| \geq \delta\eta \right\} \in \mathcal{J}.
\end{aligned}$$

Then, we have the proof.
(ii) Let $\delta > 0$. Since $\liminf_{n\to\infty} \frac{\lambda_n^\alpha}{n^\alpha} = 1$, we have $m \in N$ such that $\left| \frac{\lambda_n^\alpha}{n^\alpha} - 1 \right| < \frac{\delta}{2}$ for all $n \geq m$. For $\varepsilon > 0$,

$$\begin{aligned}
\frac{1}{n^\alpha} \left| \left\{ k \leq n : \left| \frac{\rho(x, A_k)}{\rho(x, B_k)} - L \right| \geq \varepsilon \right\} \right| &\leq \frac{n^\alpha - \lambda_n^\alpha}{n^\alpha} + \frac{1}{\lambda_n^\alpha} \left| \left\{ k \in I_n : \left| \frac{\rho(x, A_k)}{\rho(x, B_k)} - L \right| \geq \varepsilon \right\} \right| \\
&\leq 1 - (1 - \frac{1}{\delta}) + \frac{1}{\lambda_n^\alpha} \left| \left\{ k \in I_n : \left| \frac{\rho(x, A_k)}{\rho(x, B_k)} - L \right| \geq \varepsilon \right\} \right|
\end{aligned}$$

for all $n \geq m$. Hence,

$$\begin{aligned}
&\left\{ n \in \mathbb{N} : \frac{1}{n^\alpha} \left| \left\{ k \leq n : \left| \frac{\rho(x,A_k)}{\rho(x,B_k)} - L \right| \geq \varepsilon \right\} \right| \geq \delta \right\} \\
&\subseteq \left\{ n \in \mathbb{N} : \frac{1}{\lambda_n^\alpha} \left| \left\{ k \in I_n : \left| \frac{\rho(x,A_k)}{\rho(x,B_k)} - L \right| \geq \varepsilon \right\} \right| \geq \frac{\delta}{2} \right\} \cup \{1, 2, ...m\}.
\end{aligned}$$

We know that the right side belongs to the ideal because of the theorem expression. So we have the proof. □

Now let us investigate how the $p = (p_k)$ sequence affects the previous definitions. Initially, we use the constant $p = (p) = (p, p, p, p, ...)$ sequence of positive real numbers in the following two theorems.

Theorem 2. *(i) If* $\{A_k\} \overset{\left[V_\lambda^{Lp}(\mathcal{J}_W)\right]^\alpha}{\sim} \{B_k\}$ *then* $\{A_k\} \overset{\left[S_\lambda^{L}(\mathcal{J}_W)\right]^\alpha}{\sim} \{B_k\}$.

ii) If $\{A_k\},\{B_k\} \in L_\infty$ and $\{A_k\} \overset{\left[S_\lambda^{L}(\mathcal{J}_W)\right]^\alpha}{\sim} \{B_k\}$ then $\{A_k\} \overset{\left[V_\lambda^{Lp}(\mathcal{J}_W)\right]^\alpha}{\sim} \{B_k\}$.

Proof. (i) Let $\{A_k\} \overset{\left[V_\lambda^{Lp}(\mathcal{J}_W)\right]^\alpha}{\sim} \{B_k\}$ and $\varepsilon > 0$. For each $x \in X$,

$$\begin{aligned} \sum_{k \in I_n} \left| \frac{\rho(x,A_k)}{\rho(x,B_k)} - L \right|^p &\geq \sum_{\substack{k \in I_n \\ \left|\frac{\rho(x,A_k)}{\rho(x,B_k)} - L\right| \geq \varepsilon}} \left| \frac{\rho(x,A_k)}{\rho(x,B_k)} - L \right|^p \\ &\geq \varepsilon^p \left| \left\{ k \in I_n : \left| \frac{\rho(x,A_k)}{\rho(x,B_k)} - L \right| \geq \varepsilon \right\} \right| \end{aligned}$$

and so,

$$\frac{1}{\varepsilon^p} \frac{1}{\lambda_n^\alpha} \sum_{k \in I_n} \left| \frac{\rho(x, A_k)}{\rho(x, B_k)} - L \right| \geq \frac{1}{\lambda_n^\alpha} \left| \left\{ k \in I_n : \left| \frac{\rho(x, A_k)}{\rho(x, B_k)} - L \right| \geq \varepsilon \right\} \right|.$$

Then, for any $\delta > 0$ we have,

$$\begin{aligned} &\left\{ n \in \mathbb{N} : \frac{1}{\lambda_n^\alpha} \left| \left\{ k \in I_n : \left| \frac{\rho(x,A_k)}{\rho(x,B_k)} - L \right| \geq \varepsilon \right\} \right| \geq \delta \right\} \\ &\subseteq \left\{ n \in \mathbb{N} : \frac{1}{\lambda_n^\alpha} \sum_{k \in I_n} \left| \frac{\rho(x,A_k)}{\rho(x,B_k)} - L \right|^p \geq \varepsilon^p \delta \right\}. \end{aligned}$$

Therefore, $\{A_k\} \overset{\left[S_\lambda^{L}(\mathcal{J}_W)\right]^\alpha}{\sim} \{B_k\}$.

(ii) Assume that $\{A_k\},\{B_k\} \in L_\infty$ and $\{A_k\} \overset{\left[S_\lambda^{L}(\mathcal{J}_W)\right]^\alpha}{\sim} \{B_k\}$. There is an M such that $\left| \frac{\rho(x,A_k)}{\rho(x,B_k)} - L \right| \leq M$ for each $x \in X$ and all k. For each $\varepsilon > 0$,

$$\begin{aligned} \frac{1}{\lambda_n^\alpha} \sum_{k \in I_n} \left| \frac{\rho(x,A_k)}{\rho(x,B_k)} - L \right|^p &= \frac{1}{\lambda_n^\alpha} \sum_{\substack{k \in I_n \\ \left|\frac{\rho(x,A_k)}{\rho(x,B_k)} - L\right| \geq \varepsilon}} \left| \frac{\rho(x,A_k)}{\rho(x,B_k)} - L \right|^p \\ &+ \frac{1}{\lambda_n^\alpha} \sum_{\substack{k \in I_n \\ \left|\frac{\rho(x,A_k)}{\rho(x,B_k)} - L\right| < \varepsilon}} \left| \frac{\rho(x,A_k)}{\rho(x,B_k)} - L \right|^p \\ &\leq \frac{1}{\lambda_n^\alpha} M^p \left| \left\{ k \in I_n : \left| \frac{\rho(x,A_k)}{\rho(x,B_k)} - L \right| \geq \varepsilon \right\} \right| \\ &+ \frac{1}{\lambda_n^\alpha} \varepsilon^p \left| \left\{ k \in I_n : \left| \frac{\rho(x,A_k)}{\rho(x,B_k)} - L \right| < \varepsilon \right\} \right| \\ &\leq \frac{M^p}{\lambda_n^\alpha} \left| \left\{ k \in I_n : \left| \frac{\rho(x,A_k)}{\rho(x,B_k)} - L \right| \geq \varepsilon \right\} \right| + \varepsilon^p \end{aligned}$$

and then for any $\delta > 0$,

$$\left\{ n \in \mathbb{N} : \frac{1}{\lambda_n^\alpha} \sum_{k \in I_n} \left| \frac{\rho(x,A_k)}{\rho(x,B_k)} - L \right|^p \geq \varepsilon \right\}$$

$$\subseteq \left\{ n \in \mathbb{N} : \frac{1}{\lambda_n^\alpha} \left| \left\{ k \in I_n : \left| \frac{\rho(x,A_k)}{\rho(x,B_k)} - L \right| \geq \varepsilon \right\} \right| \geq \frac{\varepsilon^p}{M^p} \right\} \in \mathcal{J}.$$

□

Now let us examine the above theorems for a non-constant $p = (p_k)$ sequence of positive real numbers.

Theorem 3. *(i) Let $p = (p_k)$ be a positive real number sequence, $\inf p_k = h$ and $\sup_k p = H$. $\{A_k\} \overset{\left[V_\lambda^{L(p)}(\mathcal{J}_W)\right]^\alpha}{\sim} \{B_k\}$ implies $\{A_k\} \overset{\left[S_\lambda^{L}(\mathcal{J}_W)\right]^\alpha}{\sim} \{B_k\}$.*

(ii) Let $\{A_k\}$ and $\{B_k\}$ be bounded sequences, $\inf p_k = h$, $\sup_k p = H$ and $\varepsilon > 0$. Then, $\{A_k\} \overset{\left[S_\lambda^{L}(\mathcal{J}_W)\right]^\alpha}{\sim} \{B_k\}$ implies $\{A_k\} \overset{\left[V_\lambda^{L(p)}(\mathcal{J}_W)\right]^\alpha}{\sim} \{B_k\}$.

Proof. (i) Assume that $\{A_k\} \overset{\left[V_\lambda^{L(p)}(\mathcal{J}_W)\right]^\alpha}{\sim} \{B_k\}$ and $\varepsilon > 0$. Then, we can write

$$\begin{aligned}
\frac{1}{\lambda_n^\alpha} \sum_{k \in I_n} \left| \frac{\rho(x,A_k)}{\rho(x,B_k)} - L \right|^{p_k} &= \frac{1}{\lambda_n^\alpha} \sum_{\substack{k \in I_n \\ \left| \frac{\rho(x,A_k)}{\rho(x,B_k)} - L \right| \geq \varepsilon}} \left| \frac{\rho(x,A_k)}{\rho(x,B_k)} - L \right|^{p_k} + \frac{1}{\lambda_n^\alpha} \sum_{\substack{k \in I_n \\ \left| \frac{\rho(x,A_k)}{\rho(x,B_k)} - L \right| < \varepsilon}} \left| \frac{\rho(x,A_k)}{\rho(x,B_k)} - L \right|^{p_k} \\
&\geq \frac{1}{\lambda_n^\alpha} \sum_{\substack{k \in I_n \\ \left| \frac{\rho(x,A_k)}{\rho(x,B_k)} - L \right| \geq \varepsilon}} \left| \frac{\rho(x,A_k)}{\rho(x,B_k)} - L \right|^{p_k} \\
&\geq \frac{1}{\lambda_n^\alpha} \sum_{\substack{k \in I_n \\ \left| \frac{\rho(x,A_k)}{\rho(x,B_k)} - L \right| \geq \varepsilon}} (\varepsilon)^{p_k} \\
&\geq \frac{1}{\lambda_n^\alpha} \sum_{\substack{k \in I_n \\ \left| \frac{\rho(x,A_k)}{\rho(x,B_k)} - L \right| \geq \varepsilon}} \min \left\{ (\varepsilon)^h, (\varepsilon)^H \right\} \\
&\geq \frac{1}{\lambda_n^\alpha} \min \left\{ (\varepsilon)^h, (\varepsilon)^H \right\} \left\{ k \in I_n : \left| \frac{\rho(x,A_k)}{\rho(x,B_k)} - L \right| \geq \varepsilon \right\}
\end{aligned}$$

and so for $\delta > 0$, we have

$$\left\{ n \in \mathbb{N} : \left| \left\{ k \in I_n : \left| \frac{\rho(x,A_k)}{\rho(x,B_k)} - L \right| \geq \varepsilon \right\} \right| \geq \delta \right\}$$

$$\subseteq \left\{ n \in \mathbb{N} : \frac{1}{\lambda_n^\alpha} \sum_{k \in I_n} \left| \frac{\rho(x,A_k)}{\rho(x,B_k)} - L \right|^{p_k} \geq \delta \min \left\{ (\varepsilon)^h, (\varepsilon)^H \right\} \right\} \in \mathcal{J}.$$

(ii) From the theorem's statement there is an integer (M) such that $\left|\frac{\rho(x,A_k)}{\rho(x,B_k)} - L\right| \leq M$ for each $x \in X$ and all k. For each $\varepsilon > 0$,

$$\begin{aligned}
\frac{1}{\lambda_n^\alpha}\sum_{k\in I_n}\left|\frac{\rho(x,A_k)}{\rho(x,B_k)} - L\right|^{p_k} &= \frac{1}{\lambda_n^\alpha}\sum_{\substack{k\in I_n \\ \left|\frac{\rho(x,A_k)}{\rho(x,B_k)} - L\right|\geq\varepsilon}}\left|\frac{\rho(x,A_k)}{\rho(x,B_k)} - L\right|^{p_k} \\
&+ \frac{1}{\lambda_n^\alpha}\sum_{\substack{k\in I_n \\ \left|\frac{\rho(x,A_k)}{\rho(x,B_k)} - L\right|<\varepsilon}}\left|\frac{\rho(x,A_k)}{\rho(x,B_k)} - L\right|^{p_k} \\
&\leq \frac{1}{\lambda_n^\alpha}\left|\left\{k \in I_n : \left|\frac{\rho(x,A_k)}{\rho(x,B_k)} - L\right| \geq \frac{\varepsilon}{2}\right\}\right| \max\left\{M^h, M^H\right\} \\
&+ \frac{1}{\lambda_n^\alpha}\left|\left\{k \in I_n : \left|\frac{\rho(x,A_k)}{\rho(x,B_k)} - L\right| < \frac{\varepsilon}{2}\right\}\right| \frac{\max(\varepsilon)^{p_k}}{2} \\
&\leq \max\left\{M^h, M^H\right\}\frac{1}{\lambda_n^\alpha}\left|\left\{k \in I_n : \left|\frac{\rho(x,A_k)}{\rho(x,B_k)} - L\right| \geq \frac{\varepsilon}{2}\right\}\right| + \frac{\max\left\{(\varepsilon)^h,(\varepsilon)^H\right\}}{2}
\end{aligned}$$

and

$$\begin{aligned}
&\left\{n \in \mathbb{N} : \frac{1}{\lambda_n^\alpha}\sum_{k\in I_n}\left|\frac{\rho(x,A_k)}{\rho(x,B_k)} - L\right|^{p_k} \geq \varepsilon\right\} \\
&\subseteq \left\{n \in \mathbb{N} : \frac{1}{\lambda_n^\alpha}\left|\left\{k \in I_n : \left|\frac{\rho(x,A_k)}{\rho(x,B_k)} - L\right| \geq \frac{\varepsilon}{2}\right\}\right| \geq \frac{2\varepsilon - \max\left\{(\varepsilon)^h,(\varepsilon)^H\right\}}{2\max\left\{M^h,M^H\right\}}\right\} \in \mathcal{J}.
\end{aligned}$$

□

Finally, in the last theorem we investigate the relationship between $\mathcal{W}-$ strongly asymptotically $\mathcal{J}_\lambda-$statistical equivalence of order α and $\mathcal{W}-$strongly Cesáro asymptotically $\mathcal{J}-$statistical equivalence of order α.

Theorem 4. *If* $\{A_k\} \overset{\left[V_\lambda^{L(p)}(\mathcal{J}_W)\right]^\alpha}{\sim} \{B_k\}$, *then* $\{A_k\} \overset{\left[\sigma^{L(p)}(\mathcal{J}_W)\right]^\alpha}{\sim} \{B_k\}$.

Proof. Now, assume that $\{A_k\} \overset{\left[V_\lambda^{L(p)}(\mathcal{J}_W)\right]^\alpha}{\sim} \{B_k\}$ and $\varepsilon > 0$.

$$\begin{aligned}
\frac{1}{n^\alpha}\sum_{k=1}^{n}\left|\frac{\rho(x,A_k)}{\rho(x,B_k)} - L\right|^{p_k} &= \frac{1}{n^\alpha}\sum_{k\in I_n}\left|\frac{\rho(x,A_k)}{\rho(x,B_k)} - L\right|^{p_k} + \frac{1}{n^\alpha}\sum_{k=1}^{n-\lambda n}\left|\frac{\rho(x,A_k)}{\rho(x,B_k)} - L\right|^{p_k} \\
&\leq \frac{1}{\lambda_n^\alpha}\sum_{k\in I_n}\left|\frac{\rho(x,A_k)}{\rho(x,B_k)} - L\right|^{p_k} + \frac{1}{\lambda_n^\alpha}\sum_{k=1}^{n-\lambda n}\left|\frac{\rho(x,A_k)}{\rho(x,B_k)} - L\right|^{p_k} \\
&\leq \frac{2}{\lambda_n^\alpha}\sum_{k\in I_n}\left|\frac{\rho(x,A_k)}{\rho(x,B_k)} - L\right|^{p_k}.
\end{aligned}$$

□

According to these operations,

$$\left\{n \in \mathbb{N} : \frac{1}{n^\alpha}\sum_{k=1}^{n}\left|\frac{\rho(x,A_k)}{\rho(x,B_k)} - L\right|^{p_k} \geq \varepsilon\right\} \subseteq \left\{n \in \mathbb{N} : \frac{1}{\lambda_n^\alpha}\sum_{k\in I_n}\left|\frac{\rho(x,A_k)}{\rho(x,B_k)} - L\right|^{p_k} \geq \frac{\varepsilon}{2}\right\} \in \mathcal{J}.$$

3. Conclusions and Future Developments

In our paper, we obtained some different results by defining the $W-$asymptotically $\mathcal{J}-$statistical equivalence of order α for $\lambda = (\lambda_n)$ sequences. Later on, we generalized our results by using a positive real number sequence $p = (p_k)$. Firstly, we compared the $W-$asymptotically $\mathcal{J}-$statistical equivalence of order α and the $W-$asymptotically $\mathcal{J}_\lambda-$statistical equivalence of order α for set sequences. These results are important to understand the role of λ. In other theorems, we investigated the relations between $W-$asymptotically $\mathcal{J}_\lambda-$statistical equivalence and $W-$strongly asymptotically $\mathcal{J}_\lambda-$statistically equivalence of order α according to whether p is constant or not. Then, we searched for the relation between $W-$strongly Cesáro asymptotically $\mathcal{J}-$statistically equivalent sequences of order α and $W-$ asymptotically $\mathcal{J}_\lambda-$statistical equivalent sequences of order α.

We know that the p sequence mentioned in this article is a sequence of positive integers. It is a matter of curiosity as to how the results will be obtained if the p sequence does not provide these conditions. On the other hand, it would be interesting to compare the results obtained using a different sequence to λ with the results in this article.

Author Contributions: Conceptualization, H.G. and N.D.; Investigation, H.G. and N.D.; Writing—original draft, H.G.

Funding: This research received no external funding.

Acknowledgments: The authors are grateful to the referees and the editor for their corrections and suggestions, which have greatly improved the readability of the paper.

Conflicts of Interest: The authors declare no conflict of interest.

References

1. Fast, H. Sur la convergence statistique. *Colloq. Math.* **1951**, *2*, 241–244. [CrossRef]
2. Steinhaus, H. Sur la convergence ordinaire et la convergence asymptotique. *Colloq. Math.* **1951**, *2*, 73–74.
3. Schoenberg, I.J. The integrability of certain functions and related summability methods. *Am. Math. Mon.* **1959**, *66*, 361–375. [CrossRef]
4. Anastassiou, G.A.; Duman, O. Statistical Korovkin theory for multivariate stochastic processes. *Stoch. Anal. Appl.* **2010**, *28*, 648–661. [CrossRef]
5. Freedman, A.R.; Sember, J.; Raphael, M. Some Cesàro-type summability spaces. *Proc. Lon. Math. Soc.* **1978**, *37*, 508–520. [CrossRef]
6. Gadjiev, A.D.; Orhan, C. Some approximation theorems via statistical convergence. *Rocky Mt. J. Math.* **2002**, *32*, 129–138. [CrossRef]
7. H. I. Miller, A measure theoretical subsequence characterization of statistical convergence. *Trans. Am. Math. Soc.* **1995**, *347*, 1811–1819. [CrossRef]
8. Zygmund, A. *Trigonometric Series*; Cambridge University Press: Cambridge, UK, 1979.
9. Mursaleen, M. $\lambda-$statistical convergence. *Math. Slovaca* **2000**, *50*, 111–115.
10. Kostyrko, P.; Šalát, T.; Wilezyński, W. $\mathcal{I}-$Convergence. *Real Anal. Exch.* **2000**, *26*, 669–686.
11. Das, P.; Savaş, E.; Ghosal, S. On generalized of certain summability methods using ideals. *Appl. Math. Lett.* **2011**, *26*, 1509–1514. [CrossRef]
12. Pobyvanets, I.P. Asymptotic equivalence of some linear transformation defined by a nonnegative matrix and reduced to generalized equivalence in the sense of Cesàro and Abel. *Matematicheskaya Fizika* **1980**, *28*, 83–87.
13. Marouf, M. Asymptotic equivalence and summability. *Int. J. Math. Math. Sci.* **1993**, *16*, 755–762. [CrossRef]
14. Bilgin, T. f-Asymptotically equivalent sequences. *Acta Univ. Apulensis* **2011**, *28*, 271–278.
15. Patterson, R.F. On asymptotically statistically equivalent sequences. *Demostratio Math.* **2003**, *36*, 149–153.
16. Ümüş, H.G.; Savaş, E. On $S_\lambda^L(I)-$asymptotically statistical equivalent sequences, *Numer. Anal. Appl. Math.* **2012**, *1479*, 936–941.
17. Baronti, M.; Papini, P. *Convergence of Sequences of Sets, Methods of Functional Analysis in Approximation Theory*; Birkhauser: Basel, Switzerland, 1986; pp. 133–155.
18. Nuray, F.; Rhodes, B.E. Statistical convergence of sequences of sets. *Fasc. Math.* **2012**, *49*, 87–99.
19. Kişi, Ö.; Nuray, F. New convergence definitions for sequence of sets. *Abstr. Appl. Anal.* **2013**, *2013*. [CrossRef]
20. Hazarika, B.; Esi, A. On asymptotically Wijsman lacunary statistical convergence of set sequences in ideal context. *Filomat* **2017**, *31*, 2691–2703. [CrossRef]

21. Savas, E. Asymptotically $\mathcal{J}$—lacunary statistical equivalent of order α for sequences of sets. *J. Nonlinear Sci. Appl.* **2017**, *10*, 2860–2867. [CrossRef]
22. Hunia, S.; Das, P.; Pal, S.K. Restricting statistical convergence. *Acta Math. Hungarica* **2012**, *13*, 153–161.
23. Çolak, R. *Statistical Convergence of Order α, Modern Methods in Analysis and Its Applications*; Anamaya Publishers: New Delhi, India, 2010; pp. 121–129.
24. Çolak, R.; Bektaş, Ç.A. λ-statistical convergence of order α. *Acta Math. Sci. Ser. B Engl. Ed.* **2011**, *31*, 953–959. [CrossRef]
25. Şengül, H.; Et, M. On lacunary statistical convergence of order α. *Acta Math. Sci. Ser. B Engl. Ed.* **2014**, *34*, 473–482. [CrossRef]
26. Ghosal, S. Weighted statistical convergence of order α and its applications. *J. Egypt. Math. Soc.* **2016**, *24*, 60–67. [CrossRef]
27. Et, M.; Altın, Y.; Çolak, R. Almost statistical convergence of order α. *Acta Sci. Mar.* **2015**, *37*, 55–61. [CrossRef]
28. Das, P.; Savaş, E. On $\mathcal{I}-$statistical and $\mathcal{I}-$lacunary statistical convergence of order a. *Bull. Irani. Math. Soc.* **2014**, *40*, 459–472.

axioms

Article

Some Exact Solutions to Non-Fourier Heat Equations with Substantial Derivative

Konstantin Zhukovsky *, Dmitrii Oskolkov and Nadezhda Gubina

Faculty of Physics, M.V. Lomonosov Moscow State University, Leninskie Gory, 119991 Moscow, Russia; oskolkov.di15@physics.msu.ru (D.O.); gubina_nadya@mail.ru (N.G.)

* Correspondence: zhukovsk@physics.msu.ru; Tel.: +7-495-9393177

Received: 10 April 2018; Accepted: 13 July 2018; Published: 18 July 2018

Abstract: One-dimensional equations of telegrapher's-type (TE) and Guyer–Krumhansl-type (GK-type) with substantial derivative considered and operational solutions to them are given. The role of the exponential differential operators is discussed. The examples of their action on some initial functions are explored. Proper solutions are constructed in the integral form and some examples are studied with solutions in elementary functions. A system of hyperbolic-type inhomogeneous differential equations (DE), describing non-Fourier heat transfer with substantial derivative thin films, is considered. Exact harmonic solutions to these equations are obtained for the Cauchy and the Dirichlet conditions. The application to the ballistic heat transport in thin films is studied; the ballistic properties are accounted for by the Knudsen number. Two-speed heat propagation process is demonstrated—fast evolution of the ballistic quasi-temperature component in low-dimensional systems is elucidated and compared with slow diffusive heat-exchange process. The comparative analysis of the obtained solutions is performed.

Keywords: exponential operator; differential operator; Guyer–Krumhansl equation; moving media; non–Fourier; heat conduction; Knudsen number

---

1. Introduction

Recent progress in technology and science has driven interest to studies of heat conduction beyond common Fourier law [1]: $\partial_t T = k\partial_x^2 T$, where k is the thermal diffusivity, T is the temperature; Fourier law describes heat conduction in homogeneous matter at normal conditions well. New heat sources, such as lasers, microwaves etc, are employed in medicine, science and material processing for melting, welding, cutting, drilling, etc. Often heat source and treated media are in motion. Some of modern materials and media have one or two dimensions: ultra-thin films, layers and nano-wires. Highly inhomogeneous porous and multilayered media are also common in industry. First major deviations from Fourier law were found in liquid Helium and in some solid crystal dielectrics at low temperatures <25 ° K [2–5]. Proper phenomenon was called Second Sound [6]; its satisfactory qualitative mathematical description was proposed by Cattaneo and Vernotte [7], who supposed phonon heat transport in addition to Fourier heat diffusion. In terms of temperature it can be put as follows:

$$(\tau\partial_t^2 + \partial_t)T = k_F\nabla^2 T \quad (1)$$

where k_F is the Fourier thermal diffusivity, $\tau = k_F/C^2$ is the relaxation time of the heat waves propagation, which relates the moments of the temperature change and of the respective heat flux change. Cattaneo-Vernotte constitution implies a phase lag between the heat flux vector and the temperature gradient; in addition to heat diffusion the temperature perturbation propagates in matter like damped sound-wave at finite speed $C = \sqrt{k_F/\tau}$. The relaxation time τ is associated with th phonon–phonon interaction time; at normal conditions it is very small: $\tau \approx 10^{-13}$ s. The Cattaneo Equation (1) is the particular case of the telegrapher's equation (TE).

$$\left(\partial_t^2 + \varepsilon \partial_t\right) F(x,t) = \left(\alpha \partial_x^2 + \kappa\right) F(x,t), \; \varepsilon, \; \alpha, \; \kappa = \text{const} \tag{2}$$

which takes its name because it describes the electric signal propagation in long electric lines without radiation [8]. For heat conduction we have $\tau = 1/\varepsilon$, $k_F = \alpha/\varepsilon$, $\mu = \kappa/\varepsilon$ and $\mu = 0$ in Equation (1); the source term, $\kappa \neq 0$, describes heat exchange with the environment of low excess temperatures. However, precise quantitative description of the Second Sound with Equation (1) was not successful. There are reports on non-Fourier heat transport in highly inhomogeneous matter even at normal conditions [9–17], in fuel droplets [18], in biomaterials [19], in energy saving and insulating materials [20], in graphene, nanofibers, carbon nanotubes, silicon wires, etc. [21–27]. Heat transport in these cases was found close to that described by Guyer and Krumhansl (GK) in [28,29]. In the one-dimensional case in terms of temperature alone it has the following form:

$$\left(\partial_t^2 + \varepsilon \partial_t - \delta \partial_{t,x,x}^3\right) T(x,t) = \left(\alpha \partial_x^2 + \kappa\right) T(x,t), \; \alpha, \; \varepsilon, \; \delta, \; \kappa = \text{const}. \tag{3}$$

In its pure form the Guyer–Krumhansl law has $\kappa = 0$. The following formulation:

$$\left(\tau \partial_t^2 + \partial_t\right) T(x,t) = \left(D_b \partial_{t,x,x}^3 + k_F \partial_x^2 + \mu\right) T(x,t), \tag{4}$$

involves the parameters $\tau = 1/\varepsilon$, $\mu = \kappa/\varepsilon$, $k_F = \alpha/\varepsilon$, and $D_b = \delta/\varepsilon$, where the latter, k_b, is the ballistic type heat conductivity. The parameters in the above Equations (3) and (4) have the following dimensions: $[\tau] = s$, $[k_F] = \frac{m^2}{s}$, $[D_b] = m^2$, $[\mu] = \frac{1}{s}$, $[\kappa] = \frac{1}{s^2}$, $[\alpha] = \frac{m^2}{s^2}$, $[\delta] = \frac{m^2}{s}$, and $[\varepsilon] = \frac{1}{s}$.

Despite GK equation is usually associated with ballistic properties, which manifest when the characteristic system scale L is comparable or less than the mean free path l of the phonons $L < l$, contradictory opinions on it exist [30–32]. In particular, the term "ballistic" is questioned in the context of GK type heat conduction in macroscopic inhomogeneous materials at room temperature [10,11]. Moreover, more complicated system of equations was proposed for the ballistic heat transport in [30], while heat propagation in highly inhomogeneous materials was reported to be close to GK law [9,11].

Solutions to Hyperbolic Heat Conduction Equation (HHCE) can be obtained both analytically and numerically [33–40], although numerical methods seem to be more commonly used [41–44]. Analytical study gives deeper insight in the problem; operational analytical approach and solutions to HHE were developed in [45–50]. This method easily handles also other linear DE of high order and fractional DE [51–58]. Use of the exponential differential operators, such as the heat operator $S = e^{\partial_x^2}$ [59] allows operational solution of GK-type Equation (4) as demonstrated in [47–50].

In what follows we will obtain some exact solutions to non-Fourier heat transfer in GK-type equation with the substantial derivative, where the speed of the media is constant, and will obtain some particular operational solutions to modified GK-type equation in this case. Moreover, we will study the important case of periodic initial conditions, which occur, for example, in polymer electrolyte fuel cells (PEFCs) [60–68], as well as in thin membranes and highly inhomogeneous porous periodic structures, in printed wired heating boards [69] etc. We will analyze the solutions for the heat transport, described by the system of inhomogeneous partial derivative equation, and use the Knudsen number to account for the ballistic conditions in 1- and 2-dimensional structures [70].

2. Ballistic Heat Transport Equations with Substantial Derivative

For ultra-thin films and wires neither pure GK nor Cattaneo laws describe exactly heat transport phenomena, but their combination with Fourier law; it forms the following inhomogeneous system of differential equations for the ballistic $\theta_b(x,t)$ and diffusive component $\theta_d(x,t)$ of the complete dimensionless quasi-temperature $\theta = \theta_d + \theta_b$ (see [30]):

$$\left\{\partial_t^2 + 2\partial_t - \frac{10Kn_b^2}{3}\partial_x^2 - 3Kn_b^2\partial_{t,x,x}^3 + 1\right\}\theta_b(x,t) = 0, \tag{5}$$

$$\left(\partial_t^2 + \frac{Kn_b^2}{Kn_d^2}\partial_t - \frac{Kn_b^4}{3Kn_d^2}\partial_x^2\right)\theta_d(x,t) = \left(\partial_t + \frac{Kn_b^2}{Kn_d^2}\right)\theta_b(x,t), \tag{6}$$

where indices b and d stand for "ballistic" and "diffusive", respectively. The dimensionless quasi-temperature θ can be described as the non-dimensional energy, associated with the internal energy; the quantities θ_d and θ_b must therefore be understood to quasi-temperatures, defined as a measure of corresponding internal energy components u_d and u_b, respectively to which they are related by the simple relations $\theta_d = u_d/c\ \theta_b = u_b/c$ (see [30] for details), where c is the heat capacity.

It should be understood though, that the ballistic heat transport is effectively described by both contributions in $\theta = \theta_d + \theta_b$, $\theta_d(x,t)$ and $\theta_b(x,t)$ are distinguished for convenience. It is easy to see that the above Equations (5) and (6) are of GK-type (5) and of telegrapher's type (6); the inhomogeneous right hand side (r.h.s.) of Equation (6) is given by the solution of (5). In the above equations and in all following equations the space- and time-coordinates are obviously dimensionless as well as the proper coefficients in the equations are. The dimensionless form of equations for the temperature is useful at least because of the space-time temperature distribution is expressed in our work explicitly in terms of x and t. Proper renormalization in SI units or else how is elementary (see, for example, [30] for details) and good, especially in view of heavy notations in the analytical solutions, which we will obtain in what follows.

Low-dimensional system is characterized by the dimensionless Knudsen number $Kn = l/L$, which for the phonon heat transport describes the ratio of the free mean path of the phonons l to the characteristic scale of the system L. Knudsen number usually arises in problems, where the scale of the system and the characteristic scale of the processes in it compare with each other, such as for gas flow in ultra-narrow channels [71], etc. In distinct ballistic case in (5), $Kn = 1$, we have $\alpha = 3.333$, $\delta = 3$, $\varepsilon = 2$, $\kappa = -1$ in (3) and $k_F = 5/3$, $D_b = 3/2$, $\tau = 1/2$, $\mu = -1/2$ in (4). We assume dimensionless equations here and in what follows. In the weak ballistic case, $Kn = 0.1$, we get $\alpha = 0.03333$, $\delta = 0.03$, $\varepsilon = 2$, $\kappa = -1$ in (3) and $k_F = 0.01666$, $k_b = 0.015$, $\tau = 1/2$, $\mu = -1/2$ in (4). In the distinct ballistic case, $Kn = 1$, all heat transport terms in GK-type equation contribute more or less equally, in the weak ballistic case, $Kn = 0.1$, the Cattaneo wave-term prevails. The inhomogeneous system of PDE (5) and (6) for ballistic heat transport thin film was studied numerically in [30]; based on the periodic analytical solutions to GK-type equation [72] some solutions to (5) and (6) were obtained in [73,74].

Above Equations (5) and (6), as well as Equations (2)–(4), describe non-Fourier heat transport in stationary media. In the case when the observer moves relatively to the media with constant speed $\vec{v}$, the substantial derivative $D/Dt = \partial/\partial t + \vec{v}\vec{\nabla}$ should be used instead of the time-derivative $\partial/\partial t$ to describe the variation of quantity F along a path. Generally speaking, both the material equations and the energy, mass and momentum conservation equations in case of moving observer should be considered with account for the substantial derivative, resulting in a proper heat conduction equation for the temperature alone. This is relatively simple in the case of Fourier heat diffusion and even for relativistic heat equation [75,76], some more complicated for GK-type equation. While not intending to rigorously derive a set of equations for the heat transport in case of moving observer, accounting for the relativistic heat flux, the energy balance, etc., we will solve the extended form of the known GK-type equation with the simple substantial derivative. Derivation of a physically comprehensive one-dimensional analogue of GK-equation in case of moving observer remains beyond the scope of the present study. The following modified form of GK-type equation arises upon the use of the substantial time derivative:

$$\left(\partial_t^2 + \varepsilon\partial_t + 2v\partial_{t,x}^2 - \delta\partial_{t,x,x}^3\right)T(x,t) = \left(\left(\alpha - v^2\right)\partial_x^2 - \varepsilon v\partial_x + \delta v\partial_x^3 + \kappa\right)T(x,t). \quad (7)$$

This simply presumes that the temperature as a function of time is recorded by a floating instrument in a flow, such as a weather balloon in meteorology and oceanography, which implies the substantial derivative along the pathline traveled. With this said, we notice that the above Equation (7) differs from GK-type Equation (4) by some additional terms: the second mixed time-space derivative in the left hand side (l.h.s.) and the third- and first-order space-derivatives in the r.h.s. We need to emphasize that this equation is related to a rigid domain meanwhile the observer is moving with speed $\vec{v}$. Moreover, the coefficient for the second-order space-derivative in the r.h.s. now contains the speed v, which can compensate α, eliminating the second-order space-derivative, if $v = \sqrt{\alpha}$, or invert the sign of this term, if $v > \sqrt{\alpha}$. Telegrapher's equation accordingly modifies as follows:

$$\left(\partial_t^2 + \varepsilon\partial_t + 2v\partial_{t,x}^2\right)T(x,t) = \left(\left(\alpha - v^2\right)\partial_x^2 - \varepsilon v\partial_x + \kappa\right)T(x,t). \quad (8)$$

Differently from GK-type equation with substantial derivative (7), TE (8) has only second-order differential operators; both GK-type and TE can be solved by the operational method, employing exponential differential operators $e^{\hat{D}}$ and involving the heat operator $\hat{S} = e^{\partial_x^2}$ [59]. The operational solution to the third-order PDE (7) apparently involves exponential differential operator of the third order $e^{a\partial_x^3}$. However, in what follows we will demonstrate as this solution effectively reduces to the shift of the solution in a stationary media. Considering thin films, we have obvious relations between the coefficients in (7) and (5): $\alpha_b \leftrightarrow 10Kn_b^2/3$, $\varepsilon_b \leftrightarrow 2$, $\delta \leftrightarrow 3Kn_b^2$, $\kappa_b \leftrightarrow -1$, while for Equation (6) we see that $\alpha_d \leftrightarrow Kn_b^4/3Kn_d^2$, $\varepsilon_d \leftrightarrow Kn_b^2/Kn_d^2$. For heat transport in moving thin film, the system of DE (5) and (6) for the complete quasi-temperature $\theta = \theta_d + \theta_b$ takes the following form:

$$\left(\partial_t^2 + 2\partial_t + 2v\partial_{t,x}^2 - \left(\frac{10}{3}Kn_b^2 - v^2\right)\partial_x^2 - 3Kn_b^2\partial_{t,x,x}^3 + 2v\partial_x - 3vKn_b^2\partial_x^3 + 1\right)\theta_b(x,t) = 0, \quad (9)$$

$$\left(\partial_t^2 + \frac{Kn_b^2}{Kn_d^2}\partial_t + 2v\partial_{t,x}^2 - \left(\frac{Kn_b^4}{3Kn_d^2} - v^2\right)\partial_x^2 + v\frac{Kn_b^2}{Kn_d^2}\partial_x\right)\theta_d(x,t) = \left(\partial_t + v\partial_x + \frac{Kn_b^2}{Kn_d^2}\right)\theta_b(x,t). \quad (10)$$

In what follows we will approach the GK-type equation with substantial derivative (7) operationally, and will also provide exact analytical harmonic solution to the system of PDE (9) and (10), describing ballistic heat transfer with substantial derivative.

3. Operational Approach to Transport Equations

The above Equations (7) and (8) are the particular cases of the following DE with the coordinate-dependent operators $\hat{\varepsilon}(x)$ and $\hat{D}(x)$:

$$\left(\partial_t^2 + \hat{\varepsilon}(x)\partial_t\right)F(x,t) = \hat{D}(x)F(x,t), \quad (11)$$

which can be solved using the operational method. Equation (11) models a broad spectrum of physical phenomena. It becomes telegrapher's equation for $\hat{D}(x) = \alpha\partial_x^2 + \kappa$ and constant ε term and further reduces to Cattaneo heat equation for $\hat{D}(x) = \partial_x^2$; GK-type equation appears when $\hat{\varepsilon} = \varepsilon - \delta\partial_x^2$. Some of these equations were studied in [45–50]. Other second-order PDE and fractional DE were explored with the help of the operational approach in [51–58]. The formal converging particular operational solution to (11) reads as follows:

$$F(x,t) = e^{-\frac{t\hat{\varepsilon}(x)}{2}} e^{-\frac{t}{2}\sqrt{\hat{\varepsilon}^2(x) + 4\hat{D}(x)}}C(x), \quad (12)$$

where $C(x)$ can be obtained from the initial condition $F(x,0) = f(x)$. The particular form of the initial function will be chosen below. The other branch of the solution contains the positive argument in the exponential, $e^{\frac{t}{2}\sqrt{\hat{\varepsilon}^2(x)+4\hat{D}(x)}}$. No Laplace transforms exist for it. However, symmetry with respect to inversion, $t \to -t$, $\varepsilon \to -\varepsilon$, $\delta \to -\delta$, $v \to -v$, allows writing the other solution to Equation (7), based on the one we obtain with the help of the Laplace transforms

$$e^{-t\sqrt{V}} = \frac{t}{2\sqrt{\pi}} \int_0^\infty \frac{d\xi}{\xi\sqrt{\xi}} e^{-\frac{t^2}{4\xi}-\xi V}, \; t > 0. \tag{13}$$

The solution has the following integral form, provided it converges:

$$F(x,t) = e^{-\frac{t}{2}\hat{\varepsilon}(x)} \frac{t}{4\sqrt{\pi}} \int_0^\infty \frac{d\xi}{\xi\sqrt{\xi}} e^{-\frac{t^2}{16\xi}} e^{-\xi\hat{\varepsilon}^2(x)} e^{-4\xi\hat{D}(x)} f(x). \tag{14}$$

The ability to perform analytical integration in (14) depends on the explicit form of the operators $\hat{\varepsilon}$, $\hat{D}$ and of the initial function $f(x)$; the numerical calculation can be done though. Accounting for the explicit form of the operators $\hat{D} = (\alpha - v^2)\partial_x^2 - \varepsilon v\partial_x + \delta v\partial_x^3 + \kappa$ and $\hat{\varepsilon} = \varepsilon + 2v\partial_x - \delta\partial_x^2$ in Equation (7), we obtain from (14) the following integral:

$$F(x,t) = e^{-\frac{t}{2}(\varepsilon+2v\partial_x)} \frac{t}{4\sqrt{\pi}} \int_0^\infty \frac{d\xi}{\xi\sqrt{\xi}} e^{-\frac{t^2}{16\xi}-\xi(\varepsilon^2+4\kappa)} e^{(\frac{t}{2}\delta+2\xi(\varepsilon\delta-2\alpha))\partial_x^2} e^{-\xi\delta^2\partial_x^4} f(x), \tag{15}$$

which benefits from the use of the operational identity for $\hat{p} = \sqrt{a}\hat{D}$, $\hat{D} = \partial^2$ (see [59,77]):

$$e^{\hat{p}^2} = \frac{1}{\sqrt{\pi}} \int_{-\infty}^{\infty} \exp\left(-\xi^2 + 2\xi\hat{p}\right) d\xi, \tag{16}$$

and yields

$$e^{a\hat{D}^2} f(x) = \frac{1}{\sqrt{\pi}} \int_{-\infty}^{\infty} \exp(-\xi^2 + 2\xi\sqrt{a}\hat{D}) f(x) d\xi. \tag{17}$$

Applying formula (17) to the fourth-order exponential differential operator $e^{-\xi\delta^2\partial_x^4} f(x) = \int_{-\infty}^{\infty} e^{-\zeta^2+2i\zeta\delta\sqrt{\xi}\partial_x^2} f(x) d\zeta/\sqrt{\pi}$, we get the heat operator [59] $\hat{S} = e^{\nu\partial_x^2}$; collecting the second-order derivative terms in the exponential, we get $\int_{-\infty}^{\infty} e^{-\zeta^2+((t\delta/2)-4\xi\alpha+2\xi\varepsilon\delta+2i\zeta\delta\sqrt{\xi})\partial_x^2} f(x) d\zeta$. This yields the following particular bounded solution to GK-type heat equation with substantial derivative (7):

$$F(x,t) = \frac{e^{-\frac{t}{2}\varepsilon} t}{4\pi} \hat{\Theta} \int_0^\infty \frac{d\xi}{\xi\sqrt{\xi}} e^{-\frac{t^2}{16\xi}-\xi(\varepsilon^2+4\kappa)} \int_{-\infty}^{\infty} e^{-\zeta^2} \hat{S} f(x) d\zeta, \tag{18}$$

which involves the shift operator $\hat{\Theta} = e^{y\partial_x}$, where $y = -vt$, and the heat operator $\hat{S} = e^{\eta\partial_x^2}$, where $\eta = 2i\zeta\sqrt{\xi}\delta - \xi 2(2\alpha + \varepsilon\delta) + t\delta/2$. The shift operator produces the translation along x $e^{y(\partial_x+\alpha)} f(x) = e^{y\alpha} f(x+y)$ and the heat operator produces Gauss transforms (17) and operational relations [59], so that the solution with account for the motion contains the translation $x - vt$ of the stationary solution.

Let us consider the initial polynomial function $f(x) = \sum_k x^k$. The particular solution (18) to GK-type equation with substantial derivative (7) arises upon the application of the operational rule

$$\hat{S} e^{\gamma x} x^k = e^{\eta\partial_x^2} x^k e^{\gamma x} = e^{\gamma x + \gamma^2 \nu} H_k(x + 2\gamma\nu, \nu), \; k \in \text{Integers}, \; \gamma \in \text{Reals}, \tag{19}$$

which is easy to prove [57,58]. For $\gamma = 0$ we immediately obtain the following integral:

$$F(x,t)|_{f(x)=\sum_k x^k} = \sum_k \frac{te^{-\frac{t}{2}\varepsilon}}{4\pi} \int_0^\infty \frac{d\xi}{\xi\sqrt{\xi}} e^{-\frac{t^2}{16\xi}-\xi(\varepsilon^2+4\kappa)} \int_{-\infty}^{\infty} e^{-\zeta^2} H_k(x-vt,\eta)d\zeta, \tag{20}$$

where

$$\eta = 2\xi\varepsilon\delta - 4\xi\alpha + t\delta/2 + i\zeta 2\sqrt{\xi}\delta. \tag{21}$$

The above integral can be taken in elementary functions if we account for the explicit form of the Hermite polynomials:

$$H_k(x-vt,\eta) = k! \sum_{r=0}^{[k/2]} \frac{(x-vt)^{k-2r}\eta^r}{(k-2r)!r!} = (-i)^k \eta^{\frac{k}{2}} H_k\left(\frac{i(x-vt)}{2\sqrt{\eta}}\right). \tag{22}$$

For example, for $f(x) = x^2$ the solution reads as follows:

$$F(x,t)|_{f(x)=x^2} = e^{-\frac{t}{2}(\sqrt{V}+\varepsilon)}\left((x-vt)^2 + t\delta + \frac{t}{\sqrt{V}}(\delta\varepsilon - 2\alpha)\right), \; V = \varepsilon^2 + 4\kappa^2. \tag{23}$$

More general case of the exponential-polynomial function $f(x) = x^k e^{\gamma x}$ is cumbersome and we omit proper expressions for conciseness. However, we have performed the integration explicitly in Wolfram Mathematica program and below we give the example of the solution for $F(x,0) = x^2 e^{-x}$:

$$\begin{aligned}
&F(x,t)|_{f(x)=x^2e^{-x}} = e^{-x-\frac{t}{2}(\sqrt{r}-2v-\delta+\varepsilon)}\left(x^2 + \tfrac{t}{r^{3/2}}(a+bx) + \tfrac{t^2}{r^{3/2}}c\right),\\
&r = 4\alpha + \delta^2 - 2\delta\varepsilon + \varepsilon^2 + 4\kappa,\\
&a = \sqrt{r}\delta\left(4\alpha + (\varepsilon-\delta)^2\right) + \delta^2\left(-6\alpha - \delta^2 - 3\varepsilon(\varepsilon-\delta)\right) +\\
&\varepsilon^2(-2\alpha + \delta\varepsilon) - 4\kappa(2\alpha + \sqrt{r}\delta - 3\delta^2 + \delta\varepsilon),\\
&b = 2r\left(2\alpha + \delta^2 - \sqrt{r}v - \delta(\sqrt{r}+\varepsilon)\right),\\
&c = r^{3/2}v^2 - 2rvd + 2p + \delta^2\Delta\\
&d = 2\alpha + \delta(-\sqrt{r} + \delta - \varepsilon),\\
&p = 2\alpha^2(\sqrt{r} - 4\delta) - 2\alpha\delta\left(3\delta^2 + \sqrt{r}\varepsilon + \varepsilon^2 - 2\delta(\sqrt{r}+2\varepsilon) + 4\kappa\right),\\
&\Delta = \sqrt{r}(\varepsilon-\delta)^2 + (\varepsilon-\delta)^3 + 2\kappa\left(\sqrt{r} + 2(\varepsilon-\delta)\right).
\end{aligned} \tag{24}$$

Evidently, all the values in the above expressions are dimensionless as well as the coordinate x and the time t.

In the harmonic ansatz the evolution of the initial function $f(x) = \exp(inx)$ can be easily obtained from (18) as follows:

$$F(x,t)|_{f(x)=e^{inx}} = \exp\left(in(x-vt) - \frac{t}{2}\left(\bar{\varepsilon} + \sqrt{\bar{\varepsilon}^2 + 4(\kappa^2 - \alpha n^2)}\right)\right), \; \bar{\varepsilon} = \varepsilon + n^2\delta. \tag{25}$$

Other solutions and their detailed study will be performed in forthcoming publications. All the above solutions to equations with substantial derivative contain the shift, $x - vt$, respectively to the solution in a stationary media, which depends on x. Further examples can be easily considered with the help of the operational approach. In the following chapter we will explore in details the particular case of the harmonic solution. Evidently, it can be easily generalized for any periodic solution, expandable in Fourier series. Similarly, the Fourier integral transforms technique apply.

4. Exact Periodic Solutions to GK-Type Equation with Substantial Derivative

An exact harmonic solution $\propto e^{inx}$ for the inhomogeneous system of PDE (9) and (10) can be obtained straight from the operational solution (18) (see (25)), or by separating the variables:

$T(x,t) = X(x)y(t)$. GK-type equation with substantial derivative (7) in the harmonic ansatz, $T(x,t) = e^{inx}y(t) \equiv F(x,t)$ reduces to the following ordinary differential equation for $y(t)$:

$$y''(t) + \varepsilon_{eff}y'(t) + \lambda_{eff}y(t) = 0, \tag{26}$$

where $\varepsilon_{eff} = \bar{\varepsilon} + 2vin, \bar{\varepsilon} = \varepsilon + n^2\delta$, and $\lambda_{eff} = (\alpha - v^2)n^2 + ivn\bar{\varepsilon} - \kappa$. The function $F(x,t) \equiv T(x,t)$ is introduced here for clarity of notations. The exact solution to GK equation with substantial derivative (7) then easily follows from the solution to (26):

$$\begin{gathered} F(x,t) = e^{inx}y(t),\ y(t) = e^{-\frac{t}{2}\varepsilon_{eff}}\left(C_1 e^{-\frac{t}{2}\sqrt{U}} + C_2 e^{\frac{t}{2}\sqrt{U}}\right), \\ U = \varepsilon_{eff}^2 - 4\lambda_{eff} = \bar{\varepsilon}^2 + 4\left(\kappa - \alpha n^2\right), \end{gathered} \tag{27}$$

where the constants C_1, C_2 are determined either from the initial or boundary conditions. According to the theory of separation of variables, the whole solution has to satisfy initial or boundary conditions. Evidently, telegrapher's equation with substantial derivative in the harmonic ansatz similarly reads

$$y''(t) + \tilde{\varepsilon}y'(t) + \tilde{\lambda}y(t) = 0, \tag{28}$$

and differs from (26) by the substitutions $\varepsilon_{eff} \to \tilde{\varepsilon} = \varepsilon + 2vin$, $\bar{\varepsilon} \to \varepsilon$, $\lambda_{eff} \to \tilde{\lambda} = (\alpha - v^2)n^2 + ivn\varepsilon - \kappa$. It's solution $y(t)$ is evidently (27), where $\delta = 0$.

Consider, for example, the Cauchy initial conditions

$$F(x,0) = Ae^{inx},\ \partial_t F(x,0) = Be^{inx}. \tag{29}$$

In the stationary case, $v = 0$, the Cauchy conditions (37), where initially $\frac{\partial\theta(x,t)}{\partial t} = 0$, have the meaning of zero heat flux q if $B = 0$ for both ballistic and diffusive components of quasi-temperature [30]. If $v \neq 0$, then the dimensionless equation $\theta'_t + v\theta'_x + q = 0$ describes the one-dimensional energy balance. The initially zero heat flux q then corresponds to $B = -invA$.

Then, GK-type Equation (7) with substantial derivative in the harmonic ansatz has the solution (27), satisfying ODE (26), where the coefficients C_1, C_2 are determined from the Cauchy conditions (29): $C_1 + C_2 = A$ and $-C_1\left(\varepsilon_{eff} + \sqrt{U}\right) - C_2\left(\varepsilon_{eff} - \sqrt{U}\right) = 2B$, and read as follows:

$$C_1 = \frac{A}{2} - \frac{B + A\varepsilon_{eff}/2}{\sqrt{U}},\ C_2 = \frac{A}{2} + \frac{B + A\varepsilon_{eff}/2}{\sqrt{U}}. \tag{30}$$

TE with substantial derivative (8) has the harmonic solution, whose time-dependent part $y(t)$ satisfies ODE (28):

$$F(x,t) = e^{inx}y(t),\ y(t) = e^{-\frac{t}{2}\tilde{\varepsilon}}\left(B_1 e^{-\frac{t}{2}\sqrt{u}} + B_2 e^{\frac{t}{2}\sqrt{u}}\right),\ u = \tilde{\varepsilon}^2 - 4\tilde{\lambda} = \varepsilon^2 + 4\left(\kappa - \alpha n^2\right), \tag{31}$$

where the coefficients B_1, B_2 are obtained from the Cauchy initial conditions (29):

$$B_1 = \frac{A}{2} - \frac{B + A\tilde{\varepsilon}/2}{\sqrt{u}},\ B_2 = \frac{A}{2} + \frac{B + A\tilde{\varepsilon}/2}{\sqrt{u}}. \tag{32}$$

Now, let us consider the Dirichlet boundary conditions in the moments of time $t = 0$ and $t = T$:

$$F(x,0) = Ae^{inx},\ F(x,T) = Ge^{inx}. \tag{33}$$

The harmonic solutions (27) and (31) satisfy respectively GK Equation (7) and telegrapher's Equation (8) with substantial derivative with the coefficients, determined by Equation (33):

$$C_1 = -e^{\frac{T}{2}\varepsilon_{eff}} \frac{Ae^{-\frac{T}{2}(\varepsilon_{eff}-\sqrt{U})} - G}{e^{-\frac{T}{2}\sqrt{U}} - e^{\frac{T}{2}\sqrt{U}}}, \; C_2 = A - C_1, \tag{34}$$

and

$$B_1 = -e^{\frac{T}{2}\tilde{\varepsilon}} \frac{Ae^{-\frac{T}{2}(\tilde{\varepsilon}-\sqrt{u})} - G}{e^{-\frac{T}{2}\sqrt{u}} - e^{\frac{T}{2}\sqrt{u}}}, \; B_2 = A - B_1. \tag{35}$$

Similarly, we can consider evolution of any function, expandable in Fourier series.

Note, that for some values of the coefficients α and δ the quantities U and u can assume negative values. Albeit it is not very obvious in the form of the harmonic solution (27) to GK-type equation with substantial derivative (26) and the solution (31) to TE with substantial derivative (28), these solutions remains real at any moment of time for a real initial function, because of the complex exponential is compensated by proper complex parts in the coefficients $C_{1,2}$ and $B_{1,2}$.

5. Exact Periodic Solutions to Ballistic Heat Transport in Thin Films

Ballistic heat transport in thin films is described by the system of PDE (9) and (10). In order to obtain the exact solution to this problem we reduced GK-type Equation (9) in the harmonic ansatz to (26) and solved it (see (27)); the result naturally involves the particular solution (25). This solution now constitutes the r.h.s. of Equation (10) for θ_d. Consider the Cauchy initial problem for θ_b:

$$\theta_b(x,0) = Ae^{inx}, \; \partial_t\theta_b(x,0) = Be^{inx}, \tag{36}$$

and the Cauchy conditions for θ_d:

$$\theta_d(x,0) = Ve^{inx}, \; \partial_t\theta_d(x,0) = We^{inx}, \; V, W = \text{const}. \tag{37}$$

The quasi-temperature component $\theta_b(x,t)$ is in fact given by Equation (27):

$$\theta_b(x,t) = e^{inx}\left(C_1 e^{-\frac{t}{2}E_1} + C_2 e^{-\frac{t}{2}E_2}\right), \; E_1 = \left(\varepsilon_{eff} + \sqrt{U}\right), \; E_2 = \left(\varepsilon_{eff} - \sqrt{U}\right), \tag{38}$$

where $\varepsilon_{eff} = \bar{\varepsilon} + 2ivn$, $U = \bar{\varepsilon}^2 + 4(\kappa - \alpha_b n^2)$, $\bar{\varepsilon} = \varepsilon_b + n^2\delta$, $\alpha_b = 10Kn_b^2/3$, $\varepsilon_b = 2$, $\delta = 3Kn_b^2$, $\kappa_b = -1$, v is the speed of the media; for the Cauchy initial conditions (36) (see (29) with $F \equiv \theta_b$) the coefficients $C_{1,2}$ are given by Equation (30). The solution (38) for $\theta_b(x,t)$ contributes to the r.h.s. of the telegrapher's Equation (6) for the component $\theta_d(x,t)$. The exponential differential operators $e^{\hat{D}(x)}$ do not bring new harmonics to the initial content, $\theta_d(x,t) = \Theta_d(t)e^{inx}$, and Equation (6) reduces to the following inhomogeneous ODE for $\Theta_d(t)$:

$$\left(\frac{d^2}{dt^2} + \tilde{\varepsilon}_d \frac{d}{dt} + \tilde{\lambda}_d\right)\Theta_d(t) = +Pe^{-\frac{t}{2}E_1} + Qe^{-\frac{t}{2}E_2}, \tag{39}$$

where

$$\begin{gathered}\tilde{\varepsilon}_d = \varepsilon_d + i2vn, \; \tilde{\lambda}_d = \left(\alpha_d - v^2\right)n^2 + ivn\varepsilon_d, \; \varepsilon_d = \frac{Kn_b^2}{Kn_d^2}, \; \alpha_d = \frac{Kn_b^4}{3Kn_d^2}, \\ P = C_1\left(\varepsilon_d - \frac{\bar{\varepsilon} + \sqrt{U}}{2} + ivn\right), \; Q = C_2\left(\varepsilon_d - \frac{\bar{\varepsilon} - \sqrt{U}}{2} + ivn\right),\end{gathered} \tag{40}$$

where $C_{1,2}$ are given by Equation (30). The above Equation (39) for $\Theta_d(t)$ possesses analytical solution, which, in turn, yields the evolving in time harmonic solution for the diffusive quasi-temperature component $\theta_d(x,t)$ in the following form:

$$\theta_d(x,t) = e^{inx}\left(4\left(\frac{P}{S}e^{-\frac{t}{2}E_1} + \frac{Q}{L}e^{-\frac{t}{2}E_2}\right) + D_1 e^{-\frac{t}{2}(\widetilde{\varepsilon}_d+R)} + D_2 e^{-\frac{t}{2}(\widetilde{\varepsilon}_d-R)}\right), \tag{41}$$

$$\begin{aligned} R &= \sqrt{\widetilde{\varepsilon}_d^2 - 4\widetilde{\lambda}_d} = \sqrt{\varepsilon_d^2 - 4n^2\alpha_d}, \\ S &= E_1^2 + 4\widetilde{\lambda}_d - 2E_1\widetilde{\varepsilon}_d = \left(\overline{\varepsilon} - \varepsilon_d + \sqrt{U}\right)^2 + 4\alpha_d n^2 - \varepsilon_d^2, \\ L &= E_2^2 + 4\widetilde{\lambda}_d - 2E_2\widetilde{\varepsilon}_d = \left(\overline{\varepsilon} - \varepsilon_d - \sqrt{U}\right)^2 + 4\alpha_d n^2 - \varepsilon_d^2, \end{aligned} \tag{42}$$

where the constants $D_{1,2}$ are determined from the Cauchy initial conditions (37) for $\theta_d(x,t)$:

$$D_1 = \frac{V}{2}\left(1 - \frac{\widetilde{\varepsilon}_d}{R}\right) - \frac{1}{R}\left(W + \frac{2P}{S}\left(\overline{\varepsilon} - \varepsilon_d + \sqrt{U} + R\right) + \frac{2Q}{L}\left(\overline{\varepsilon} - \varepsilon_d - \sqrt{U} + R\right)\right), \tag{43}$$

$$D_2 = \frac{V}{2}\left(1 + \frac{\widetilde{\varepsilon}_d}{R}\right) + \frac{1}{R}\left(W + \frac{2P}{S}\left(\overline{\varepsilon} - \varepsilon_d + \sqrt{U} - R\right) + \frac{2Q}{L}\left(\overline{\varepsilon} - \varepsilon_d - \sqrt{U} - R\right)\right). \tag{44}$$

The exact analytical harmonic solutions for the $\theta_b(x,t)$ is given by (38) and (30), and for $\theta_d(x,t)$, it is given by (41)–(44); their sum represents the complete solution for the ballistic heat transfer in thin films [30] with the Cauchy initial conditions (36), (37), valid for arbitrary values of the Knudsen numbers Kn_d, Kn_b. Note, that a real initial distribution of quasi-temperature $\theta(x,0)$ evolves in real domain even for imaginary exponentials in the solutions due to complex coefficients C_1, C_2 and D_1, D_2.

Now let us consider the Dirichlet conditions at the boundaries

$$\theta_b(x,0) = Ae^{inx},\ \theta_b(x,T) = Ge^{inx},\ \theta_d(x,0) = Ve^{inx},\ \theta_d(x,T) = We^{inx}, \tag{45}$$

where $A, G, V, W = \text{const}$ (see also Equation (33)). The evolution of $\theta_b(x,t)$ is given by Equation (38), where $C_{1,2}$ are set by Equation (34). The evolution of the counterpart $\theta_d(x,t)$ is governed by Equations (39) and (40). The r.h.s. of Equation (39) is set by Equation (38) for the ballistic component $\theta_b(x,t)$, where $C_{1,2}$ are given by (34). The explicit solution $\theta_d(x,t)$ to inhomogeneous DE (39) is possible, although it is much more cumbersome than in the Cauchy case:

$$\theta_d(x,t) = e^{inx}\left(\frac{b_2}{w_1}e^{-E_2 t} + \frac{b_1}{w_2}e^{-E_1 t} + \frac{q_1}{w_1 w_2}e^{-\frac{1}{2}t(\varepsilon_d+R)} - \frac{q_2}{w_1 w_2}e^{-\frac{1}{2}t(\varepsilon_d-R)}\right), \tag{46}$$

where

$$\begin{gathered} w_1 = E_2^2 - E_2\widetilde{\varepsilon}_d + \widetilde{\lambda}_d,\ w_2 = E_1^2 - E_1\widetilde{\varepsilon}_d + \widetilde{\lambda}_d, \\ q_1 = \frac{1}{e^{RT}-1}\left(e^{-E_2 T} w_1 b_1 d_2 + e^{-E_2 T} b_2 d_2 w_2 + d_1 w_1 w_2\right), \\ q_2 = \frac{1}{e^{RT}-1}\left(e^{-E_2 T}\left(\widetilde{\lambda}_d - w_2\right)(b_2 d_6 - w_1 d_4) + e^{-E_1 T}\left(w_1\left(\widetilde{\lambda}_d d_3 - b_1 d_5\right) + \widetilde{\lambda}_d b_2 d_7\right)\right), \\ b_1 = C_1\left(\widetilde{\varepsilon}_d - \frac{\varepsilon_{eff} + \sqrt{U}}{2} + ivn\right),\ b_2 = C_2\left(\widetilde{\varepsilon}_d - \frac{\varepsilon_{eff} - \sqrt{U}}{2} + ivn\right),\ d_1 = Ve^{RT} - We^{\frac{T}{2}(\widetilde{\varepsilon}_d+R)}, \\ d_2 = e^{\frac{T}{2}(\widetilde{\varepsilon}_d+R)} - e^{T(E_1+R)},\ d_3 = Ve^{E_1 T} - We^{(2E_1+\widetilde{\varepsilon}_d+R)\frac{T}{2}},\ d_4 = Ve^{E_2 T} - We^{(2E_2+\widetilde{\varepsilon}_d+R)\frac{T}{2}}, \\ d_5 = e^{E_1 T} - e^{(\widetilde{\varepsilon}_d+R)\frac{T}{2}},\ d_6 = e^{E_2 T} - e^{(\widetilde{\varepsilon}_d+R)\frac{T}{2}},\ d_7 = -e^{E_1 T} + e^{(\widetilde{\varepsilon}_d+R+2(E_1-E_2))\frac{T}{2}}. \end{gathered} \tag{47}$$

The solution for $\theta_d(x,t)$ in the form (46), (47) explicitly involves the Dirichlet conditions and together with the solution for $\theta_b(x,t)$, given in the Dirichlet case by (38) and (34), they describe the evolution of the complete harmonic quasi-temperature $\theta = \theta_d + \theta_b$.

Let us now consider some examples of the harmonic solutions to GK-type heat equation with substantial derivative. Consider, for example, the Cauchy problem with initial conditions $\theta_b(x,0) = \theta_d(x,0) = e^{ix}$, $\partial_t\theta_b(x,0) = \partial_t\theta_d(x,0) = 0$. Let's first consider $Kn_b = Kn_d = 1$. Proper solutions for $\theta_b(x,t)$, $\theta_d(x,t)$ and $\theta(x,t)$ in the stationary media are presented in Figure 1.

It can be seen in the top left plot in Figure 1 as the ballistic component $\theta_b(x,t)$ rapidly relaxes to the stationary state. The diffusive component $\theta_d(x,t)$ behaves similarly, but the relaxation process takes much more time. The behavior of ballistic and diffusive components in the domain is more distinct for $Kn < 1$. The complete quasi-temperature $\theta(x,t)$ (see bottom plot in Figure 1) follows the behavior of both the ballistic $\theta_b(x,t)$ and diffusive $\theta_d(x,t)$ constituents. However, the latter, being the solution of Equation (39), is strongly influenced by the solution (38) for the ballistic component $\theta_b(x,t)$, which sets the inhomogeneous part in Equation (39). Thus, the evolution of the quasi-temperature $\theta(x,t)$ in thin films has two speeds: the ballistic constituent $\theta_b(x,t)$ develops rapidly and sets the r.h.s. in Equation (39) for the diffusive constituent $\theta_d(x,t)$. The evolution of $\theta_d(x,t)$ is much slower: after the first instances it proceeds under the influence of already relaxed true ballistic constituent $\theta_b(x,t)$.

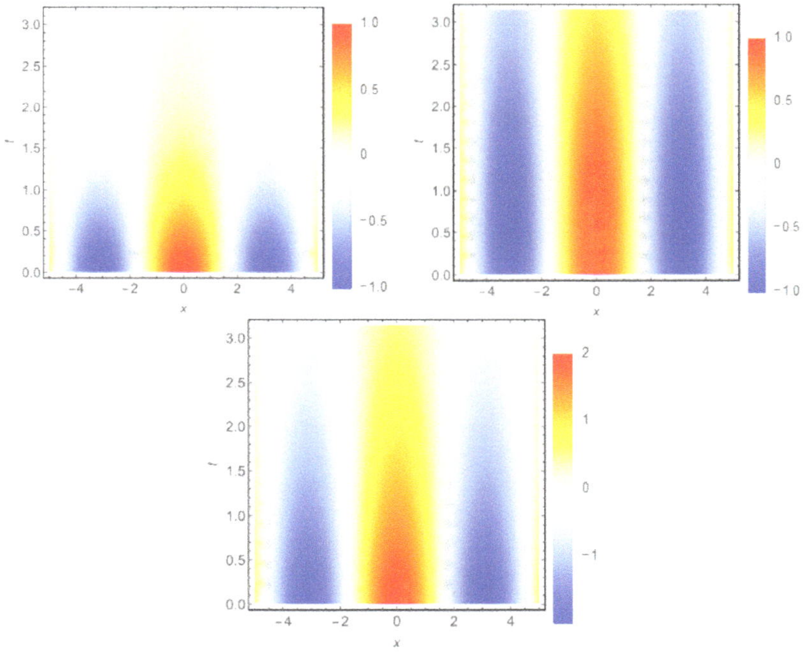

Figure 1. Evolution of quasi-temperature components in a thin stationary film with $Kn_b = Kn_d = 1$ for $v = 0$: $\theta_b(x,t)$—top left plot, $\theta_d(x,t)$—top right plot and $\theta(x,t) = \theta_b(x,t) + \theta_d(x,t)$—bottom plot. The Cauchy conditions for PDEs (9) and (10) system are $\theta_b(x,0) = \theta_d(x,0) = e^{inx}$, $\partial_t\theta_b(x,0) = \partial_t\theta_d(x,0) = 0$.

In the case when the rate of change of temperature for a given position in a field depends both on the instantaneous rate of change of temperature at that location $(\partial/\partial t)$ as well as on the rate at which the temperature is convected to that location by the fluid motion, the behaviors of the solutions change significantly. The solutions for $Kn_d = Kn_b = 1$ and $v = 10$, are presented in Figure 2, for $v = -10$ in Figure 3.

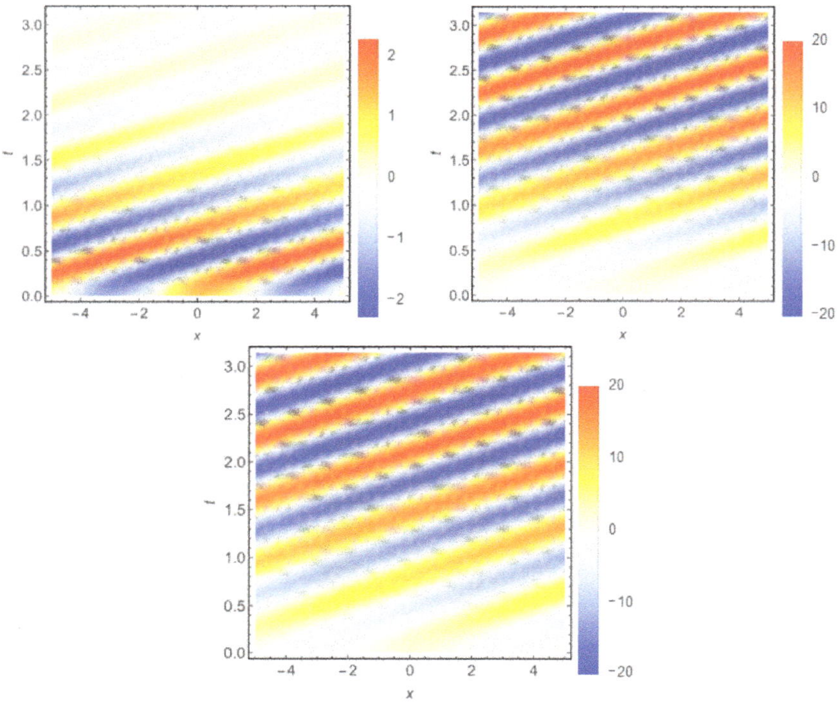

Figure 2. Evolution of quasi-temperature components in a thin film with $Kn_b = Kn_d = 1$ for $v = +10$: $\theta_b(x,t)$—top left plot, $\theta_d(x,t)$—top right plot and $\theta(x,t) = \theta_b(x,t) + \theta_d(x,t)$—bottom plot. The Cauchy conditions for PDEs (9) and (10) system are $\theta_b(x,0) = \theta_d(x,0) = e^{inx}$, $\partial_t\theta_b(x,0) = \partial_t\theta_d(x,0) = 0$.

The ballistic constituent θ_b rapidly vanishes with time; the direction of the colorful waves in the plot depends on the sign of the speed v (see top left plots in Figures 2 and 3). Diffusive counterpart of the complete quasi-temperature θ_d has waves of increasing amplitude as seen in top right plots in Figures 2 and 3. The complete quasi-temperature $\theta = \theta_d + \theta_b$ has obvious non-Fourier behavior, which largely follows that of θ_d.

For smaller values of Knudsen number, $Kn = 0.1$, the behavior of the harmonic solutions to ballistic heat propagation in thin films, obeying the Eqs. system (9) and (10) with Cauchy conditions (36) (37), has less distinctive wave interference with slower fade of the solution for θ_b and weaker amplitude increase speed for θ_d; proper solutions for $v = +10$ are presented in Figure 4.

It is important to note, that the whole system is symmetric under the speed inversion, which becomes obvious with account for formulae (38)–(44).

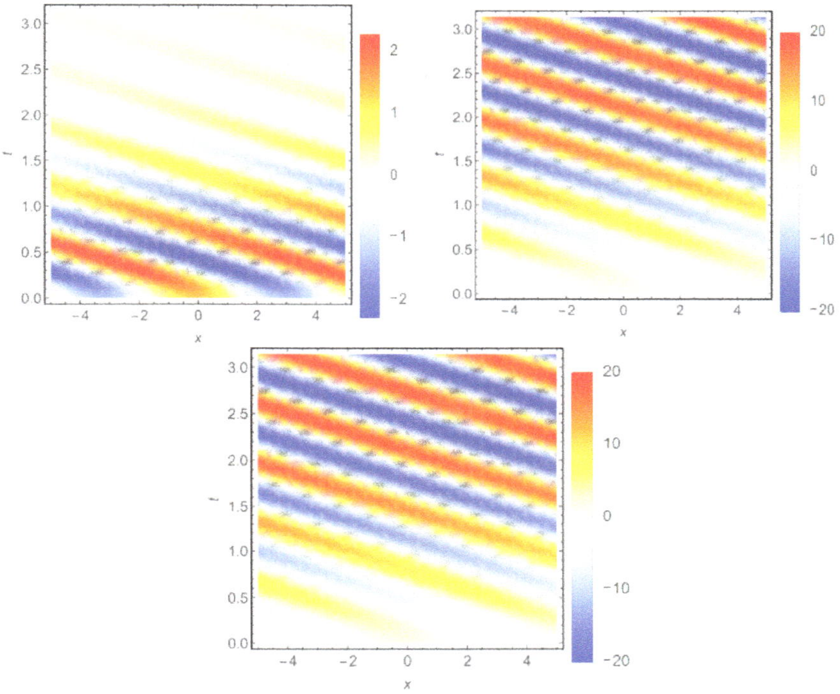

Figure 3. Evolution of quasi-temperature components in a thin film with $Kn_b = Kn_d = 1$ for $v = -10$: $\theta_b(x,t)$—top left plot, $\theta_d(x,t)$—top right plot and $\theta(x,t) = \theta_b(x,t) + \theta_d(x,t)$—bottom plot. The Cauchy conditions for PDEs (9) and (10) system are $\theta_b(x,0) = \theta_d(x,0) = e^{inx}$, $\partial_t\theta_b(x,0) = \partial_t\theta_d(x,0) = 0$.

Note, that for $Kn_b = Kn_d = 0.1$ $\theta_b(x,t)$ and $\theta_d(x,t)$ constituents behave differently (see Figure 4). The diffusive constituent slowly grows as the ballistic constituent shows gradual amplitude decrease, as shown in Figure 5. Moreover, for $v = 10$ the complete quasi-temperature $\theta(x,t) = \theta_b(x,t) + \theta_d(x,t)$ monotonously increases, following the behavior of the diffusive constituent as shown in Figure 5. The maximum principle, established though for parabolic equations, is violated for the diffusive component and complete quasi-temperature with Cauchy conditions for non-Fourier ballistic heat transport in thin films. In all of the cases, the evolution of $\theta_b(x,t)$ occurs faster than that of $\theta_d(x,t)$; the two-speed heat propagation process can be seen in every set of plots in Figures 1–5.

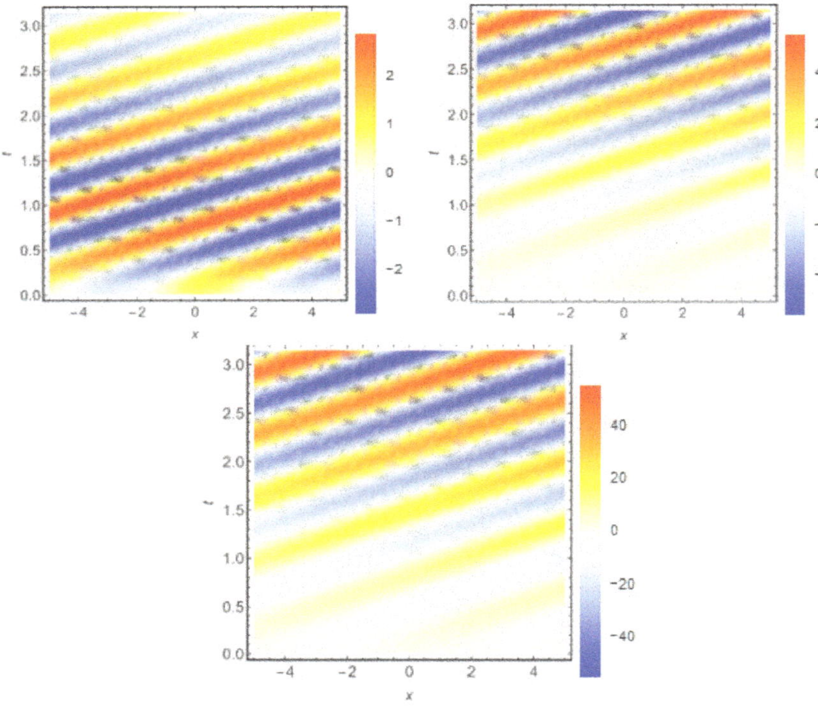

Figure 4. Evolution of quasi-temperature components in a thin film with $Kn_b = Kn_d = 0.1$ for $v = +10$: $\theta_b(x,t)$—top left plot, $\theta_d(x,t)$—top right plot and $\theta(x,t) = \theta_b(x,t) + \theta_d(x,t)$—bottom plot. The Cauchy conditions for PDEs (9) and (10) system are $\theta_b(x,0) = \theta_d(x,0) = e^{inx}$, $\partial_t\theta_b(x,0) = \partial_t\theta_d(x,0) = 0$.

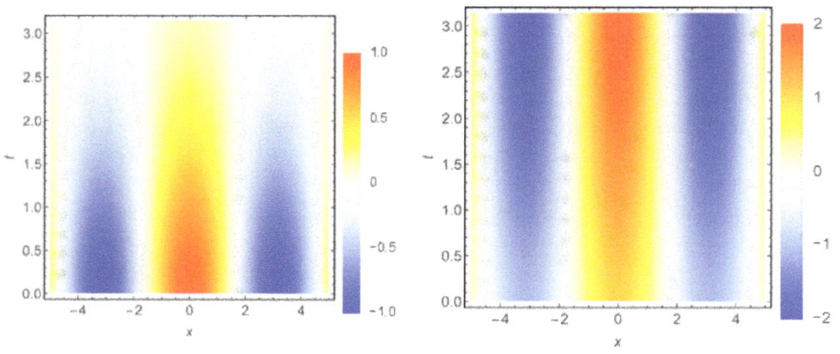

Figure 5. Evolution of quasi-temperature components in a stationary thin film with $Kn_b = Kn_d = 0.1$ for $v = 0$: $\theta_b(x,t)$—left plot, $\theta_d(x,t)$—right plot. The Cauchy conditions for PDEs (9) and (10) system are $\theta_b(x,0) = \theta_d(x,0) = e^{inx}$, $\partial_t\theta_b(x,0) = \partial_t\theta_d(x,0) = 0$.

The exploration of the solutions for Dirichlet conditions will be done elsewhere.

It should be noted that the one-dimensional energy balance equation yields zero value for the initial heat flux, if $B = -invA$. Above we have obtained the solutions, which allow nonzero initial conditions on the first time derivative; we have investigated some examples while writing this paper.

For example, for the case of $Kn_b = Kn_d = 1, v = 10$ the maximum principle is violated for both zero and nonzero initial conditions imposed on the first time-derivative of the ballistic component; however, there was no significant qualitative difference in the behaviors of the diffusive component in this case.

Explicit study for the solution in thin films with zero initial heat flux and non-zero velocity v will be done in forthcoming publications.

6. Conclusions

In the present work we have studied and analytically solved one-dimensional heat transport equations of Cattaneo- and Guyer–Krumhansl-type with substantial derivative. Exponential differential operators and the operational method were used to obtain integral forms of exact particular solutions to heat transport equations. The media speed is accounted for by the exponential differential operator $e^{-vt\partial_x}$, which produces a shift $x - vt$. Together with the heat operator $\hat{S}$ they transform the initial temperature profile and determine its evolution. Their action on the initial polynomial $\sum_n x^n$ yields sums of the Hermite polynomials $H_n(x, y)$; the action on the monomial-exponential function yields more complicated sums of Hermite polynomials. Proper exact operational solutions were obtained in the sums of converging integrals of elementary functions. Several examples were considered, the solutions for $F(x, 0) = x^2$, $F(x, 0) = x^2 e^{-x}$ and $F(x, 0) = e^{inx}$ were given. The solutions exactly satisfy the heat transport equations, which has been proven by direct substitution.

The equations with substantial derivative of Guyer-Krumhansl-type and of telegrapher's-type are demonstrated to have similar structure in the harmonic ansatz. Their exact harmonic solutions have been obtained by both operational method and the separation of variables. The latter reduces these PDEs to ODEs. We have shown that the harmonic solutions may have local and global extremums in the domain due to the interference of the Cattaneo heat waves; proper second-order time-derivative term also insures finite speed of the temperature perturbation propagation. We have shown that in a GK-type equation with dominant ballistic term δ, the Cattaneo heat waves are damped.

We have obtained the exact analytical solution to the ballistic heat transport in thin films by solving the system of inhomogeneous PDEs [30] in the harmonic ansatz; both Cauchy and Dirichlet conditions were explored. We found that true ballistic constituent of quasi-temperature, θ_b, evolves much faster than the diffusive counterpart of quasi-temperature, θ_d. We have studied the evolution of the complete quasi-temperature $\theta(x, t) = \theta_b(x, t) + \theta_d(x, t)$ under the ballistic conditions; the latter apply when the phonon mean free path l is comparable with the system scale L. The quasi-temperature θ first senses fast evolving true ballistic part θ_b. The contribution of θ_b sets the r.h.s. in the inhomogeneous DE for the component θ_d. After relatively fast relaxation of θ_b, further evolution of the complete quasi-temperature θ follows the diffusive constituent θ_d. Thus the initial fast-developing true ballistic component θ_b determines the evolution of the complete quasi-temperature of the system θ by imposing distribution for the diffusive quasi-temperature θ_d at the beginning of the process.

The obtained harmonic solutions for the Cauchy problem (9), (10), (36), (37) were applied for the study of the temperature patterns in thin films for the Knudsen number values $Kn = 0.1$ and $Kn = 1$. In the stationary case $v = 0$, the true ballistic component θ_b monotonously relaxes (see Figures 1 and 5); its diffusive counterpart θ_d behavior shows noticeable dependence on the value of the Knudsen numbers: for $Kn = 1$ it slowly relaxes (see Figure 5) and for $Kn = 0.1$ it droningly grows (see Figure 1). The effect of the speed of the observer on the perceived heat transport in thin films is distinguished for the pure ballistic and diffusive quasi-temperature constituents. In the presence of the motion, the extremum of the ballistic component θ_b occurs inside the domain after the initial moment $t = 0$ (see Figures 2–4). The diffusive quasi-temperature component monotonously grows in the domain both in the weak ballistic (see Figure 4) and strong ballistic cases (see Figures 2 and 3). This behavior contrasts common Fourier heat diffusion, for which the maximum principle holds. It is also noticeable that the system shows symmetric behavior while the value of the speed is reversed, which does not contradict the Second law. The Knudsen number variation from 0.1 to 1 in the presence of the speed $v = 10$, are practically not sensed by the system. Thus the speed of the observer v influences the registered temperature rather than the ballistic heat transport contribution.

Author Contributions: K.Z. has made the major work on the paper. N.G. has made a contribution in the Section 3. D.O. has made the contribution in Sections 4 and 5.

Acknowledgments: The authors are grateful to Anatolii Borisov, Andrey Lobanov for the discussions of the heat equations; to Laszlo I. Kiss (UQAC, Department of Applied Sciences, Quebec University) and Yasar Demirel (Department of Chemical and Biological Engineering, University of Nebraska, Lincoln) for discussions of the heat evolution patterns.

Conflicts of Interest: The authors declare no conflict of interest.

References

1. Fourier, J.P.J. *The Analytical Theory of Heat*; Cambridge University Press: London, UK, 1878.
2. Ackerman, C.C.; Guyer, R.A. Temperature pulses in dielectric solids. *Ann. Phys.* **1968**, *50*, 128–185. [CrossRef]
3. Ackerman, C.C.; Overton, W.C. Second sound in solid helium-3. *Phys. Rev. Lett.* **1969**, *22*, 764. [CrossRef]
4. McNelly, T.F.; Rogers, S.J.; Channin, D.J.; Rollefson, R.; Goubau, W.M.; Schmidt, G.E.; Krumhansl, J.A.; Pohl, R.O. Heat pulses in NaF: Onset of second sound. *Phys. Rev. Lett.* **1970**, *24*, 100. [CrossRef]
5. Narayanamurti, V.; Dynes, R.D. Observation of second sound in Bismuth. *Phys. Rev. Lett.* **1972**, *26*, 1461–1465. [CrossRef]
6. Peshkov, V. Second sound in Helium II. *J. Phys. (Moscow)* **1944**, *8*, 381.
7. Cattaneo, C. Sur une forme de l'equation de la chaleur eliminant le paradoxe d'une propagation instantanee. *C. R. l'Acad. Sci. Paris* **1958**, *247*, 431–433.
8. Terman, F.E. *Radio Engineers' Handbook*, 1st ed.; McGraw-Hill: New York, NY, USA, 1943.
9. Both, S.; Czél, B.; Fülöp, T.; Ván, P.; Verhás, J. Deviation from the Fourier law in room-temperature heat pulse experiments. *J. Non-Equilib. Thermodyn.* **2016**, *41*, 41–48. [CrossRef]
10. Kovács, R.; Ván, P. Models of ballistic propagation of heat at low temperatures. *Int. J. Thermophys.* **2016**, *37*, 95. [CrossRef]
11. Ván, P.; Berezovski, A.; Fülöp, T.; Gróf, G.; Kovács, R.; Lovas, Á.; Verhás, J. Guyer-Krumhansl-type heat conduction at room temperature. *EPL* **2017**, *118*, 50005. [CrossRef]
12. Cahill, D.G. Thermal conductivity measurement from 30 to 750 K: The 3ω method. *Rev. Sci. Instrum.* **1990**, *61*, 802–808. [CrossRef]
13. Tang, D.W.; Araki, N. Non-Fourier heat conduction behaviour in finite mediums under pulse surface heating. *Mater. Sci. Eng. A* **2000**, *292*, 173–178. [CrossRef]
14. Kaminski, W. Hyperbolic heat conduction equations for materials with a nonhomogeneous inner structure. *J. Heat Transf.* **1990**, *112*, 555–560. [CrossRef]
15. Mitra, K.; Kumar, S.; Vedavarz, A.; Moallemi, M.K. Experimental evidence of hyperbolic heat conduction in processed meat. *J. Heat Transf.* **1995**, *117*, 568–573. [CrossRef]
16. Herwig, H.; Beckert, K. Fourier versus non-Fourier heat conduction in materials with a nonhomogeneous inner structure. *J. Heat Transf.* **2000**, *122*, 363–365. [CrossRef]
17. Roetzel, W.; Putra, N.; Das, S.K. Experiment and analysis for non-Fourier conduction in materials with non-homogeneous inner structure. *Int. J. Therm. Sci.* **2003**, *42*, 541–552. [CrossRef]
18. Sazhin, S.S. Modelling of fuel droplet heating and evaporation: Recent results and unsolved problems. *Fuel* **2017**, *196*, 69–101. [CrossRef]
19. Scott, E.P.; Tilahun, M.; Vick, B. The question of thermal waves in heterogeneous and biological materials. *J. Biomech. Eng.* **2009**, *131*, 074518. [CrossRef] [PubMed]
20. Ricciu, R.; Besalduch, L.A.; Galatioto, A.; Ciulla, G. Thermal characterization of insulating materials. *Renew. Sustain. Energy Rev.* **2018**, *82*, 1765–1773. [CrossRef]
21. Porrà, J.M.; Masoliver, J.; Weiss, G.H. When the telegrapher's equation furnishes a better approximation to the transport equation than the diffusion approximation. *Phys. Rev. E* **1997**, *55*, 7771–7774. [CrossRef]
22. Shiomi, J.; Maruyama, S. Non-Fourier heat conduction in a single-walled carbon nanotube: Classical molecular dynamics simulations. *Phys. Rev. B* **2006**, *73*, 205420. [CrossRef]

23. Baringhaus, J.; Ruan, M.; Edler, F.; Tejeda, A.; Sicot, M.; Taleb-Ibrahimi, A.; Li, A.P.; Jiang, Z.; Conrad, E.H.; Berger, C.; et al. Exceptional ballistic transport in epitaxial graphene nanoribbons. *Nature* **2014**, *506*, 349–354. [CrossRef] [PubMed]
24. Hochbaum, A.I.; Chen, R.; Delgado, R.D.; Liang, W.; Garnett, E.C.; Najarian, M.; Majumdar, A.; Yang, P. Enhanced thermoelectric performance of rough silicon nanowires. *Nature (London)* **2008**, *451*, 163–167. [CrossRef] [PubMed]
25. Boukai, A.I.; Bunimovich, Y.; Tahir-Kheli, J.; Yu, J.-K.; Goddard, W.A.; Heath, J.R. Silicon nanowires as efficient thermoelectric materials. *Nature (London)* **2008**, *451*, 168. [CrossRef] [PubMed]
26. Paddock, C.A.; Eesley, G.L. Transient thermoreflectance from thin metal films. *J. Appl. Phys.* **1986**, *60*, 285–290. [CrossRef]
27. Maldovan, M. Transition between ballistic and diffusive heat transport regimes in silicon materials. *Appl. Phys. Lett.* **2012**, *101*, 113110. [CrossRef]
28. Guyer, R.A.; Krumhansl, J.A. Solution of the linearized phonon Boltzmann equation. *Phys. Rev.* **1966**, *148*, 766–778. [CrossRef]
29. Guyer, R.A.; Krumhansl, J.A. Thermal conductivity, second sound and phonon hydrodynamic phenomena in non-metallic crystals. *Phys. Rev.* **1966**, *148*, 778–788. [CrossRef]
30. Lebon, G.; Machrafi, H.; Gremela, M.; Dubois, C. An extended thermodynamic model of transient heat conduction at sub-continuum scales. *Proc. R. Soc. A* **2011**, *467*, 3241–3256. [CrossRef]
31. Müller, I.; Ruggeri, T. *Rational Extended Thermodynamics*; Springer: Berlin/Heidelberg, Germany, 1998.
32. Rogolino, P.; Kovács, R.; Ván, P.; Cimmelli, V.A. Generalized heat-transport equations: Parabolic and hyperbolic models. *Contin. Mech. Thermodyn.* **2018**. [CrossRef]
33. Moosaie, A. Non-Fourier heat conduction in a finite medium with insulated boundaries and arbitrary initial conditions. *Int. Commun. Heat Mass Transf.* **2008**, *35*, 103–111. [CrossRef]
34. Ahmadikia1, H.; Rismanian, M. Analytical solution of non-Fourier heat conduction problem on a fin under periodic boundary conditions. *J. Mech. Sci. Technol.* **2011**, *25*, 2919–2926. [CrossRef]
35. Yen, C.C.; Wu, C.Y. Modelling hyperbolic heat conduction in a finite medium with periodic thermal disturbance and surface radiation. *Appl. Math. Model.* **2003**, *27*, 397–408. [CrossRef]
36. Lewandowska, M. Hyperbolic heat conduction in the semi-infinite body with a time-dependent laser heat source. *Heat Mass Transf.* **2001**, *37*, 333–342. [CrossRef]
37. Lewandowska, M.; Malinowski, L. An analytical solution of the hyperbolic heat conduction equation for the case of a finite medium symmetrically heated on both sides. *Int. Commun. Heat Mass Transf.* **2006**, *33*, 61–69. [CrossRef]
38. Saedodin, S.; Torabi, M. Analytical solution of non-Fourier heat conduction in cylindrical coordinates. *Int. Rev. Mech. Eng.* **2009**, *3*, 726–732.
39. Challamel, N.; Grazide, C.; Picandet, V.; Perrot, A.; Zhang, Y. A nonlocal Fourier's law and its application to the heat conduction of one-dimensional and two-dimensional thermal lattices. *C. R. Mec.* **2016**, *344*, 388–401. [CrossRef]
40. Saedodin, S.; Torabi, M. Algebraically explicit analytical solution of three-dimensional hyperbolic heat conduction equation. *Adv. Theor. Appl. Mech.* **2010**, *3*, 369–383.
41. Chen, G. Ballistic-diffusive heat-conduction equations. *Phys. Rev. Lett.* **2001**, *86*, 2297–2300. [CrossRef] [PubMed]
42. Hsiao, T.K.; Chang, H.K.; Liou, S.C.; Chu, M.W.; Lee, S.C.; Chang, C.W. Observation of room-temperature ballistic thermal conduction persisting over 8.3 mm in SiGe nanowires. *Nat. Nanotechnol.* **2013**, *8*, 534–538. [CrossRef] [PubMed]
43. Zhang, Y.; Ye, W. Modified ballistic–diffusive equations for transient non-continuum heat conduction. *Int. J. Heat Mass Transf.* **2015**, *83*, 51–63. [CrossRef]
44. Kovacs, R.; Van, P. Generalized heat conduction in heat pulse experiments. *Int. J. Heat Mass Transf.* **2015**, *83*, 613–620. [CrossRef]
45. KZhukovsky, V. Operational method of solution of linear non-integer ordinary and partial differential equations. *SpringerPlus* **2016**, *5*, 119. [CrossRef] [PubMed]
46. Zhukovsky, K. Operational approach and solutions of hyperbolic heat conduction equations. *Axioms* **2016**, *5*, 28. [CrossRef]

47. Zhukovsky, K.V.; Srivastava, H.M. Analytical solutions for heat diffusion beyond Fourier law. *Appl. Math. Comput.* **2017**, *293*, 423–437. [CrossRef]
48. Zhukovsky, K.V. Violation of the maximum principle and negative solutions with pulse propagation in Guyer–Krumhansl model. *Int. J. Heat Mass Transf.* **2016**, *98*, 523–529. [CrossRef]
49. Zhukovsky, K.V. Exact solution of Guyer–Krumhansl type heat equation by operational method. *Int. J. Heat Mass Transf.* **2016**, *96*, 132–144. [CrossRef]
50. Zhukovsky, K. Exact negative solutions for Guyer–Krumhansl type equation and the violation of the maximum principle. *Entropy* **2017**, *19*, 440. [CrossRef]
51. Zhukovsky, K. Operational solution for some types of second order differential equations and for relevant physical problems. *J. Math. Anal. Appl.* **2017**, *446*, 628–647. [CrossRef]
52. Zhukovsky, K.V. A method of inverse differential operators using ortogonal polynomials and special functions for solving some types of differential equations and physical problems. *Mosc. Univ. Phys. Bull.* **2015**, *70*, 93–100. [CrossRef]
53. Zhukovsky, K. Solution of some types of differential equations: Operational calculus and inverse differential operators. *Sci. World J.* **2014**, *2014*. [CrossRef] [PubMed]
54. Zhukovsky, K.V. Solving evolutionary-type differential equations and physical problems using the operator method. *Theor. Math. Phys.* **2017**, *190*, 52–68. [CrossRef]
55. Dattoli, G.; Srivastava, H.M.; Zhukovsky, K.V. Operational methods and Differential Equations with applications to initial-value problems. *Appl. Math. Comput.* **2007**, *184*, 979–1001. [CrossRef]
56. Zhukovsky, K.V. Operational solution of differential equations with derivatives of non-integer order, Black–Scholes type and heat conduction. *Mosc. Univ. Phys. Bull.* **2016**, *71*, 237–244. [CrossRef]
57. Dattoli, G.; Srivastava, H.M.; Zhukovsky, K.V. Orthogonality properties of the Hermite and related polynomials. *J. Comput. Appl. Math.* **2005**, *182*, 165–172. [CrossRef]
58. Dattoli, G.; Srivastava, H.M.; Zhukovsky, K.V. A new family of integral transforms and their applications. *Integral Transforms Spec. Funct.* **2006**, *17*, 31–37. [CrossRef]
59. Srivastava, H.M.; Manocha, H.L. *A Treatise on Generating Functions;* Halsted Press (Ellis Horwood Limited): Chichester, UK; John Wiley and Sons: New York, NY, USA; Chichester, UK; Brisbane, Australia; Toronto, ON, Canada, 1984.
60. Boucetta, A.; Ghodbane, H.; Ayad, M.Y.; Bahri, M. A review on the performance and modelling of proton exchange membrane fuel cells. *AIP Conf. Proc.* **2016**, *1758*, 030019.
61. Arato, E.; Pinna, M.; Mazzoccoli, M.; Bosio, B. Gas-phase mass-transfer resistances at polymeric electrolyte membrane fuel cells electrodes: Theoretical analysis on the effectiveness of interdigitated and serpentine flow arrangements. *Energies* **2016**, *9*, 229. [CrossRef]
62. Maidhily, M.; Rajalakshmi, N.; Dhathathreyan, K.S. Electrochemical impedance spectroscopy as a diagnostic tool for the evaluation of flow field geometry in polymer electrolyte membrane fuel cells. *Renew. Energy* **2013**, *51*, 79–84. [CrossRef]
63. St-Pierre, J. Hydrogen mass transport in fuel cell gas diffusion electrodes. *Fuel Cells* **2011**, *11*, 263–273. [CrossRef]
64. Zhukovsky, K.; Pozio, A. Maximum current limitations of the PEM fuel cell with serpentine gas supply channels. *J. Power Sources* **2004**, *130*, 95–105. [CrossRef]
65. Zhukovsky, K.V. Three Dimensional model of gas transport in a porous diffuser of a polymer electrolyte fuel cell. *AIChE J.* **2003**, *49*, 3029–3036. [CrossRef]
66. Zhukovsky, K. Modeling of the Current Limitations of PEFC. *AIChE J.* **2006**, *52*, 2356–2366. [CrossRef]
67. Weber, A.Z.; Newman, J. Modeling transport in polymer-electrolyte fuel cells. *Chem. Rev.* **2004**, *104*, 4679–4726. [CrossRef] [PubMed]
68. Kawase, M.; Sato, K.; Mitsui, R.; Asonuma, H.; Kageyama, M.; Yamaguchi, K.; Inoue, G. Electrochemical reaction engineering of polymer electrolyte fuel cell. *AIChE J.* **2017**, *63*, 249–256. [CrossRef]
69. Rogié, B.; Monier-Vinard, E.; Nguyen, M.-N.; Bissuel, V.; Laraqi, N. Practical analytical modeling of 3D multi-layer Printed Wired Board with buried volumetric heating sources. *Int. J. Therm. Sci.* **2018**, *129*, 404–415. [CrossRef]
70. Calvo-Schwarzwälder, M.; Hennessy, M.G.; Torres, P.; Myers, T.G.; Alvarez, F.X. A slip-based model for the size-dependent effective thermal conductivity of nanowires. *Int. Commun. Heat Mass Transf.* **2018**, *91*, 57–63. [CrossRef]
71. Zhukovskij, K.V. Gas flow in long microchannels. *Vestn. Mosk. Univ. Ser. 3 Fiz. Astron.* **2001**, *3*, 49–54.

72. Zhukovsky, K.V. A harmonic solution for the hyperbolic heat conduction equation and its relationship to the Guyer–Krumhansl Equation. *Mosc. Univ. Phys. Bull.* **2018**, *73*, 45–52. [CrossRef]
73. Zhukovsky, K. Exact harmonic solution to ballistic type heat propagation in thin films and wires. *Int. J. Heat Mass Transf.* **2018**, *120*, 944–955. [CrossRef]
74. Zhukovsky, K.; Oskolkov, D. Exact harmonic solutions to Guyer–Krumhansl-type equation and application to heat transport in thin films. *Contin. Mech. Thermodyn.* **2018**. [CrossRef]
75. Ali, Y.M.; Zhang, L.C. Relativistic heat conduction. *Int. J. Heat Mass Transf.* **2005**, *48*, 2397–2406. [CrossRef]
76. Al-Khairy, R.T.; Al-Ofey, Z.M. Analytical solution of the hyperbolic heat conduction equation for moving semi-infinite medium under the effect of time-dependent laser heat source. *J. Appl. Math.* **2009**, *2009*. [CrossRef]
77. Wolf, K.B. *Integral Transforms in Science and Engineering*; New York Plenum Press: New York, NY, USA, 1979.

axiom

Article

Some Summation Theorems for Generalized Hypergeometric Functions

Mohammad Masjed-Jamei [1,2,*] and Wolfram Koepf [1]

1 Department of Mathematics, University of Kassel, Heinrich-Plett-Str. 40, D-34132 Kassel, Germany; koepf@mathematik.uni-kassel.de

2 Department of Mathematics, K. N. Toosi University of Technology, P.O. Box 16315-1618, 11369 Tehran, Iran

* Correspondence: mmjamei@kntu.ac.ir or mmjamei@yahoo.com

Received: 2 May 2018; Accepted: 6 June 2018; Published: 8 June 2018

Abstract: Essentially, whenever a generalized hypergeometric series can be summed in terms of gamma functions, the result will be important as only a few such summation theorems are available in the literature. In this paper, we apply two identities of generalized hypergeometric series in order to extend some classical summation theorems of hypergeometric functions such as Gauss, Kummer, Dixon, Watson, Whipple, Pfaff–Saalschütz and Dougall formulas and also obtain some new summation theorems in the sequel.

Keywords: generalized hypergeometric series; Gauss and confluent hypergeometric functions; classical summation theorems of hypergeometric functions

MSC: 33C20; 33C05; 65B10

1. Introduction

Let $\mathbb{R}$ and $\mathbb{C}$ denote the sets of real and complex numbers and z be a complex variable. For real or complex parameters a and b, the generalized binomial coefficient:

$$\binom{a}{b} = \frac{\Gamma(a+1)}{\Gamma(b+1)\Gamma(a-b+1)} = \binom{a}{a-b} \qquad (a, b \in \mathbb{C}),$$

in which:

$$\Gamma(z) = \int_0^\infty x^{z-1} e^{-x} dx,$$

denotes the well-known gamma function for $Re(z) > 0$, can be reduced to the particular case:

$$\binom{a}{n} = \frac{(-1)^n (-a)_n}{n!},$$

where $(a)_b$ denotes the Pochhammer symbol [1] given by:

$$(a)_b = \frac{\Gamma(a+b)}{\Gamma(a)} = \begin{cases} 1 & (b = 0,\ a \in \mathbb{C}\backslash\{0\}), \\ a(a+1)...(a+n-1) & (b \in \mathbb{N},\ a \in \mathbb{C}). \end{cases} \tag{1}$$

Based on Pochhammer's symbol (1), the generalized hypergeometric functions [2]:

$${}_pF_q\left(\begin{matrix} a_1, & \dots & , a_p \\ b_1, & \dots & , b_q \end{matrix} \middle| z \right) = \sum_{k=0}^{\infty} \frac{(a_1)_k \dots (a_p)_k}{(b_1)_k \dots (b_q)_k} \frac{z^k}{k!}, \tag{2}$$

are indeed a Taylor series expansion for a function, say f, as $\sum_{k=0}^{\infty} c_k^* z^k$ with $c_k^* = f^{(k)}(0)/k!$, for which the ratio of successive terms can be written as:

$$\frac{c_{k+1}^*}{c_k^*} = \frac{(k+a_1)(k+a_2)...(k+a_p)}{(k+b_1)(k+b_2)...(k+b_q)(k+1)}.$$

According to the ratio test [3,4], the series (2) is convergent for any $p \leq q+1$. In fact, it converges in $|z| < 1$ for $p = q+1$, converges everywhere for $p < q+1$ and converges nowhere ($z \neq 0$) for $p > q+1$. Moreover, for $p = q+1$, it absolutely converges for $|z| = 1$ if the condition:

$$A^* = \mathrm{Re}\left(\sum_{j=1}^{q} b_j - \sum_{j=1}^{q+1} a_j\right) > 0,$$

holds and is conditionally convergent for $|z| = 1$ and $z \neq 1$ if $-1 < A^* \leq 0$ and is finally divergent for $|z| = 1$ and $z \neq 1$ if $A^* \leq -1$.

There are two important cases of the series (2) arising in many physics problems [5,6]. The first case (convergent in $|z| \leq 1$) is the Gauss hypergeometric function:

$$y = {}_2F_1\left(\begin{array}{c} a,\ b \\ c \end{array} \middle| z\right) = \sum_{k=0}^{\infty} \frac{(a)_k (b)_k}{(c)_k} \frac{z^k}{k!},$$

with the integral representation:

$${}_2F_1\left(\begin{array}{c} a,\ b \\ c \end{array} \middle| z\right) = \frac{\Gamma(c)}{\Gamma(b)\Gamma(c-b)} \int_0^1 t^{b-1}(1-t)^{c-b-1}(1-tz)^{-a} dt,$$

$$(\mathrm{Re}\, c > \mathrm{Re}\, b > 0;\ |\arg(1-z)| < \pi). \quad (3)$$

Replacing $z = 1$ in (3) directly leads to the well-known Gauss identity [1]:

$${}_2F_1\left(\begin{array}{c} a,\ b \\ c \end{array} \middle| 1\right) = \frac{\Gamma(c)\Gamma(c-a-b)}{\Gamma(c-a)\Gamma(c-b)} \qquad \mathrm{Re}(c-a-b) > 0\,. \quad (4)$$

The second case, which converges everywhere, is the Kummer confluent hypergeometric function:

$$y = {}_1F_1\left(\begin{array}{c} b \\ c \end{array} \middle| z\right) = \sum_{k=0}^{\infty} \frac{(b)_k}{(c)_k} \frac{z^k}{k!},$$

with the integral representation:

$${}_1F_1\left(\begin{array}{c} b \\ c \end{array} \middle| z\right) = \frac{\Gamma(c)}{\Gamma(b)\Gamma(c-b)} \int_0^1 t^{b-1}(1-t)^{c-b-1} e^{zt} dt,$$

$$(\mathrm{Re}\, c > \mathrm{Re}\, b > 0;\ |\arg(1-z)| < \pi).$$

Essentially, whenever a generalized hypergeometric series can be summed in terms of gamma functions, the result will be important as only a few such summation theorems are available in the literature; see, e.g., [7–13]. In this sense, the classical summation theorems such as Kummer and Gauss for ${}_2F_1$, Dixon, Watson, Whipple and Pfaff–Saalschütz for ${}_3F_2$, Whipple for ${}_4F_3$, Dougall for ${}_5F_4$ and Dougall for ${}_7F_6$ are well known [1,14]. In this paper, we apply two identities of generalized hypergeometric functions in order to obtain some new summation theorems and extend the above-mentioned classical theorems. For this purpose, we should first recall the classical theorems as follows.

* Kummer's theorem ([1], p. 108):

$$ {}_2F_1\left(\begin{array}{c} a,\ b \\ 1+a-b \end{array} \middle| -1 \right) = \frac{\Gamma(1+a-b)\Gamma(1+(a/2))}{\Gamma(1-b+(a/2))\Gamma(1+a)}. \tag{5} $$

* Second Gauss theorem ([1], p. 108):

$$ {}_2F_1\left(\begin{array}{c} a,\ b \\ (a+b+1)/2 \end{array} \middle| \frac{1}{2} \right) = \frac{\sqrt{\pi}\,\Gamma((a+b+1)/2)}{\Gamma((a+1)/2)\Gamma((b+1)/2)}. \tag{6} $$

* Bailey's theorem ([1], p. 108):

$$ {}_2F_1\left(\begin{array}{c} a,\ 1-a \\ b \end{array} \middle| \frac{1}{2} \right) = \frac{\Gamma(b/2)\Gamma((b+1)/2)}{\Gamma((a+b)/2)\Gamma((b-a+1)/2)}. \tag{7} $$

* Dixon's theorem ([1], p. 108):

$$ {}_3F_2\left(\begin{array}{c} a,\ b,\ c \\ 1+a-b,\ 1+a-c \end{array} \middle| 1 \right) = \frac{\Gamma(1+a/2)\Gamma(1+a-b)\Gamma(1+a-c)\Gamma(1-b-c+a/2)}{\Gamma(1+a)\Gamma(1-b+a/2)\Gamma(1-c+a/2)\Gamma(1+a-b-c)}. \tag{8} $$

* Watson's theorem ([1], p. 108):

$$ {}_3F_2\left(\begin{array}{c} a,\ b,\ c \\ (a+b+1)/2,\ 2c \end{array} \middle| 1 \right) = \frac{\sqrt{\pi}\,\Gamma(1+c/2)\Gamma((a+b+1)/2)\Gamma(c-(a+b-1)/2)}{\Gamma((a+1)/2)\Gamma((b+1)/2)\Gamma(c-(a-1)/2)\Gamma(c-(b-1)/2)}. \tag{9} $$

* Whipple's theorem ([1], p. 108):

$$ {}_3F_2\left(\begin{array}{c} a,\ 1-a,\ c \\ c,\ 2b-c+1 \end{array} \middle| 1 \right) = \frac{\pi\, 2^{1-2b}\Gamma(c)\Gamma(2b-c+1)}{\Gamma((a+c)/2)\Gamma(b+(a-c+1)/2)\Gamma((1-a+c)/2)\Gamma(b+1-(a+c)/2)}. \tag{10} $$

* Pfaff–Saalschütz theorem ([1], p. 108):

$$ {}_3F_2\left(\begin{array}{c} a,\ b,\ -n \\ c,\ 1+a+b-c-n \end{array} \middle| 1 \right) = \frac{(c-a)_n(c-b)_n}{(c)_n(c-a-b)_n}. \tag{11} $$

* Second Whipple theorem ([1], p. 108):

$$ {}_4F_3\left(\begin{array}{c} a,\ 1+a/2,\ b,\ c \\ a/2,\ a-b+1,\ a-c+1 \end{array} \middle| -1 \right) = \frac{\Gamma(a-b+1)\Gamma(a-c+1)}{\Gamma(a+1)\Gamma(a-b-c+1)}. \tag{12} $$

* Dougall's theorem ([1], p. 108):

$$ {}_5F_4\left(\begin{array}{c} a,\ 1+a/2,\ c,\ d,\ e \\ a/2,\ a-c+1,\ a-d+1,\ a-e+1 \end{array} \middle| 1 \right) $$

$$= \frac{\Gamma(a-c+1)\Gamma(a-d+1)\Gamma(a-e+1)\Gamma(a-c-d-e+1)}{\Gamma(a+1)\Gamma(a-d-e+1)\Gamma(a-c-e+1)\Gamma(a-c-d+1)}. \quad (13)$$

* Second Dougall theorem ([1], p. 108):

$${}_7F_6\left(\begin{array}{c} a,\ 1+a/2,\ b,\ c,\ d,\ 1+2a-b-c+n,\ -n \\ a/2,\ a-b+1,\ a-c+1,\ a-d+1,\ b+c+d-a-n,\ a+1+n \end{array}\middle|\, 1\right)$$

$$= \frac{(a+1)_n(a-b-c+1)_n(a-b-d+1)_n(a-c-d+1)_n}{(a+1-b)_n(a+1-c)_n(a+1-d)_n(a+1-b-c-d)_n}. \quad (14)$$

In order to derive the first identity and only for simplicity, we will use the following symbol for representing finite sums of hypergeometric series:

$${}_p\overset{(m)}{F}_q\left(\begin{array}{c} a_1, \ \dots \ , a_p \\ b_1, \ \dots \ , b_q \end{array}\middle|\, z\right) = \sum_{k=0}^{m} \frac{(a_1)_k \dots (a_p)_k}{(b_1)_k \dots (b_q)_k} \frac{z^k}{k!}.$$

For instance, we have:

$${}_p\overset{(-1)}{F}_q(z) = 0, \quad {}_p\overset{(0)}{F}_q(z) = 1 \quad \text{and} \quad {}_p\overset{(1)}{F}_q(z) = 1 + \frac{a_1 \dots a_p}{b_1 \dots b_q} z.$$

2. First Hypergeometric Identity

Let m, n be two natural numbers so that $n \leq m$. By referring to Relation (1), since:

$$\frac{(n)_k}{(m)_k} = \frac{\Gamma(k+n)\Gamma(m)}{\Gamma(k+m)\Gamma(n)} = \frac{\Gamma(m)}{\Gamma(n)} \frac{1}{(k+n)(k+n+1)\dots(k+m-1)}, \quad (15)$$

substituting (15) in a special case of (2) yields:

$${}_pF_q\left(\begin{array}{c} a_1, \ \dots \ , a_{p-1}, \ \ n \\ b_1, \ \dots \ , b_{q-1}, \ \ m \end{array}\middle|\, z\right) = \frac{\Gamma(m)}{\Gamma(n)} \sum_{k=0}^{\infty} \frac{(a_1)_k \dots (a_{p-1})_k}{(b_1)_k \dots (b_{q-1})_k} \frac{(k+1)(k+2)\dots(k+n-1)\, z^k}{(k+m-1)!}$$

$$= \frac{\Gamma(m)}{\Gamma(n)} \sum_{j=m-1}^{\infty} \frac{(a_1)_{j-m+1} \dots (a_{p-1})_{j-m+1}}{(b_1)_{j-m+1} \dots (b_{q-1})_{j-m+1}} \frac{(j+2-m)(j+3-m)\dots(j-m+n)\, z^{j-m+1}}{j!}. \quad (16)$$

Relation (16) shows that we encounter a complicated computational problem that cannot be easily evaluated. However, some particular cases such as $n = 1$ and $n = 2$ can be directly computed. We leave other cases as open problems.

The case $n = 1$ leads to a known result in the literature [14], because:

$$
{}_pF_q\left(\begin{matrix} a_1, \ldots, a_{p-1}, & 1 \\ b_1, \ldots, b_{q-1}, & m \end{matrix} \middle| z\right) = \Gamma(m) \sum_{j=m-1}^{\infty} \frac{(a_1)_{j-m+1} \cdots (a_{p-1})_{j-m+1}}{(b_1)_{j-m+1} \cdots (b_{q-1})_{j-m+1}} \frac{z^{j-m+1}}{j!}
$$
$$
= \Gamma(m) \left(\sum_{j=0}^{\infty} \frac{(a_1)_{j-m+1} \cdots (a_{p-1})_{j-m+1}}{(b_1)_{j-m+1} \cdots (b_{q-1})_{j-m+1}} \frac{z^{j-m+1}}{j!} - \sum_{j=0}^{m-2} \frac{(a_1)_{j-m+1} \cdots (a_{p-1})_{j-m+1}}{(b_1)_{j-m+1} \cdots (b_{q-1})_{j-m+1}} \frac{z^{j-m+1}}{j!} \right), \quad (17)
$$

and since:

$$
(a)_{j-m+1} = \frac{\Gamma(a-m+1)}{\Gamma(a)} (a-m+1)_j,
$$

Relation (17) is simplified as:

$$
{}_pF_q\left(\begin{matrix} a_1, \ldots, a_{p-1}, & 1 \\ b_1, \ldots, b_{q-1}, & m \end{matrix} \middle| z\right)
$$
$$
= \frac{\Gamma(b_1) \ldots \Gamma(b_{q-1})}{\Gamma(a_1) \ldots \Gamma(a_{p-1})} \frac{\Gamma(a_1-m+1) \ldots \Gamma(a_{p-1}-m+1)}{\Gamma(b_1-m+1) \ldots \Gamma(b_{q-1}-m+1)} \frac{(m-1)!}{z^{m-1}}
$$
$$
\times \left({}_{p-1}F_{q-1}\left(\begin{matrix} a_1-m+1, \ldots, a_{p-1}-m+1 \\ b_1-m+1, \ldots, b_{q-1}-m+1 \end{matrix} \middle| z\right) \right.
$$
$$
\left. - {}_{p-1}^{(m-2)}F_{q-1}\left(\begin{matrix} a_1-m+1, \ldots, a_{p-1}-m+1 \\ b_1-m+1, \ldots, b_{q-1}-m+1 \end{matrix} \middle| z\right) \right). \quad (18)
$$

However, the interesting point is that using Relation (18), we can obtain various special cases that extend all classical summation theorems as follows.

Special Case 1. When $p = 3, q = 2$ and $x = 1$, Relation (18) is simplified as:

$$
{}_3F_2\left(\begin{matrix} a, & b, & 1 \\ & c, & m \end{matrix} \middle| 1\right) = \frac{\Gamma(m)\Gamma(c)\Gamma(a-m+1)\Gamma(b-m+1)}{\Gamma(a)\Gamma(b)\Gamma(c-m+1)}
$$
$$
\times \left(\frac{\Gamma(c-m+1)\Gamma(c-a-b+m-1)}{\Gamma(c-a)\Gamma(c-b)} - {}_2^{(m-2)}F_1\left(\begin{matrix} a-m+1, \ b-m+1 \\ c-m+1 \end{matrix} \middle| 1\right) \right). \quad (19)
$$

For $m = 1$, Relation (19) exactly gives Formula (4), while for $m = 2, 3$, we have:

$$
{}_3F_2\left(\begin{matrix} a, & b, & 1 \\ & c, & 2 \end{matrix} \middle| 1\right) = \frac{c-1}{(a-1)(b-1)} \left(\frac{\Gamma(c-1)\Gamma(c-a-b+1)}{\Gamma(c-a)\Gamma(c-b)} - 1 \right),
$$

and:

$$
{}_3F_2\left(\begin{matrix} a, & b, & 1 \\ & c, & 3 \end{matrix} \middle| 1\right) = \frac{2(c-2)_2}{(a-2)_2(b-2)_2}
$$
$$
\times \left(\frac{\Gamma(c-2)\Gamma(c-a-b+2)}{\Gamma(c-a)\Gamma(c-b)} - \frac{ab+c-2a-2b+2}{c-2} \right).
$$

These two formulas are given in [14].

Special Case 2. When $p = 3, q = 2$ and $x = -1$, by noting the Kummer theorem (5), Relation (18) is simplified as:

$$ {}_3F_2\left(\begin{array}{c} a,\ b,\ 1 \\ a-b+m,\ m \end{array} \middle| -1\right) = (-1)^{m-1}\frac{\Gamma(m)\Gamma(a-b+m)\Gamma(a-m+1)\Gamma(b-m+1)}{\Gamma(a)\Gamma(b)\Gamma(a-b+1)} \times \left(\frac{\Gamma(a-b+1)\Gamma(1+(a-m+1)/2)}{\Gamma(2+a-m)\Gamma(m-b+(a-m+1)/2)} - {}_2F_1^{(m-2)}\left(\begin{array}{c} a-m+1,\ b-m+1 \\ a-b+1 \end{array} \middle| -1\right)\right). \quad (20) $$

For $m=1$, Relation (20) exactly gives the Kummer formula, while for $m=2,3$, we have:

$$ {}_3F_2\left(\begin{array}{c} a,\ b,\ 1 \\ a-b+2,\ 2 \end{array} \middle| -1\right) = \frac{a-b+1}{(a-1)(b-1)}\left(1-\frac{\Gamma(a-b+1)\Gamma(1+(a-1)/2)}{\Gamma(a)\Gamma(-b+2+(a-1)/2)}\right), $$

and:

$$ {}_3F_2\left(\begin{array}{c} a,\ b,\ 1 \\ a-b+3,\ 3 \end{array} \middle| -1\right) = \frac{2(a-b+1)_2}{(a-2)_2(b-2)_2}\left(\frac{\Gamma(a-b+1)\Gamma(a/2)}{\Gamma(a-1)\Gamma(-b+2+a/2)} - \frac{3a+b-ab-3}{a-b+1}\right). $$

Special Case 3. When $p=3, q=2$ and $x=1/2$, by noting the second kind of Gauss Formula (6), Relation (18) is simplified as:

$$ {}_3F_2\left(\begin{array}{c} a,\ b,\ 1 \\ (a+b+1)/2,\ m \end{array} \middle| \frac{1}{2}\right) = (2)^{m-1} \times \frac{\Gamma(m)\Gamma((a+b+1)/2)\Gamma(a-m+1)\Gamma(b-m+1)}{\Gamma(a)\Gamma(b)\Gamma(-m+1+(a+b+1)/2)} \times \left(\frac{\sqrt{\pi}\,\Gamma(-m+1+(a+b+1)/2)}{\Gamma(1+(a-m)/2)\,\Gamma(1+(b-m)/2)} - {}_2F_1^{(m-2)}\left(\begin{array}{c} a-m+1,\ b-m+1 \\ -m+1+(a+b+1)/2 \end{array} \middle| \frac{1}{2}\right)\right). \quad (21) $$

For $m=1$, Relation (21) exactly gives the second kind of Gauss formula, while for $m=2,3$, we have:

$$ {}_3F_2\left(\begin{array}{c} a,\ b,\ 1 \\ (a+b+1)/2,\ 2 \end{array} \middle| \frac{1}{2}\right) = \frac{a+b-1}{(a-1)(b-1)}\left(\frac{\sqrt{\pi}\,\Gamma(-1+(a+b+1)/2)}{\Gamma(a/2)\,\Gamma(b/2)} - 1\right), $$

and:

$$ {}_3F_2\left(\begin{array}{c} a,\ b,\ 1 \\ (a+b+1)/2,\ 3 \end{array} \middle| \frac{1}{2}\right) = \frac{2(a+b-1)(a+b-3)}{(a-2)_2(b-2)_2} \times \left(\frac{\sqrt{\pi}\,\Gamma((a+b-3)/2)}{\Gamma((a-1)/2)\,\Gamma((b-1)/2)} - \frac{ab-a-b+1}{a+b-3}\right). $$

Special Case 4. When $p=3, q=2$ and $x=1/2$, by noting the Bailey theorem (7), Relation (18) is simplified as:

$$
{}_3F_2\left(\begin{matrix} a,\ 2m-a-1,\ 1 \\ b,\ m \end{matrix}\ \middle|\ \frac{1}{2}\right) = (2)^{m-1}\frac{\Gamma(m)\Gamma(b)\Gamma(a-m+1)\Gamma(m-a)}{\Gamma(a)\Gamma(2m-a-1)\Gamma(b-m+1)}
$$

$$
\times\left(\frac{\Gamma((b-m+1)/2)\Gamma((b-m+2)/2)}{\Gamma(-m+1+(a+b)/2)\,\Gamma((b-a+1)/2)} - {}^{(m-2)}_{\ \ 2}F_1\left(\begin{matrix} a-m+1,\ m-a \\ b-m+1 \end{matrix}\ \middle|\ \frac{1}{2}\right)\right). \quad (22)
$$

For $m = 1$, Relation (22) exactly gives the Bailey formula, while for $m = 2, 3$, we have:

$$
{}_3F_2\left(\begin{matrix} a,\ 3-a,\ 1 \\ b,\ 2 \end{matrix}\ \middle|\ \frac{1}{2}\right) = \frac{2(1-b)}{(1-a)_2}\left(\frac{\Gamma((b-1)/2)\Gamma(b/2)}{\Gamma(-1+(a+b)/2)\,\Gamma((b-a+1)/2)} - 1\right),
$$

and:

$$
{}_3F_2\left(\begin{matrix} a,\ 5-a,\ 1 \\ b,\ 3 \end{matrix}\ \middle|\ \frac{1}{2}\right) = \frac{8(b-2)_2}{(a-4)_4}
$$

$$
\times\left(\frac{\Gamma((b-1)/2)\Gamma((b-2)/2)}{\Gamma(-2+(a+b)/2)\,\Gamma((b-a+1)/2)} - \frac{5a-a^2+2b-10}{2(b-2)}\right).
$$

Special Case 5. When $p = 4, q = 3$ and $x = 1$, by noting the Dixon theorem (8), Relation (18) is simplified as:

$$
{}_4F_3\left(\begin{matrix} a,\ b,\ c,\ 1 \\ a-b+m,\ a-c+m,\ m \end{matrix}\ \middle|\ 1\right)
$$

$$
= \frac{\Gamma(m)\Gamma(a-b+m)\Gamma(a-c+m)\Gamma(a+1-m)\Gamma(b+1-m)\Gamma(c+1-m)}{\Gamma(a)\Gamma(b)\Gamma(c)\Gamma(a-b+1)\Gamma(a-c+1)}
$$

$$
\times\left(\begin{matrix} \frac{\Gamma((a+3-m)/2)\Gamma(a-b+1)\Gamma(a-c+1)\Gamma(-b-c+(a+3m-1)/2)}{\Gamma(a+2-m)\Gamma(-b+(a+m+1)/2)\Gamma(-c+(a+m+1)/2)\Gamma(a-b-c+m)} \\ - {}^{(m-2)}_{\ \ 3}F_2\left(\begin{matrix} a-m+1,\ b-m+1,\ c-m+1 \\ a-b+1,\ a-c+1 \end{matrix}\ \middle|\ 1\right) \end{matrix}\right). \quad (23)
$$

For $m = 1$, Relation (23) exactly gives the Dixon formula, while for $m = 2, 3$, we have:

$$
{}_4F_3\left(\begin{matrix} a,\ b,\ c,\ 1 \\ a-b+2,\ a-c+2,\ 2 \end{matrix}\ \middle|\ 1\right) = \frac{(a-b+1)(a-c+1)}{(a-1)(b-1)(c-1)}
$$

$$
\times\left(\frac{\Gamma((a+1)/2)\Gamma(a-b+1)\Gamma(a-c+1)\Gamma(-b-c+(a+5)/2)}{\Gamma(a)\Gamma(-b+(a+3)/2)\Gamma(-c+(a+3)/2)\Gamma(a-b-c+2)} - 1\right),
$$

and:

$$ {}_4F_3\left(\begin{matrix} a,\ b,\ c,\ 1 \\ a-b+3,\ a-c+3,\ 3 \end{matrix} \,\middle|\, 1\right) = \frac{2(a-b+1)_2(a-c+1)_2}{(a-2)_2(b-2)_2(c-2)_2} $$

$$ \times\left(\frac{\Gamma(a/2)\Gamma(a-b+1)\Gamma(a-c+1)\Gamma((a/2)-b-c+4)}{\Gamma(a-1)\Gamma((a/2)-b+2)\Gamma((a/2)-c+2)\Gamma(a-b-c+3)} - \frac{(a-2)(b-2)(c-2)}{(a-b+1)(a-c+1)} - 1\right). $$

Special Case 6. When $p = 4, q = 3$ and $x = 1$, by noting the Watson theorem (9), Relation (18) is simplified as:

$$ {}_4F_3\left(\begin{matrix} a,\ b,\ c,\ 1 \\ (a+b+1)/2,\ 2c+1-m,\ m \end{matrix} \,\middle|\, 1\right) $$

$$ = \frac{\Gamma(m)\Gamma((a+b+1)/2)\Gamma(2c+1-m)\Gamma(a+1-m)\Gamma(b+1-m)\Gamma(c+1-m)}{\Gamma(a)\Gamma(b)\Gamma(c)\Gamma(-m+(a+b+3)/2)\Gamma(2c-2m+2)} $$

$$ \times\left(\begin{matrix} \frac{\sqrt{\pi}\,\Gamma(c-m+(3/2))\Gamma(-m+(a+b+3)/2)\Gamma(c-(a+b-1)/2)}{\Gamma(1+(a-m)/2)\Gamma(1+(b-m)/2)\Gamma(c+1-(a+m)/2)\Gamma(c+1-(b+m)/2)} \\ -\ {}_3F_2^{(m-2)}\left(\begin{matrix} a-m+1,\ b-m+1,\ c-m+1 \\ -m+1+(a+b+1)/2,\ 2c-2m+2 \end{matrix} \,\middle|\, 1\right) \end{matrix}\right). \tag{24} $$

For $m = 1$, Relation (24) exactly gives the Watson formula, while for $m = 2, 3$, we have:

$$ {}_4F_3\left(\begin{matrix} a,\ b,\ c,\ 1 \\ (a+b+1)/2,\ 2c-1,\ 2 \end{matrix} \,\middle|\, 1\right) = \frac{a+b-1}{(a-1)(b-1)} $$

$$ \times\left(\frac{\sqrt{\pi}\,\Gamma(c-(1/2))\Gamma((a+b-1)/2)\Gamma(c-(a+b-1)/2)}{\Gamma(a/2)\Gamma(b/2)\Gamma(c-(a/2))\Gamma(c-(b/2))} - 1\right), $$

and:

$$ {}_4F_3\left(\begin{matrix} a,\ b,\ c,\ 1 \\ (a+b+1)/2,\ 2c-2,\ 3 \end{matrix} \,\middle|\, 1\right) = \frac{(2c-3)(a+b-1)(a+b-3)}{(a-2)_2(b-2)_2(c-1)} $$

$$ \times\left(\frac{\sqrt{\pi}\,\Gamma(c-(3/2))\Gamma((a+b-3)/2)\Gamma(c-(a+b-1)/2)}{\Gamma((a-1)/2)\Gamma((b-1)/2)\Gamma(c-(a+1)/2)\Gamma(c-(b+1)/2)} - \frac{(a-2)(b-2)}{a+b-3} - 1\right). $$

Special Case 7. When $p = 4, q = 3$ and $x = 1$, by noting the Whipple theorem (10), Relation (18) is simplified as:

$$
{}_4F_3\left(\begin{array}{c} a,\ 2m-1-a,\ b,\ 1 \\ c,\ 2b-c+1,\ m \end{array}\,\middle|\, 1\right)
$$

$$
= \frac{\Gamma(m)\Gamma(c)\Gamma(2b-c+1)\Gamma(a+1-m)\Gamma(m-a)\Gamma(b+1-m)}{\Gamma(a)\Gamma(2m-1-a)\Gamma(b)\Gamma(c+1-m)\Gamma(2b-c-m+2)}
$$

$$
\times \left(\begin{array}{c} \frac{\pi\, 2^{2m-2b-1}\Gamma(c-m+1)\Gamma(2b-c+2-m)}{\Gamma(-m+1+(a+c)/2)\Gamma(-m+1+b+(a-c+1)/2)\Gamma((1-a+c)/2)\Gamma(b+1-(a+c)/2)} \\ -\ {}_3^{(m-2)}F_2\left(\begin{array}{c} a-m+1,\ m-a,\ b-m+1 \\ c-m+1,\ 2b-c-m+2 \end{array}\,\middle|\, 1\right) \end{array}\right). \quad (25)
$$

For $m=1$, Relation (25) exactly gives the Whipple formula, while for $m=2,3$, we have:

$$
{}_4F_3\left(\begin{array}{c} a,\ 3-a,\ b,\ 1 \\ c,\ 2b-c+1,\ 2 \end{array}\,\middle|\, 1\right) = \frac{(c-1)(c-2b)}{(a-2)_2(b-1)}\times
\left(\frac{\pi\, 2^{3-2b}\Gamma(c-1)\Gamma(2b-c)}{\Gamma(-1+(a+c)/2)\Gamma(b+(a-c-1)/2)\Gamma((1-a+c)/2)\Gamma(b+1-(a+c)/2)}-1\right),
$$

and:

$$
{}_4F_3\left(\begin{array}{c} a,\ 5-a,\ b,\ 1 \\ c,\ 2b-c+1,\ 3 \end{array}\,\middle|\, 1\right) = \frac{2(c-2)_2(2b-c-1)_2}{(a-4)_4(b-2)_2}
\times\left(\frac{\pi\, 2^{5-2b}\Gamma(c-2)\Gamma(2b-c-1)}{\Gamma(-2+(a+c)/2)\Gamma(b+(a-c-3)/2)\Gamma((1-a+c)/2)\Gamma(b+1-(a+c)/2)} - \frac{(a-2)(3-a)(b-2)}{(c-2)(2b-c-1)}-1\right).
$$

Special Case 8. When $p=4, q=3$ and $x=1$, by noting the Pfaff–Saalschütz theorem (11), Relation (18) is simplified as:

$$
{}_4F_3\left(\begin{array}{c} a,\ b,\ -n+m-1,\ 1 \\ c,\ 1+a+b-c-n,\ m \end{array}\,\middle|\, 1\right) = \frac{(m-1)!(1-c)_{m-1}(c-a-b+n)_{m-1}}{(1-a)_{m-1}(1-b)_{m-1}(n+2-m)_{m-1}}
$$

$$
\times\left(\frac{(c-a)_n(c-b)_n}{(c+1-m)_n(c-a-b+m-1)_n} - {}_3^{(m-2)}F_2\left(\begin{array}{c} a-m+1,\ b-m+1,\ -n \\ c-m+1,\ 2+a+b-c-m-n \end{array}\,\middle|\, 1\right)\right). \quad (26)
$$

For $m=1$, Relation (26) exactly gives the Pfaff–Saalschütz formula, while for $m=2,3$, we have:

$$
{}_4F_3\left(\begin{array}{c} a,\ b,\ -n+1,\ 1 \\ c,\ 1+a+b-c-n,\ 2 \end{array}\,\middle|\, 1\right) = \frac{(1-c)(c-a-b+n)}{n\,(1-a)(1-b)}
\times\left(\frac{(c-a)_n(c-b)_n}{(c-1)_n(c-a-b+1)_n}-1\right),
$$

and:

$$ {}_4F_3\left(\begin{matrix} a,\ b,\ -n+2,\ 1 \\ c,\ 1+a+b-c-n,\ 3 \end{matrix}\,\middle|\, 1\right) = \frac{2(1-c)_2(c-a-b+n)_2}{n(1-a)_2(1-b)_2(n-1)_2} \times \left(\frac{(c-a)_n(c-b)_n}{(c-2)_n(c-a-b+2)_n} + \frac{n(a-2)(b-2)}{(c-2)(a+b-c-n-1)} - 1\right). $$

Special Case 9. When $p = 5, q = 4$ and $x = -1$, by noting the second theorem of Whipple (12), Relation (18) is simplified as:

$$ {}_5F_4\left(\begin{matrix} a,\ (a+m+1)/2,\ b,\ c,\ 1 \\ (a+m-1)/2, a-b+m,\ a-c+m,\ m \end{matrix}\,\middle|\, -1\right) = (-1)^{m-1}\times $$

$$ \frac{\Gamma(m)\Gamma((a+m-1)/2)\Gamma(a-b+m)\Gamma(a-c+m)\Gamma(a-m+1)\Gamma((a-m+3)/2)\Gamma(b-m+1)\Gamma(c-m+1)}{\Gamma(a)\Gamma((a+m+1)/2)\Gamma(b)\Gamma(c)\Gamma((a-m+1)/2)\Gamma(a-b+1)\Gamma(a-c+1)} $$

$$ \times\left(\frac{\Gamma(1+a-b)\Gamma(1+a-c)}{\Gamma(2-m+a)\Gamma(m+a-b-c)} - {}_4F_3^{(m-2)}\left(\begin{matrix} a-m+1,\ (a-m+3)/2,\ b-m+1,\ c-m+1 \\ (a-m+1)/2,\ a-b+1,\ a-c+1 \end{matrix}\,\middle|\, -1\right)\right). \tag{27} $$

For $m = 1$, Relation (27) exactly gives the Whipple formula, while for $m = 2, 3$, we have:

$$ {}_5F_4\left(\begin{matrix} a,\ (a+3)/2,\ b,\ c,\ 1 \\ (a+1)/2, a-b+2,\ a-c+2,\ 2 \end{matrix}\,\middle|\, -1\right) $$

$$ = \frac{4(a-b+1)(a-c+1)}{(a^2-1)(a-1)(b-1)(c-1)}\left(1 - \frac{\Gamma(1+a-b)\Gamma(1+a-c)}{\Gamma(a)\Gamma(2+a-b-c)}\right), $$

and:

$$ {}_5F_4\left(\begin{matrix} a,\ (a+4)/2,\ b,\ c,\ 1 \\ (a+2)/2, a-b+3,\ a-c+3,\ 3 \end{matrix}\,\middle|\, -1\right) = \frac{2(a-b+1)_2(a-c+1)_2}{(a+2)(a-1)(b-2)_2(c-2)_2} $$

$$ \times\left(\frac{\Gamma(1+a-b)\Gamma(1+a-c)}{\Gamma(a-1)\Gamma(3+a-b-c)} + \frac{a(b-2)(c-2)}{(a-b+1)(a-c+1)} - 1\right). $$

Special Case 10. When $p = 6, q = 5$ and $x = 1$, by noting the Dougall theorem (13), Relation (18) is simplified as:

$$ {}_6F_5\left(\begin{matrix} a,\ (a+m+1)/2,\ c,\ d,\ e,\ 1 \\ (a+m-1)/2, a-c+m,\ a-d+m,\ a-e+m,\ m \end{matrix}\,\middle|\, 1\right) $$

$$ = \Gamma(m)\Gamma((a+m-1)/2)\Gamma(a-c+m) $$

$$ \times\frac{\Gamma(a-d+m)\Gamma(a-e+m)\Gamma(a-m+1)\Gamma((a-m+3)/2)\Gamma(c-m+1)\Gamma(d-m+1)\Gamma(e-m+1)}{\Gamma(a)\Gamma((a+m+1)/2)\Gamma(c)\Gamma(d)\Gamma(e)\Gamma((a-m+1)/2)\Gamma(a-c+1)\Gamma(a-d+1)\Gamma(a-e+1)} $$

$$ \times\left(\begin{matrix} \frac{\Gamma(a-c+1)\Gamma(a-d+1)\Gamma(a-e+1)\Gamma(a-c-d-e+2m-1)}{\Gamma(a+2-m)\Gamma(a-d-e+m)\Gamma(a-c-e+m)\Gamma(a-c-d+m)} \\ -\ {}_5F_4^{(m-2)}\left(\begin{matrix} a-m+1,\ (a-m+3)/2,\ c-m+1, d-m+1,\ e-m+1 \\ (a-m+1)/2,\ a-c+1,\ a-d+1,\ a-e+1 \end{matrix}\,\middle|\, 1\right) \end{matrix}\right). \tag{28} $$

For $m = 1$, Relation (28) exactly gives the Dougall formula, while for $m = 2, 3$, we have:

$$
{}_6F_5\left(\left. \begin{matrix} a,\ (a+3)/2,\ c,\ d,\ e,\ 1 \\ (a+1)/2, a-c+2,\ a-d+2,\ a-e+2,\ 2 \end{matrix} \right| 1 \right)
$$

$$
= \frac{(a-c+1)(a-d+1)(a-e+1)}{(c-1)(d-1)(e-1)(a+1)}
$$

$$
\times \left(\frac{\Gamma(a-c+1)\Gamma(a-d+1)\Gamma(a-e+1)\Gamma(a-c-d-e+3)}{\Gamma(a)\Gamma(a-d-e+2)\Gamma(a-c-e+2)\Gamma(a-c-d+2)} - 1 \right),
$$

and:

$$
{}_6F_5\left(\left. \begin{matrix} a,\ (a+4)/2,\ c,\ d,\ e,\ 1 \\ (a+2)/2, a-c+3,\ a-d+3,\ a-e+3,\ 3 \end{matrix} \right| 1 \right)
$$

$$
= \frac{2\,(a-c+1)_2(a-d+1)_2(a-e+1)_2}{(a-1)(a+2)(c-2)_2(d-2)_2(e-2)_2}
$$

$$
\times \left(\frac{\Gamma(a-c+1)\Gamma(a-d+1)\Gamma(a-e+1)\Gamma(a-c-d-e+5)}{\Gamma(a-1)\Gamma(a-d-e+3)\Gamma(a-c-e+3)\Gamma(a-c-d+3)} \right.
$$

$$
\left. - \frac{a(c-2)(d-2)(e-2)}{(a-c+1)(a-d+1)(a-e+1)} - 1 \right).
$$

Special Case 11. When $p = 8, q = 7$ and $x = 1$, by noting the second theorem of Dougall (14), Relation (18) is simplified as:

$$
{}_8F_7\left(\left. \begin{matrix} a,\ (a+m+1)/2,\ b, c,\ d,\ 2a-b-c-d+2m-1+n,\ m-n-1,\ 1 \\ (a+m-1)/2,\ a-b+m,\ a-c+m,\ a-d+m,\ b+c+d-a+1-m-n, a+n+1,\ m \end{matrix} \right| 1 \right)
$$

$$
= (-1)^{m-1}(m-1)! \times
$$

$$
\frac{((3-a-m)/2)_{m-1}(1-a+b-m)_{m-1}(1-a+c-m)_{m-1}(1-a+d-m)_{m-1}(m+n+a-b-c-d)_{m-1}(-a-n)_{m-1}}{((1-a-m)/2)_{m-1}(1-a)_{m-1}(1-b)_{m-1}(1-c)_{m-1}(1-d)_{m-1}(b+c+d-2a+2-2m-n)_{m-1}(n+2-m)_{m-1}}
$$

$$
\times \left(\begin{matrix} \frac{(a-m+2)_n(a-b-c+m)_n(a-b-d+m)_n(a-c-d+m)_n}{(a-b+1)_n(a-c+1)_n(a-d+1)_n(a-b-c-d+2m-1)_n} \\ - {}^{(m-2)}_{\quad 7}F_6\left(\left. \begin{matrix} a-m+1,\ (a-m+3)/2,\ b-m+1, c-m+1,\ d-m+1,\ 2a-b-c-d+m+n, -n \\ (a-m+1)/2, a-b+1,\ a-c+1,\ a-d+1,\ b+c+d-a+2-2m-n, a-m+n+2 \end{matrix} \right| 1 \right) \end{matrix} \right). \tag{29}
$$

For $m = 1$, Relation (29) exactly gives the Dougall formula, while for $m = 2, 3$, we have:

$$
{}_8F_7\left(\begin{array}{c} a,\ (a+3)/2,\ b,c,\ d,\ 2a-b-c-d+3+n,\ -n+1,\ 1 \\ (a+1)/2,\ a-b+2,\ a-c+2,\ a-d+2,\ b+c+d-a-1-n, a+n+1,\ 2 \end{array} \middle| \ 1\right)
$$

$$
= \frac{(-a+b-1)(-a+c-1)(-a+d-1)(n+2+a-b-c-d)(a+n)}{n\,(1+a)(1-b)(1-c)(1-d)(b+c+d-2a-2-n)}
$$

$$
\times \left(1-\frac{(a)_n(a-b-c+2)_n(a-b-d+2)_n(a-c-d+2)_n}{(a-b+1)_n(a-c+1)_n(a-d+1)_n(a-b-c-d+3)_n}\right),
$$

and:

$$
{}_8F_7\left(\begin{array}{c} a,\ (a+4)/2,\ b,c,\ d,\ 2a-b-c-d+5+n,\ -n+2,\ 1 \\ (a+2)/2,\ a-b+3,\ a-c+3,\ a-d+3,\ b+c+d-a-2-n, a+n+1,\ 3 \end{array} \middle| \ 1\right)
$$

$$
= \frac{(a-2)(-a+b-2)_2(-a+c-2)_2(-a+d-2)_2(3+n+a-b-c-d)_2(-a-n)_2}{(a+2)(1-a)_2(1-b)_2(1-c)_2(1-d)_2(b+c+d-2a-4-n)_2(n-1)_2}
$$

$$
\times \left(\begin{array}{c} \frac{(a-1)_n(a-b-c+3)_n(a-b-d+3)_n(a-c-d+3)_n}{(a-b+1)_n(a-c+1)_n(a-d+1)_n(a-b-c-d+5)_n} \\ +\frac{na(b-2)(c-2)(d-2)(2a-b-c-d+3+n)}{(a-b+1)(a-c+1)(a-d+1)(b+c+d-a-n-4)(n+a-1)} - 1 \end{array}\right).
$$

Remark 1. *There are two further special cases, which however do not belong to classical summation theorems. When $p = q = 1$, Relation (18) is simplified as:*

$$
{}_1F_1\left(\begin{array}{c} 1 \\ m \end{array} \middle| \ z\right) = \frac{(m-1)!}{z^{m-1}}\left(e^z - \sum_{j=0}^{m-2}\frac{z^j}{j!}\right),
$$

and when $p = q+1 = 2$, it yields:

$$
{}_2F_1\left(\begin{array}{c} a,\ 1 \\ m \end{array} \middle| \ z\right) = \frac{(m-1)!}{z^{m-1}}\frac{\Gamma(a-m+1)}{\Gamma(a)}\left((1-z)^{m-a-1} - \sum_{j=0}^{m-2}(a-m+1)_j\frac{z^j}{j!}\right).
$$

Similarly, for the case $n = 2$, Relation (16) changes to:

$$
{}_pF_q\left(\begin{array}{c} a_1,\ \dots a_{p-1},\ \ 2 \\ b_1,\ \dots b_{q-1},\ \ m \end{array} \middle| \ z\right) = \frac{\Gamma(m)}{z^{m-1}}\sum_{j=m-1}^{\infty}\frac{(a_1)_{j-m+1}\cdots(a_{p-1})_{j-m+1}}{(b_1)_{j-m+1}\cdots(b_{q-1})_{j-m+1}}\frac{(j+2-m)\,z^j}{j!}
$$

$$
= \frac{\Gamma(m)}{z^{m-1}}\left(z\sum_{r=m-2}^{\infty}\frac{(a_1)_{j-m+2}\cdots(a_{p-1})_{j-m+2}}{(b_1)_{j-m+2}\cdots(b_{q-1})_{j-m+2}}\frac{z^r}{r!} + (2-m)\sum_{j=m-1}^{\infty}\frac{(a_1)_{j-m+1}\cdots(a_{p-1})_{j-m+1}}{(b_1)_{j-m+1}\cdots(b_{q-1})_{j-m+1}}\frac{z^j}{j!}\right)
$$

$$
= \frac{(m-1)!}{z^{m-2}}\frac{\Gamma(b_1)\dots\Gamma(b_{q-1})}{\Gamma(a_1)\dots\Gamma(a_{p-1})}\frac{\Gamma(a_1-m+2)\dots\Gamma(a_{p-1}-m+2)}{\Gamma(b_1-m+2)\dots\Gamma(b_{q-1}-m+2)}
$$

$$\times \left({}_{p-1}F_{q-1}\left(\begin{array}{c} a_1-m+2,\ ...,\ a_{p-1}-m+2 \\ b_1-m+2,\ ...,\ b_{q-1}-m+2 \end{array} \middle| z \right) - {}_{p-1}^{(m-3)}F_{q-1}\left(\begin{array}{c} a_1-m+2,\ ...,\ a_{p-1}-m+2 \\ b_1-m+2,\ ...,\ b_{q-1}-m+2 \end{array} \middle| z \right) \right)$$

$$+ (2-m)\frac{(m-1)!}{z^{m-1}} \frac{\Gamma(b_1)...\Gamma(b_{q-1})}{\Gamma(a_1)...\Gamma(a_{p-1})} \frac{\Gamma(a_1-m+1)...\Gamma(a_{p-1}-m+1)}{\Gamma(b_1-m+1)...\Gamma(b_{q-1}-m+1)}$$

$$\times \left({}_{p-1}F_{q-1}\left(\begin{array}{c} a_1-m+1,\ ...,\ a_{p-1}-m+1 \\ b_1-m+1,\ ...,\ b_{q-1}-m+1 \end{array} \middle| z \right) - {}_{p-1}^{(m-2)}F_{q-1}\left(\begin{array}{c} a_1-m+1,\ ...,\ a_{p-1}-m+1 \\ b_1-m+1,\ ...,\ b_{q-1}-m+1 \end{array} \middle| z \right) \right).$$

Therefore:

$$ {}_pF_q\left(\begin{array}{c} a_1,\ ...\ a_{p-1},\quad 2 \\ b_1,\ ...\ b_{q-1},\quad m \end{array} \middle| z \right) = \frac{(m-1)!}{z^{m-1}} \frac{\Gamma(b_1)...\Gamma(b_{q-1})}{\Gamma(a_1)...\Gamma(a_{p-1})} \frac{\Gamma(a_1-m+1)...\Gamma(a_{p-1}-m+1)}{\Gamma(b_1-m+1)...\Gamma(b_{q-1}-m+1)}$$
$$\times \left(\begin{array}{l} \frac{(a_1-m+1)...(a_{p-1}-m+1)}{(b_1-m+1)...(b_{q-1}-m+1)} z \\ \times \left({}_{p-1}F_{q-1}\left(\begin{array}{c} a_1-m+2,\ ...,\ a_{p-1}-m+2 \\ b_1-m+2,\ ...,\ b_{q-1}-m+2 \end{array} \middle| z \right) - {}_{p-1}^{(m-3)}F_{q-1}\left(\begin{array}{c} a_1-m+2,\ ...,\ a_{p-1}-m+2 \\ b_1-m+2,\ ...,\ b_{q-1}-m+2 \end{array} \middle| z \right) \right) \\ -(m-2)\left({}_{p-1}F_{q-1}\left(\begin{array}{c} a_1-m+1,\ ...,\ a_{p-1}-m+1 \\ b_1-m+1,\ ...,\ b_{q-1}-m+1 \end{array} \middle| z \right) - {}_{p-1}^{(m-2)}F_{q-1}\left(\begin{array}{c} a_1-m+1,\ ...,\ a_{p-1}-m+1 \\ b_1-m+1,\ ...,\ b_{q-1}-m+1 \end{array} \middle| z \right) \right) \end{array} \right). \quad (30)$$

For instance, if $m = 3$, Relation (30) reads as:

$$ {}_pF_q\left(\begin{array}{c} a_1,\ ...\ ,a_{p-1},\quad 2 \\ b_1,\ ...\ ,b_{q-1},\quad 3 \end{array} \middle| z \right) = \frac{2}{z^2} \frac{\Gamma(b_1)...\Gamma(b_{q-1})}{\Gamma(a_1)...\Gamma(a_{p-1})} \frac{\Gamma(a_1-2)...\Gamma(a_{p-1}-2)}{\Gamma(b_1-2)...\Gamma(b_{q-1}-2)}$$
$$\times \left(\frac{(a_1-2)...(a_{p-1}-2)}{(b_1-2)...(b_{q-1}-2)} z\, {}_{p-1}F_{q-1}\left(\begin{array}{c} a_1-1,\ ...,\ a_{p-1}-1 \\ b_1-1,\ ...,\ b_{q-1}-1 \end{array} \middle| z \right) \right.$$
$$\left. - {}_{p-1}F_{q-1}\left(\begin{array}{c} a_1-2,\ ...,\ a_{p-1}-2 \\ b_1-2,\ ...,\ b_{q-1}-2 \end{array} \middle| z \right) + 1 \right).$$

Hence, for $p = q + 1 = 3$ and $z = 1$, we have:

$$ {}_3F_2\left(\begin{array}{c} a,\ b,\ 2 \\ c,\ 3 \end{array} \middle| 1 \right) = \frac{2}{(a-2)_2(b-2)_2}$$
$$\times \left((c-2)_2 + \frac{\Gamma(c)\Gamma(c-a-b+1)}{\Gamma(c-a)\Gamma(c-b)}(ab-a-b-c+3) \right).$$

3. Second Hypergeometric Identity

By noting Relation (1), first it is not difficult to verify that:

$$(a+m)_k = \frac{(a)_{k+m}}{(a)_m}. \quad (31)$$

Now, if the identity (31) is applied in a special case of (2), we obtain:

$$
{}_pF_q\left(\begin{array}{ccc} a_1+m, & \ldots\, a_{p-1}+m, & 1 \\ b_1+m, & \ldots & ,b_q+m \end{array}\middle| z\right) = \sum_{k=0}^{\infty} \frac{(a_1+m)_k \ldots (a_{p-1}+m)_k}{(b_1+m)_k \ldots (b_q+m)_k} z^k
$$

$$
= \frac{(b_1)_m \ldots (b_q)_m}{(a_1)_m \ldots (a_{p-1})_m} \sum_{k=0}^{\infty} \frac{(a_1)_{k+m} \ldots (a_{p-1})_{k+m}}{(b_1)_{k+m} \ldots (b_{q-1})_{k+m}} z^k
$$

$$
= \frac{(b_1)_m \ldots (b_q)_m}{(a_1)_m \ldots (a_{p-1})_m} \sum_{j=m}^{\infty} \frac{(a_1)_j \ldots (a_{p-1})_j}{(b_1)_j \ldots (b_{q-1})_j} z^{j-m}
$$

$$
= \frac{(b_1)_m \ldots (b_q)_m}{(a_1)_m \ldots (a_{p-1})_m} z^{-m} \left(\sum_{j=0}^{\infty} \frac{(a_1)_j \ldots (a_{p-1})_j}{(b_1)_j \ldots (b_{q-1})_j} z^j - \sum_{j=0}^{m-1} \frac{(a_1)_j \ldots (a_{p-1})_j}{(b_1)_j \ldots (b_{q-1})_j} z^j \right),
$$

leading to the second identity:

$$
{}_pF_q\left(\begin{array}{ccc} a_1+m, & \ldots\, a_{p-1}+m, & 1 \\ b_1+m, & \ldots & ,b_q+m \end{array}\middle| z\right) = \frac{(b_1)_m \ldots (b_q)_m}{(a_1)_m \ldots (a_{p-1})_m} z^{-m}
$$

$$
\times \left({}_pF_q\left(\begin{array}{ccc} a_1, & \ldots\, a_{p-1}, & 1 \\ b_1, & \ldots & ,b_q \end{array}\middle| z\right) - {}_pF_q^{(m-1)}\left(\begin{array}{ccc} a_1, & \ldots\, a_{p-1}, & 1 \\ b_1, & \ldots & ,b_q \end{array}\middle| z\right)\right), \quad (32)
$$

which is equivalent to:

$$
{}_pF_q\left(\begin{array}{ccc} a_1, & \ldots\, a_{p-1}, & 1 \\ b_1, & \ldots & ,b_q \end{array}\middle| z\right) = \frac{(b_1-m)_m \ldots (b_q-m)_m}{(a_1-m)_m \ldots (a_{p-1}-m)_m} z^{-m}
$$

$$
\times \left({}_pF_q\left(\begin{array}{ccc} a_1-m, & \ldots\, a_{p-1}-m, & 1 \\ b_1-m, & \ldots & ,b_q-m \end{array}\middle| z\right)\right.
$$

$$
\left. - {}_pF_q^{(m-1)}\left(\begin{array}{ccc} a_1-m, & \ldots\, a_{p-1}-m, & 1 \\ b_1-m, & \ldots & ,b_q-m \end{array}\middle| z\right)\right). \quad (33)
$$

Once again, the interesting point is that by using Relation (32) or (33), various special cases can be considered as follows.

Special Case 12. When $p = 2, q = 1$ and $x = -1$, by noting the Kummer formula and Relation (32), we get:

$$
{}_2F_1\left(\begin{array}{c} b+m,\ 1 \\ 2-b+m \end{array}\middle| -1\right)
$$

$$
= (-1)^m \frac{(2-b)_m}{(b)_m} \left(\frac{\sqrt{\pi}}{2} \frac{\Gamma(2-b)}{\Gamma(-b+(3/2))} - {}_2F_1^{(m-1)}\left(\begin{array}{c} b,\ 1 \\ 2-b \end{array}\middle| -1\right)\right). \quad (34)
$$

For instance, if $m = 1, 2$, Relation (34) is simplified as:

$$ {}_2F_1\left(\begin{array}{c} b+1,\ 1 \\ 3-b \end{array} \middle| -1\right) = -\frac{\sqrt{\pi}}{2}\frac{\Gamma(3-b)}{\Gamma((3/2)-b)} + \frac{2}{b} - 1, $$

and:

$$ {}_2F_1\left(\begin{array}{c} b+2,\ 1 \\ 4-b \end{array} \middle| -1\right) = \frac{\sqrt{\pi}}{2b(b+1)}\frac{\Gamma(4-b)}{\Gamma((3/2)-b)} - \frac{2(b-1)(b-3)}{b(b+1)}. $$

Special Case 13. When $p = 2, q = 1$ and $x = 1/2$, by noting the second kind of Gauss formula and Relation (32), we get:

$$ {}_2F_1\left(\begin{array}{c} a+m,\ 1 \\ (a/2)+m+1 \end{array} \middle| \frac{1}{2}\right) = \frac{(1+(a/2))_m}{(a)_m}2^m\left(\frac{\sqrt{\pi}\,\Gamma(1+(a/2))}{\Gamma((a+1)/2)} - {}_2F_1^{(m-1)}\left(\begin{array}{c} a,\ 1 \\ (a/2)+1 \end{array} \middle| \frac{1}{2}\right)\right). \quad (35) $$

For instance, if $m = 1, 2$, Relation (35) is simplified as:

$$ {}_2F_1\left(\begin{array}{c} a+1,\ 1 \\ (a/2)+2 \end{array} \middle| \frac{1}{2}\right) = \frac{a+2}{a}\left(\sqrt{\pi}\frac{\Gamma(1+(a/2))}{\Gamma((a+1)/2)} - 1\right), $$

and:

$$ {}_2F_1\left(\begin{array}{c} a+2,\ 1 \\ (a/2)+3 \end{array} \middle| \frac{1}{2}\right) = \sqrt{\pi}\frac{(a+2)(a+4)}{a(a+1)}\frac{\Gamma(1+(a/2))}{\Gamma((a+1)/2)} - 2(1+\frac{4}{a}). $$

Special Case 14. When $p = 3, q = 2$ and $x = 1$, by noting the Dixon formula and Relation (32), we get:

$$ {}_3F_2\left(\begin{array}{c} b+m,\ c+m,\ 1 \\ 2-b+m,\ 2-c+m \end{array} \middle| 1\right) = \frac{(2-b)_m(2-c)_m}{(b)_m(c)_m} $$

$$ \times\left(\frac{\sqrt{\pi}}{2}\frac{\Gamma(2-b)\Gamma(2-c)\Gamma(-b-c+(3/2))}{\Gamma(-b+(3/2))\Gamma(-c+(3/2))\Gamma(2-b-c)} - {}_3F_2^{(m-1)}\left(\begin{array}{c} b,\ c,\ 1 \\ 2-b,\ 2-c \end{array} \middle| 1\right)\right). \quad (36) $$

For instance, if $m = 1, 2$, Relation (36) is simplified as:

$$ {}_3F_2\left(\begin{array}{c} b+1,\ c+1,\ 1 \\ 3-b,\ 3-c \end{array} \middle| 1\right) = \frac{\sqrt{\pi}}{2bc}\frac{\Gamma(3-b)\Gamma(3-c)\Gamma((3/2)-b-c)}{\Gamma((3/2)-b)\Gamma((3/2)-c)\Gamma(2-b-c)} - \frac{(b-2)(c-2)}{bc}, $$

and:

$$ {}_3F_2\left(\begin{array}{c} b+2,\ c+2,\ 1 \\ 4-b,\ 4-c \end{array} \middle| 1\right) = \frac{\sqrt{\pi}}{2(b)_2(c)_2}\frac{\Gamma(4-b)\Gamma(4-c)\Gamma((3/2)-b-c)}{\Gamma((3/2)-b)\Gamma((3/2)-c)\Gamma(2-b-c)} - \frac{2(bc-b-c-2)(b-3)(c-3)}{(b)_2(c)_2}. $$

Special Case 15. When $p = 3, q = 2$ and $x = 1$, by noting the Watson formula and Relation (32), we get:

$$ {}_3F_2\left(\begin{array}{c} b+m,\ c+m,\ 1 \\ (b/2)+m+1,\ 2c+m \end{array} \middle| 1\right) = \frac{(1+(b/2))_m(2c)_m}{(b)_m(c)_m} $$

$$ \times\left(\frac{\sqrt{\pi}\,\Gamma(1+(c/2))\Gamma(1+(b/2))\Gamma(c-(b/2))}{\Gamma((b+1)/2)\Gamma(c)\Gamma(c-(b-1)/2)} - {}_3F_2^{(m-1)}\left(\begin{array}{c} b,\ c,\ 1 \\ (b/2)+1,\ 2c \end{array} \middle| 1\right)\right). \quad (37) $$

For instance, if $m = 1, 2$, Relation (37) is simplified as:

$$ {}_3F_2\left(\begin{array}{c} b+1,\ c+1,\ 1 \\ (b/2)+2,\ 2c+1 \end{array} \middle| 1\right) = (1+\frac{2}{b})(\frac{\sqrt{\pi}\,\Gamma(1+(c/2))\Gamma(1+(b/2))\Gamma(c-(b/2))}{\Gamma((b+1)/2)\Gamma(c)\Gamma(c-(b-1)/2)} - 1), $$

and:

$$ {}_3F_2\left(\begin{array}{c} b+2,\ c+2,\ 1 \\ (b/2)+3,\ 2c+2 \end{array} \middle| 1\right) $$

$$ = \frac{2\sqrt{\pi}c(2c+1)}{b(b+1)}\,\frac{\Gamma((c/2)+1)\Gamma((b/2)+3)\Gamma(c-(b/2))}{\Gamma((b+1)/2)\Gamma(c+2)\Gamma(c-(b-1)/2)} - \frac{(b+4)(2c+1)}{b(c+1)}. $$

Special Case 16. When $p = 3, q = 2$ and $x = 1$, by noting the Whipple formula and Relation (32), we get:

$$ {}_3F_2\left(\begin{array}{c} a+m,\ 1-a+m,\ 1 \\ c+m,\ 3-c+m \end{array} \middle| 1\right) = \frac{(c)_m(3-c)_m}{(a)_m(1-a)_m} $$

$$ \times\left(\frac{\pi\,\Gamma(c)\Gamma(3-c)}{2\,\Gamma((a+c)/2)\Gamma((a-c+3)/2)\Gamma((1-a+c)/2)\Gamma(2-(a+c)/2)}\right. $$

$$ \left. - {}_3F_2^{(m-1)}\left(\begin{array}{c} a,\ 1-a,\ 1 \\ c,\ 3-c \end{array} \middle| 1\right)\right). \quad (38) $$

For instance, if $m = 1, 2$, Relation (38) is simplified as:

$$ {}_3F_2\left(\begin{array}{c} a+1,\ 2-a,\ 1 \\ c+1,\ 4-c \end{array} \middle| 1\right) $$

$$ = \frac{\pi}{2a(1-a)}\,\frac{\Gamma(c+1)\Gamma(4-c)}{\Gamma((a+c)/2)\Gamma((a-c+3)/2)\Gamma((1-a+c)/2)\Gamma(2-(a+c)/2)} - \frac{c(c-3)}{a(a-1)}, $$

and:

$$ {}_3F_2\left(\begin{array}{c} a+2,\ 3-a,\ 1 \\ c+2,\ 5-c \end{array} \middle| 1\right) $$

$$ = \frac{\pi}{2(a-2)_4}\,\frac{\Gamma(c+2)\Gamma(5-c)}{\Gamma((a+c)/2)\Gamma((a-c+3)/2)\Gamma((1-a+c)/2)\Gamma(2-(a+c)/2)} $$

$$-\frac{(c+1)(4-c)(c(3-c)+a(1-a))}{(a-2)_4}.$$

Special Case 17. When $p = 3, q = 2$ and $x = 1$, by noting the Pfaff–Saalschutz formula and Relation (32), we get:

$$ {}_3F_2\left(\begin{matrix} b+m, \ -n+m, \ 1 \\ c+m, \ 2+b-c-n+m \end{matrix} \middle| \ 1\right) $$

$$ = \frac{(c)_m(2+b-c-n)_m}{(b)_m(-n)_m}\left(\frac{(c-1)(c-1-b+n)}{(c-1-b)(c-1+n)} - {}_3F_2^{(m-1)}\left(\begin{matrix} b, \ -n, \ 1 \\ c, \ 2+b-c-n \end{matrix} \middle| \ 1\right)\right). \quad (39)$$

For instance, if $m = 1, 2$, Relation (39) is simplified as:

$$ {}_3F_2\left(\begin{matrix} b+1, \ -n+1, \ 1 \\ c+1, \ 3+b-c-n \end{matrix} \middle| \ 1\right) = \frac{c(c-2-b+n)}{nb}\left(\frac{(c-1)(c-1-b+n)}{(c-1-b)(c-1+n)} - 1\right), $$

and:

$$ {}_3F_2\left(\begin{matrix} b+2, \ -n+2, \ 1 \\ c+2, \ 4+b-c-n \end{matrix} \middle| \ 1\right) $$

$$ = \frac{(c-1)c(c+1)(c-1-b+n)(c-2-b+n)(c-3-b+n)}{n(n-1)b(b+1)(c-1-b)(c-1+n)} $$

$$ -\frac{(c+1)(c-3-b+n)(nb+c(c-2-b+n))}{n(n-1)b(b+1)}. $$

Special Case 18. When $p = 4, q = 3$ and $x = -1$, by noting the Whipple formula and Relation (32), we get:

$$ {}_4F_3\left(\begin{matrix} m+(3/2), \ b+m, \ c+m, \ 1 \\ m+(1/2), \ 2-b+m, \ 2-c+m \end{matrix} \middle| -1\right) = \frac{(-1)^m}{2m+1}\frac{(2-b)_m(2-c)_m}{(b)_m(c)_m} $$

$$ \times\left(\frac{\Gamma(2-b)\Gamma(2-c)}{\Gamma(2-b-c)} - {}_4F_3^{(m-1)}\left(\begin{matrix} 3/2, \ b, \ c, \ 1 \\ 1/2, \ 2-b, \ 2-c \end{matrix} \middle| -1\right)\right). \quad (40)$$

For instance, if $m = 1, 2$, Relation (40) is simplified as:

$$ {}_4F_3\left(\begin{matrix} 5/2, \ b+1, \ c+1, \ 1 \\ 3/2, \ 3-b, \ 3-c \end{matrix} \middle| -1\right) = \frac{1}{3bc}((b-2)(c-2) - \frac{\Gamma(3-b)\Gamma(3-c)}{\Gamma(2-b-c)}), $$

and:

$$ {}_4F_3\left(\begin{matrix} 7/2, \ b+2, \ c+2, \ 1 \\ 5/2, \ 4-b, \ 4-c \end{matrix} \middle| -1\right) = \frac{(2-b)_2(2-c)_2}{5\,(b)_2(c)_2} $$

$$\times\left(\frac{\Gamma(2-b)\Gamma(2-c)}{\Gamma(2-b-c)}+2\frac{bc+b+c-2}{(b-2)(c-2)}\right).$$

Special Case 19. When $p=5, q=4$ and $x=1$, by noting the Dougall formula and Relation (32), we get:

$$ {}_5F_4\left(\begin{matrix} m+(3/2),\ c+m,\ d+m,\ e+m,\ 1 \\ m+(1/2),\ 2-c+m,\ 2-d+m,\ 2-e+m \end{matrix}\ \middle|\ 1\right)$$

$$=\frac{1}{2m+1}\frac{(2-c)_m(2-d)_m(2-e)_m}{(c)_m(d)_m(e)_m}$$

$$\times\left(\frac{\Gamma(2-c)\Gamma(2-d)\Gamma(2-e)\Gamma(2-c-d-e)}{\Gamma(2-d-e)\Gamma(2-c-e)\Gamma(2-c-d)}-\,{}_5F_4^{(m-1)}\left(\begin{matrix} 3/2,\ c,\ d,\ e,\ 1 \\ 1/2,\ 2-c,\ 2-d,\ 2-e \end{matrix}\ \middle|\ 1\right)\right). \quad (41)$$

For instance, if $m=1,2$, Relation (41) is simplified as:

$$ {}_5F_4\left(\begin{matrix} 5/2,\ c+1,\ d+1,\ e+1,\ 1 \\ 3/2,\ 3-c,\ 3-d,\ 3-e \end{matrix}\ \middle|\ 1\right)=\frac{(2-c)(2-d)(2-e)}{3\,cde}$$

$$\times\left(\frac{\Gamma(2-c)\Gamma(2-d)\Gamma(2-e)\Gamma(2-c-d-e)}{\Gamma(2-d-e)\Gamma(2-c-e)\Gamma(2-c-d)}-1\right),$$

and:

$$ {}_5F_4\left(\begin{matrix} 7/2,\ c+2,\ d+2,\ e+2,\ 1 \\ 5/2,\ 4-c,\ 4-d,\ 4-e \end{matrix}\ \middle|\ 1\right)=\frac{(2-c)_2(2-d)_2(2-e)_2}{5\,(c)_2(d)_2(e)_2}$$

$$\times\left(\frac{\Gamma(2-c)\Gamma(2-d)\Gamma(2-e)\Gamma(2-c-d-e)}{\Gamma(2-d-e)\Gamma(2-c-e)\Gamma(2-c-d)}-\frac{3\,cde}{(2-c)(2-d)(2-e)}-1\right).$$

Special Case 20. When $p=7, q=6$ and $x=1$, by noting the Dougall formula and Relation (32), we get:

$$ {}_7F_6\left(\begin{matrix} m+(3/2),\ b+m, c+m,\ d+m,\ 3-b-c+n+m,\ m-n,\ 1 \\ m+(1/2), 2-b+m, 2-c+m, 2-d+m,\ b+c+d-n-1+m, n+2+m \end{matrix}\ \middle|\ 1\right)$$

$$=\frac{1}{2m+1}\frac{(2-b)_m(2-c)_m(2-d)_m(b+c+d-n-1)_m(n+2)_m}{(b)_m(c)_m(d)_m(3-b-c+n)_m(-n)_m}$$

$$\times\left(\frac{(2)_n(2-b-c)_n(2-b-d)_n(2-c-d)_n}{(2-b)_n(2-c)_n(2-d)_n(2-b-c-d)_n}\right.$$

$$\left.-\,{}_7F_6^{(m-1)}\left(\begin{matrix} 3/2,\ b, c,\ d,\ 3-b-c+n,\ -n,\ 1 \\ 1/2, 2-b, 2-c, 2-d,\ b+c+d-n-1, n+2 \end{matrix}\ \middle|\ 1\right)\right). \quad (42)$$

For instance, if $m = 1, 2$, Relation (42) is simplified as:

$$
{}_7F_6\left(\begin{array}{c} 5/2,\ b+1, c+1,\ d+1,\ 3-b-c+n+1, 1-n,\ 1 \\ 3/2, 3-b, 3-c, 3-d,\ b+c+d-n, n+3 \end{array}\middle|\ 1\right)
$$

$$
= \frac{(2-b)(2-c)(2-d)(b+c+d-n-1)(n+2)}{3\,nbcd(3-b-c+n)}
$$

$$
\times\left(1-\frac{(2)_n(2-b-c)_n(2-b-d)_n(2-c-d)_n}{(2-b)_n(2-c)_n(2-d)_n(2-b-c-d)_n}\right),
$$

and:

$$
{}_7F_6\left(\begin{array}{c} 7/2,\ b+2, c+2,\ d+2,\ 5-b-c+n,\ 2-n,\ 1 \\ 5/2, 4-b, 4-c, 4-d,\ b+c+d-n+1, n+4 \end{array}\middle|\ 1\right)
$$

$$
= \frac{(2-b)_2(2-c)_2(2-d)_2(b+c+d-n-1)_2(n+2)_2}{5\,(b)_2(c)_2(d)_2(3-b-c+n)_2(-n)_2}
$$

$$
\times\left(\frac{(2)_n(2-b-c)_n(2-b-d)_n(2-c-d)_n}{(2-b)_n(2-c)_n(2-d)_n(2-b-c-d)_n}\right.
$$

$$
\left.+\frac{3\,bcd\,n\,(3-b-c+n)}{(2-b)(2-c)(2-d)(b+c+d-n-1)(n+2)}-1\right).
$$

4. Conclusions

In this paper, we applied two identities for generalized hypergeometric series in order to extend some classical summation theorems of hypergeometric functions such as Gauss, Kummer, Dixon, Watson, Whipple, Pfaff–Saalschütz and Dougall formulas and then obtained some new summation theorems using the second introduced hypergeometric identity.

Author Contributions: Both authors have the same contribution in all sections.

Acknowledgments: This research was funded by the Alexander von Humboldt Foundation under Grant Number Ref 3.4-IRN-1128637-GF-E.

Conflicts of Interest: The authors declare no conflict of interest.

References

1. Koepf, W. *Hypergeometric Summation: An Algorithmic Approach to Summation and Special Function Identities*, 2nd ed.; Springer: London, UK, 2014.
2. Slater, L.J. *Generalized Hypergeometric Functions*; Cambridge University Press: Cambridge, UK, 1966.
3. Andrews, G.E.; Askey, R.; Roy, R. Special Functions. In *Encyclopedia of Mathematics and Its Applications*; Cambridge University Press: Cambridge, UK, 1999; Volume 71.
4. Arfken, G. *Mathematical Methods for Physicists*; Academic Press: New York, NY, USA, 1985.
5. Mathai, A.M.; Saxena, R.K. *Generalized Hypergeometric Functions with Applications in Statistics and Physical Sciences*; Lecture Notes in Mathematics; Springer: Berlin/Heidelberg, Germany; New York, NY, USA, 1973; Volume 348.
6. Nikiforov, A.F.; Uvarov, V.B. *Special Functions of Mathematical Physics. A Unified Introduction with Applications*; Birkhauser: Basel, Switzerland, 1988.

7. Bailey, W.N. Products of generalized hypergeometric series. *Proc. Lond. Math. Soc.* **1928**, *2*, 242–254. [CrossRef]
8. Karlsson, P.W. Hypergeometric functions with integral parameter differences. *J. Math. Phys.* **1971**, *12*, 270–271. [CrossRef]
9. Miller, A.R. Certain summation and transformation formulas for generalized hypergeometric series. *J. Comput. Appl. Math.* **2009**, *231*, 964–972. [CrossRef]
10. Miller, A.R.; Srivastava, H.M. Karlsson-Minton summation theorems for the generalized hypergeometric series of unit argument. *Integral Transform. Spec. Funct.* **2010**, *21*, 603–612. [CrossRef]
11. Minton, B.M. Generalized hypergeometric function of unit argument. *J. Math. Phys.* **1970**, *11*, 1375–1376. [CrossRef]
12. Rosengren, H. Karlsson-Minton type hypergeometric functions on the root system C_n. *J. Math. Anal. Appl.* **2003**, *281*, 332–345. [CrossRef]
13. Srivastava, H.M. Generalized hypergeometric functions with integral parameter differences. *Indag. Math.* **1973**, *76*, 38–40. [CrossRef]
14. Prudnikov, A.P.; Brychkov, Y.A.; Marichev, O.I. *More Special Functions;* Integrals and Series; Gordon and Breach Science Publishers: Amsterdam, The Netherlands, 1990; Volume 3.

axioms

MDPI

Article

Pre-Metric Spaces Along with Different Types of Triangle Inequalities

Hsien-Chung Wu

Department of Mathematics, National Kaohsiung Normal University, Kaohsiung 802, Taiwan; hcwu@nknucc.nknu.edu.tw

Received: 25 April 2018; Accepted: 21 May 2018; Published: 24 May 2018

Abstract: The T_1-spaces induced by the pre-metric spaces along with many forms of triangle inequalities are investigated in this paper. The limits in pre-metric spaces are also studied to demonstrate the consistency of limit concept in the induced topologies.

Keywords: quasi-metric space; Hausdorff space; T_1-space; triangle inequalities

MSC: 54E35; 54E55

1. Introduction

Let X be a nonempty universal set, and let $d : X \times X \to \mathbb{R}_+$ be a nonnegative real-valued function defined on the product set $X \times X$. We say that (X, d) is a metric space if and only if the following conditions are satisfied:

- for any $x, y \in M$, $d(x, y) = 0$ implies $x = y$;
- (self-distance condition) for any $x \in M$, $d(x, x) = 0$;
- (symmetric condition) for any $x, y \in M$, $d(x, y) = d(y, x)$;
- (triangle inequality) for any $x, y, z \in M$, $d(x, z) \leq d(x, y) + d(y, z)$.

In the literature, different kinds of spaces are considered by weakening the above conditions. Wilson [1] says that (X, d) is a quasi-metric space when the symmetric condition is not satisfied; that is, the following conditions are satisfied:

- for any $x, y \in M$, $d(x, y) = 0$ if and only if $x = y$;
- for any $x, y, z \in M$, $d(x, z) \leq d(x, y) + d(y, z)$.

After that, many authors (referring to [2–15] and the references therein) also defined the quasi-metric space as follows:

- for any $x, y \in M$, $d(x, y) = 0 = d(y, x)$ if and only if $x = y$;
- for any $x, y, z \in M$, $d(x, z) \leq d(x, y) + d(y, z)$.

However, these two definitions are not equivalent. The reason is that $d(x, y) = 0$ does not necessarily imply $d(y, x) = 0$, since the symmetric condition is not satisfied. It is clear to see that, in the Wilson's sense, we also have $d(y, x) = 0$ if and only if $y = x$.

Wilson [16] also says that (X, d) is a semi-metric space when the triangle inequality is not satisfied; that is, the following conditions are satisfied:

- for any $x, y \in M$, $d(x, y) = 0$ if and only if $x = y$;
- for any $x, y \in M$, $d(x, y) = d(y, x)$.

On the other hand, Matthews [11] says that (X, d) is a partial metric space if and only if the following conditions are satisfied:

- for any $x, y \in M$, $x = y$ if and only if $d(x,x) = d(x,y) = d(y,y)$;
- for any $x, y \in M$, $d(x,x) \leq d(x,y)$;
- for any $x, y \in M$, $d(x,y) = d(y,x)$.
- for any $x, y, z \in M$, $d(x,z) \leq d(x,y) + d(y,z) - d(y,y)$.

The partial metric space does not assume the self-distance condition $d(x,x) = 0$.

In this paper, we shall consider a so-called pre-metric space in which we just assume that $d(x,y) = 0$ implies $x = y$ for any $x, y \in X$. In other words, the pre-metric space does not assume the self-distance condition and symmetric condition. Since the triangle inequality plays a very important role, without considering the symmetric condition, the triangle inequality can be considered in four forms, which was not discussed in the literature. Based on the four different kinds of triangle inequalities, we can induce the T_1-space space from the pre-metric space under some suitable conditions.

This paper is organized as follows. In Section 2 , we propose the so-called pre-metric space in which four forms of triangle inequalities are considered and studied. Many basic properties are also obtained for further investigation. In Section 3, we induce the T_1-space from a given pre-metric space under some suitable assumptions. In Section 4, the limits in pre-metric space are also studied. We present the consistency of limit concepts in the pre-metric space and the induced topologies.

2. Definitions and Properties

In this section, we shall introduce the concept of pre-metric space, and the four concepts of triangle inequalities. We also derive some interesting properties that will be used in the further study. Without considering the symmetric condition, we first introduce four types of triangle inequality as follows.

Definition 1. *Let X be a nonempty universal set, and let d be a mapping defined on $X \times X$ into $\mathbb{R}_+$.*

- *We say that d satisfies the* $\bowtie$-triangle inequality *if and only if the following inequality is satisfied:*

$$d(x,y) + d(y,z) \geq d(x,z) \text{ for all } x, y, z \in X.$$

- *We say that d satisfies the* $\triangleright$-triangle inequality *if and only if the following inequality is satisfied:*

$$d(x,y) + d(z,y) \geq d(x,z) \text{ for all } x, y, z \in X.$$

- *We say that d satisfies the* $\triangleleft$-triangle inequality *if and only if the following inequality is satisfied:*

$$d(y,x) + d(y,z) \geq d(x,z) \text{ for all } x, y, z \in X.$$

- *We say that d satisfies the* $\diamond$-triangle inequality *if and only if the following inequality is satisfied:*

$$d(y,x) + d(z,y) \geq d(x,z) \text{ for all } x, y, z \in X.$$

It is obvious that if d satisfies the symmetric condition, then the concepts of $\bowtie$-triangle inequality, $\triangleright$-triangle inequality, $\triangleleft$-triangle inequality and $\diamond$-triangle inequality are all equivalent.

Example 1. *We define a function $d : \mathbb{R}_+ \times \mathbb{R}_+ \to \mathbb{R}_+$ by $d(x,y) = \max\{x,y\}$. Then $d(x,x) = x$ for any $x \geq 0$, which also says that $d(x,x)$ is not always zero. It is not hard to check*

$$\max\{x,y\} + \max\{y,z\} \geq \max\{x,z\},$$

which also says that

$$d(x,y) + d(y,z) \geq d(x,z).$$

This shows that d satisfies the $\bowtie$-triangle inequality. Since d also satisfies the symmetric condition, it means that all the four forms of triangle inequalities are equivalent. However, since $d(x,x) > 0$ for $x > 0$, it says that $(\mathbb{R}_+, d)$ is still not a metric space.

Example 2. *We define a function $d : \mathbb{R}_+ \times \mathbb{R}_+ \to \mathbb{R}_+$ by*

$$d(x,y) = \begin{cases} x & \text{if } x \geq y \\ 2y - x & \text{if } x < y. \end{cases}$$

Then $d(x,x) = x$ for any $x \geq 0$, which also says that $d(x,x)$ is not always zero. For $x > y$, we see that $d(x,y) = x$ and $d(y,x) = 2x - y$, which says that $d(x,y) \neq d(y,x)$ in general; that is, the symmetric condition is not satisfied. It is not hard to check

$$d(x,y) + d(y,z) \geq d(x,z).$$

This shows that d also satisfies the $\bowtie$-triangle inequality.

Examples 1 and 2 say that $d(x,x) \neq 0$ for $x \neq 0$. Therefore, we propose the following definition.

Definition 2. *Let X be a nonempty universal set, and let d be a mapping defined on $X \times X$ into $\mathbb{R}_+$. We say that (X,d) is a* pre-metric space *if and only if $d(x,y) = 0$ implies $x = y$ for any $x, y \in X$.*

We see that (X,d) is a *quasi-metric space* if and only if (X,d) is a pre-metric space satisfying the $\bowtie$-triangle inequality and $d(x,x) = 0$ for all $x \in X$.

Example 3. *Examples 1 and 2 are pre-metric spaces, since it is not hard to check that $d(x,y) = 0$ implies $x = y = 0$ based on the nonnegativity.*

Remark 1. *Let (X,d) be a pre-metric space. Then $d(x,y) = 0$ implies $x = y$, which also implies $d(x,y) = 0 = d(x,x) = d(y,x)$ without needing the symmetric condition. We remark that this symmetric situation can only happen when $d(x,y) = 0$ or $d(y,x) = 0$. Therefore, if $d(x,y) > 0$ then we cannot have $d(x,y) = d(y,x)$ in general. On the other hand, we also see that $d(x,y) = 0$ or $d(y,x) = 0$ implies $d(x,x) = 0$. However, this situation does not say $d(x,x) = 0$ for all $x \in X$. We can just say that $d(x,x) = 0$ when $d(x,y) = 0$ for some $x, y \in X$. In other words, we can just say that $d(x,x) = 0$ for some $x \in X$. This situation can also be realized from Example 2.*

Proposition 1. *Let X be a nonempty universal set, and let d be a mapping defined on $X \times X$ into $\mathbb{R}_+$. Suppose that the following conditions are satisfied:*

- *$d(x,x) = 0$ for all $x \in X$;*
- *d satisfies the $\triangleright$-triangle inequality or the $\triangleleft$-triangle inequality or the $\diamond$-triangle inequality.*

Then d satisfies the symmetric condition.

Proof. Suppose that d satisfies the $\triangleright$-triangle inequality. Then, given any $x, y \in X$, we have

$$d(x,y) \leq d(x,x) + d(y,x) = d(y,x).$$

By interchanging the roles of x and y, we can also obtain $d(y,x) \leq d(x,y)$. This shows that $d(x,y) = d(y,x)$. The other cases of satisfying the $\triangleleft$-triangle inequality and the $\diamond$-triangle inequality can be similarly obtained. This completes the proof. □

Remark 2. *Suppose that $d(x,x) = 0$ for all $x \in X$, and that d satisfies the $\circ$-triangle inequality for some $\circ \in \{\triangleright, \triangleleft, \diamond\}$. Then, using Proposition 1, we see that all the four forms of triangle inequalities are equivalent.*

3. T_1-Space

We want to show that the pre-metric space along with the different kinds of triangle inequalities can induce the T_1-Space based on the concepts of open balls defined below.

Definition 3. *Let (X,d) be a pre-metric space. Given $r > 0$, the* open balls *centered at x are denoted and defined by*

$$B^{\triangleleft}(x;r) = \{y \in X : d(x,y) < r\}$$

and

$$B^{\triangleright}(x;r) = \{y \in X : d(y,x) < r\}.$$

Let $\mathcal{B}^{\triangleleft}$ denote the family of all open balls $B^{\triangleleft}(x;r)$, and let $\mathcal{B}^{\triangleright}$ denote the family of all open balls $B^{\triangleright}(x;r)$.

In the sequel, we also assume that the open balls $B^{\triangleleft}(x;r)$ and $B^{\triangleright}(x;r)$ are nonempty for each $x \in X$ and $r > 0$. In other words, given any $x \in X$ and $r > 0$, we assume that there exist y_1 and y_2 such that $d(x,y_1) < r$ and $d(x,y_2) < r$, respectively. It is also clear that if d satisfies the symmetric condition, then

$$B^{\triangleleft}(x;r) = B^{\triangleright}(x;r).$$

In this case, we simply write $B(x;r)$ to denote the open balls centered at x, and write $\mathcal{B}$ to denote the family of all open balls $B(x;r)$.

Proposition 2. *Let (X,d) be a pre-metric space.*

(i) Given any $x \in X$, we have the following properties.
   - *Suppose that $d(x,x) = 0$. Then $x \in B^{\triangleleft}(x;r) \in \mathcal{B}^{\triangleleft}$ and $x \in B^{\triangleright}(x;r) \in \mathcal{B}^{\triangleright}$ for all $r > 0$.*
   - *Suppose that $x \in B^{\triangleleft}(x;r)$ for all $r > 0$, or that $x \in B^{\triangleright}(x;r)$ for all $r > 0$. Then $d(x,x) = 0$.*

(ii) If $x \neq y$, then there exist $r_1 > 0$ and $r_2 > 0$ such that $y \notin B^{\triangleleft}(x;r_1)$ and $y \notin B^{\triangleright}(x;r_2)$.

(iii) For each $x \in X$, we have the following properties.
   - *Given any $B^{\triangleleft}(x;r) \in \mathcal{B}^{\triangleleft}$, there exists $n \in \mathbb{N}$ such that $B^{\triangleleft}(x;\frac{1}{n}) \subseteq B^{\triangleleft}(x;r)$.*
   - *Given any $B^{\triangleright}(x;r) \in \mathcal{B}^{\triangleright}$, there exists $n \in \mathbb{N}$ such that $B^{\triangleright}(x;\frac{1}{n}) \subseteq B^{\triangleright}(x;r)$.*

Proof. The first statement of part (i) is obvious. To prove the second statement of part (i), we take a sequence $\{r_n\}_{n=1}^{\infty}$ of positive numbers such that it is decreasing to zero. Then we have $d(x,x) < r_n$ for all n, which implies $d(x,x) = 0$ by taking $n \to \infty$. To prove part (ii), since $x \neq y$, it follows that $d(x,y) > 0$ and $d(y,x) > 0$ by the definition of pre-metric space. Using the denseness of $\mathbb{R}$, there exists $r_1 > 0$ such that $0 < r_1 < d(x,y)$, which also says that $y \notin B^{\triangleleft}(x;r_1)$. We also have $y \notin B^{\triangleright}(x;r_2)$ for some $r_2 > 0$ satisfying $0 < r_2 < d(y,x)$. Part (iii) follows from the existence of a positive integer n with $1/n < r$. This completes the proof. □

Proposition 3. *Let (X,d) be a pre-metric space. Then we have the following inclusions.*

(i) Suppose that d satisfies the $\bowtie$-triangle inequality.
   - *Given any $y \in B^{\triangleleft}(x;r)$, there exists $\bar{r} > 0$ such that $B^{\triangleleft}(y;\bar{r}) \subseteq B^{\triangleleft}(x;r)$.*
   - *Given any $y \in B^{\triangleright}(x;r)$, there exists $\bar{r} > 0$ such that $B^{\triangleright}(y;\bar{r}) \subseteq B^{\triangleright}(x;r)$.*

(ii) Suppose that d satisfies the $\triangleright$-triangle inequality. Given any $y \in B^{\triangleleft}(x;r)$, there exists $\bar{r} > 0$ such that $B^{\triangleright}(y;\bar{r}) \subseteq B^{\triangleleft}(x;r)$ and $B^{\triangleright}(y;\bar{r}) \subseteq B^{\triangleright}(x;r)$.

(iii) Suppose that d satisfies the $\triangleleft$-triangle inequality. Given any $y \in B^{\triangleright}(x;r)$, there exists $\bar{r} > 0$ such that $B^{\triangleleft}(y;\bar{r}) \subseteq B^{\triangleright}(x;r)$ and $B^{\triangleleft}(y;\bar{r}) \subseteq B^{\triangleleft}(x;r)$.

(iv) Suppose that d satisfies the $\diamond$-triangle inequality.

- *Given any* $y \in B^{\triangleleft}(x;r)$*, there exists* $\bar{r} > 0$ *such that* $B^{\triangleleft}(y;\bar{r}) \subseteq B^{\triangleright}(x;r)$.
- *Given any* $y \in B^{\triangleright}(x;r)$*, there exists* $\bar{r} > 0$ *such that* $B^{\triangleright}(y;\bar{r}) \subseteq B^{\triangleleft}(x;r)$.

(v) Suppose that d satisfies the $\triangleright$*-triangle inequality and the* $\triangleleft$*-triangle inequality.*

- *Given any* $y \in B^{\triangleleft}(x;r)$*, there exists* $\bar{r} > 0$ *such that* $B^{\triangleleft}(y;\bar{r}) \subseteq B^{\triangleleft}(x;r)$.
- *Given any* $y \in B^{\triangleright}(x;r)$*, there exists* $\bar{r} > 0$ *such that* $B^{\triangleright}(y;\bar{r}) \subseteq B^{\triangleright}(x;r)$.

Proof. To prove part (i), for $y \in B^{\triangleleft}(x;r)$ and $z \in B^{\triangleleft}(y;\bar{r})$, let $\bar{r} \leq r - d(x,y)$. Using the $\bowtie$-triangle inequality, we have

$$d(x,z) \leq d(x,y) + d(y,z) < d(x,y) + \bar{r} \leq d(x,y) + r - d(x,y) = r,$$

which says that $z \in B^{\triangleleft}(x;r)$. Therefore we obtain the inclusion $B^{\triangleleft}(y;\bar{r}) \subseteq B^{\triangleleft}(x;r)$. For $y \in B^{\triangleright}(x;r)$ and $z \in B^{\triangleright}(y;\bar{r})$, let $\bar{r} \leq r - d(y,x)$. Then we can similarly obtain the inclusion $B^{\triangleright}(y;\bar{r}) \subseteq B^{\triangleright}(x;r)$.

To prove part (ii), for $y \in B^{\triangleleft}(x;r)$ and $z \in B^{\triangleright}(y;\bar{r})$, let $\bar{r} \leq r - d(x,y)$. Using the $\triangleright$-triangle inequality, we have

$$d(x,z) \leq d(x,y) + d(z,y) < d(x,y) + \bar{r} \leq d(x,y) + r - d(x,y) = r,$$

which says that $z \in B^{\triangleleft}(x;r)$. Therefore we obtain the inclusion $B^{\triangleright}(y;\bar{r}) \subseteq B^{\triangleleft}(x;r)$. We can similarly obtain the inclusion $B^{\triangleright}(y;\bar{r}) \subseteq B^{\triangleright}(x;r)$.

Parts (iii) and (iv) can be similarly obtained. To prove the first statement of part (v), using part (ii), we can take $\bar{r}^* > 0$ such that $B^{\triangleright}(y;\bar{r}^*) \subseteq B^{\triangleleft}(x;r)$. Using part (iii), we can also take $\bar{r} > 0$ such that $B^{\triangleleft}(y;\bar{r}) \subseteq B^{\triangleright}(y;\bar{r}^*)$. This shows that $B^{\triangleleft}(y;\bar{r}) \subseteq B^{\triangleleft}(x;r)$. The second statement of part (v) can be similarly obtained. This completes the proof. □

Proposition 4. *Let* (X,d) *be a pre-metric space. Then we have the following inclusions.*

(i) Suppose that d satisfies the $\bowtie$*-triangle inequality.*

- *If* $x \in B^{\triangleleft}(x_1,r_1) \cap B^{\triangleleft}(x_2,r_2)$*, then there exists* $r_3 > 0$ *such that*

$$B^{\triangleleft}(x,r_3) \subseteq B^{\triangleleft}(x_1,r_1) \cap B^{\triangleleft}(x_2,r_2).$$

- *If* $x \in B^{\triangleright}(x_1,r_1) \cap B^{\triangleright}(x_2,r_2)$*, then there exists* $r_3 > 0$ *such that*

$$B^{\triangleright}(x,r_3) \subseteq B^{\triangleright}(x_1,r_1) \cap B^{\triangleright}(x_2,r_2).$$

(ii) Suppose that d satisfies the $\triangleright$*-triangle inequality. If* $x \in B^{\triangleleft}(x_1,r_1) \cap B^{\triangleleft}(x_2,r_2)$*, then there exists* $r_3 > 0$ *such that*

$$B^{\triangleright}(x,r_3) \subseteq B^{\triangleleft}(x_1,r_1) \cap B^{\triangleleft}(x_2,r_2) \text{ and } B^{\triangleright}(x,r_3) \subseteq B^{\triangleright}(x_1,r_1) \cap B^{\triangleright}(x_2,r_2).$$

(iii) Suppose that d satisfies the $\triangleleft$*-triangle inequality. If* $x \in B^{\triangleright}(x_1,r_1) \cap B^{\triangleright}(x_2,r_2)$*, then there exists* $r_3 > 0$ *such that*

$$B^{\triangleleft}(x,r_3) \subseteq B^{\triangleright}(x_1,r_1) \cap B^{\triangleright}(x_2,r_2) \text{ and } B^{\triangleleft}(x,r_3) \subseteq B^{\triangleleft}(x_1,r_1) \cap B^{\triangleleft}(x_2,r_2).$$

(iv) Suppose that d satisfies the $\diamond$*-triangle inequality.*

- *If* $x \in B^{\triangleleft}(x_1,r_1) \cap B^{\triangleleft}(x_2,r_2)$*, then there exists* $r_3 > 0$ *such that*

$$B^{\triangleleft}(x,r_3) \subseteq B^{\triangleright}(x_1,r_1) \cap B^{\triangleright}(x_2,r_2).$$

- *If* $x \in B^{\triangleright}(x_1,r_1) \cap B^{\triangleright}(x_2,r_2)$*, then there exists* $r_3 > 0$ *such that*

$$B^{\triangleright}(x,r_3) \subseteq B^{\triangleleft}(x_1,r_1) \cap B^{\triangleleft}(x_2,r_2).$$

(v) *Suppose that d satisfies the* $\triangleright$*-triangle inequality and the* $\triangleleft$*-triangle inequality. We have the following inclusions.*

- *If* $x \in B^{\triangleleft}(x_1, r_1) \cap B^{\triangleleft}(x_2, r_2)$, *then there exists* $r_4 > 0$ *such that*
$$B^{\triangleleft}(x, r_4) \subseteq B^{\triangleleft}(x_1, r_1) \cap B^{\triangleleft}(x_2, r_2).$$
- *If* $x \in B^{\triangleright}(x_1, r_1) \cap B^{\triangleright}(x_2, r_2)$, *then there exists* $r_4 > 0$ *such that*
$$B^{\triangleright}(x, r_4) \subseteq B^{\triangleright}(x_1, r_1) \cap B^{\triangleright}(x_2, r_2).$$

Proof. To prove part (i), for $x \in B^{\triangleleft}(x_1, r_1)$ and $x \in B^{\triangleleft}(x_2, r_2)$, using part (i) of Proposition 3, there exist $\bar{r}_1$ and $\bar{r}_2$ such that
$$B^{\triangleleft}(x, \bar{r}_1) \subseteq B^{\triangleleft}(x_1, r_1) \text{ and } B^{\triangleleft}(x, \bar{r}_2) \subseteq B^{\triangleleft}(x_2, r_2).$$
We take $r_3 = \min\{\bar{r}_1, \bar{r}_2\}$. Then
$$B^{\triangleleft}(x, r_3) \subseteq B^{\triangleleft}(x, \bar{r}_1) \cap B^{\triangleleft}(x, \bar{r}_2) \subseteq B^{\triangleleft}(x_1, r_1) \cap B^{\triangleleft}(x_2, r_2).$$
Therefore we obtain the first inclusion. The second inclusion can be similarly obtained.

To prove part (ii), for $x \in B^{\triangleleft}(x_1, r_1)$ and $x \in B^{\triangleleft}(x_2, r_2)$, using part (ii) of Proposition 3, there exist $\bar{r}_1$ and $\bar{r}_2$ such that
$$B^{\triangleright}(x, \bar{r}_1) \subseteq B^{\triangleleft}(x_1, r_1) \text{ and } B^{\triangleright}(x, \bar{r}_1) \subseteq B^{\triangleright}(x_1, r_1).$$
and
$$B^{\triangleright}(x, \bar{r}_2) \subseteq B^{\triangleleft}(x_2, r_2) \text{ and } B^{\triangleright}(x, \bar{r}_2) \subseteq B^{\triangleright}(x_2, r_2).$$
Let $r_3 = \min\{\bar{r}_1, \bar{r}_2\}$. Then
$$B^{\triangleright}(x, r_3) \subseteq B^{\triangleright}(x, \bar{r}_1) \cap B^{\triangleright}(x, \bar{r}_2) \subseteq B^{\triangleleft}(x_1, r_1) \cap B^{\triangleleft}(x_2, r_2)$$
and
$$B^{\triangleright}(x, r_3) \subseteq B^{\triangleright}(x, \bar{r}_1) \cap B^{\triangleright}(x, \bar{r}_2) \subseteq B^{\triangleright}(x_1, r_1) \cap B^{\triangleright}(x_2, r_2).$$
Therefore we obtain the desired inclusions.

Parts (iii) and (iv) can be similarly obtained. To prove the first statement of part (v), using part (iii) of Proposition 3 and part (ii) of this proposition, we can find $r_4 > 0$ such that
$$B^{\triangleleft}(x, r_4) \subseteq B^{\triangleright}(x, r_3) \subseteq B^{\triangleleft}(x_1, r_1) \cap B^{\triangleleft}(x_2, r_2).$$
The second statement can be similarly obtained. This completes the proof. □

Proposition 5. *Let* (X, d) *be a pre-metric space. Suppose that* $x \neq y$. *Then we have the following properties.*

(i) *Suppose that d satisfies the* $\bowtie$*-triangle inequality or the* $\diamond$*-triangle inequality. Then* $B^{\triangleleft}(x; r) \cap B^{\triangleright}(y; r) = \emptyset$ *and* $B^{\triangleright}(x; r) \cap B^{\triangleleft}(y; r) = \emptyset$ *for some* $r > 0$.
(ii) *Suppose that d satisfies the* $\triangleright$*-triangle inequality. Then* $B^{\triangleleft}(x; r) \cap B^{\triangleleft}(y; r) = \emptyset$ *for some* $r > 0$.
(iii) *Suppose that d satisfies the* $\triangleleft$*-triangle inequality. Then* $B^{\triangleright}(x; r) \cap B^{\triangleright}(y; r) = \emptyset$ *for some* $r > 0$.

Proof. Since $x \neq y$, it says that $d(x, y) > 0$ and $d(y, x) > 0$. We consider the following cases.

- Suppose that d satisfies the $\triangleright$-triangle inequality. Let $r \leq d(x, y)/2$. We are going to prove $B^{\triangleleft}(x; r) \cap B^{\triangleleft}(y; r) = \emptyset$ by contradiction. Suppose that $z \in B^{\triangleleft}(x; r) \cap B^{\triangleleft}(y; r)$. Since d satisfies the $\triangleright$-triangle inequality, it follows that
$$d(x, y) \leq d(x, z) + d(y, z) < r + r = 2r \leq d(x, y),$$

which is a contradiction. Suppose that d satisfies the $\triangleleft$-triangle inequality. Then we can similarly obtain the desired result.

- Suppose that d satisfies the $\bowtie$-triangle inequality. Let $r \leq d(x,y)/2$. For $z \in B^{\triangleleft}(x;r) \cap B^{\triangleright}(y;r)$, it follows that
$$d(x,y) \leq d(x,z) + d(z,y) < r + r = 2r \leq d(x,y),$$
which is a contradiction. On the other hand, let $r \leq d(y,x)/2$, for $z \in B^{\triangleright}(x;r) \cap B^{\triangleleft}(y;r)$, it follows that
$$d(y,x) \leq d(y,z) + d(z,x) < r + r = 2r \leq d(y,x),$$
which is a contradiction. Suppose that d satisfies the $\diamond$-triangle inequality. Then we can similarly obtain the desired result.

This completes the proof. □

Theorem 1. *Let (X,d) be a pre-metric space. Define*
$$\tau^{\triangleleft} = \left\{O^{\triangleleft} \subseteq X : x \in O^{\triangleleft} \text{ if and only if there exist } r > 0 \text{ such that } x \in B^{\triangleleft}(x;r) \subseteq O^{\triangleleft}\right\}. \tag{1}$$
and
$$\tau^{\triangleright} = \left\{O^{\triangleright} \subseteq X : x \in O^{\triangleright} \text{ if and only if there exist } r > 0 \text{ such that } x \in B^{\triangleright}(x;r) \subseteq O^{\triangleright}\right\}. \tag{2}$$
Suppose that d satisfies the $\bowtie$-triangle inequality. Then we have the following results.

- *Assume additionally that $d(x,x) = 0$ for all $x \in X$, or that $x \in B^{\triangleleft}(x;r)$ for all $x \in X$ and $r > 0$. Then $(X,\tau^{\triangleleft})$ is a T_1-space such that $\mathcal{B}^{\triangleleft}$ is a base for the topology $\tau^{\triangleleft}$.*
- *Assume additionally that $d(x,x) = 0$ for all $x \in X$, or that $x \in B^{\triangleright}(x;r)$ for all $x \in X$ and $r > 0$. Then $(X,\tau^{\triangleright})$ is a T_1-space such that $\mathcal{B}^{\triangleright}$ is a base for the topology $\tau^{\triangleright}$.*

The T_1-spaces $(X,\tau^{\triangleleft})$ and $(X,\tau^{\triangleright})$ also satisfy the first axiom of countability. Moreover, $B^{\triangleleft}(x;r)$ is a $\tau^{\triangleleft}$-open set and $B^{\triangleright}(x;r)$ is a $\tau^{\triangleright}$-open set.

Proof. Using part (i) of Proposition 2 and part (i) of Proposition 4, we see that $\tau^{\triangleleft}$ is a topology such that $\mathcal{B}^{\triangleleft}$ is a base for $\tau^{\triangleleft}$. Part (ii) of Proposition 2 says that $(X,\tau^{\triangleleft})$ is a T_1-space. Part (iii) of Proposition 2 says that there exists a countable local base at each $x \in X$ for $\tau^{\triangleleft}$, which also says that $\tau^{\triangleleft}$ satisfies the first axiom of countability. Regarding $\tau^{\triangleright}$, we can similarly obtain the desired results. Finally, part (i) of Proposition 3 says that $B^{\triangleleft}(x;r)$ is a $\tau^{\triangleleft}$-open set and $B^{\triangleright}(x;r)$ is a $\tau^{\triangleright}$-open set. This completes the proof. □

We remark that, in Theorem 1, although we assume $d(x,x) = 0$ for all $x \in X$, (X,d) is not necessarily a metric space, since the symmetric condition is still not satisfied. The following example provides this observation.

Example 4. *We define a function $d : \mathbb{R}_+ \times \mathbb{R}_+ \to \mathbb{R}_+$ by*
$$d(x,y) = \begin{cases} x & \text{if } x > y \\ 0 & \text{if } x = y \\ 2y - x & \text{if } x < y. \end{cases}$$
Then $d(x,x) = 0$ for all $x \in X$. By referring to Example 2, we also see that the symmetric condition is not satisfied, and that d satisfies the $\bowtie$-triangle inequality. Using Theorem 1, we can induce two T_1-spaces $(\mathbb{R}_+,\tau^{\triangleleft})$ and $(\mathbb{R}_+,\tau^{\triangleright})$. Moreover, the spaces $(\mathbb{R}_+,\tau^{\triangleleft})$ and $(\mathbb{R}_+,\tau^{\triangleright})$ also satisfy the first axiom of countability.

4. Limits in Pre-Metric Space

Let (X,d) be a pre-metric space. Since the symmetric condition is not necessarily satisfied, the different concepts of limit are proposed below.

Definition 4. *Let (X,d) be a pre-metric space, and let $\{x_n\}_{n=1}^{\infty}$ be a sequence in X.*

- *We write $x_n \xrightarrow{d^{\triangleright}} x$ as $n \to \infty$ if and only if $d(x_n,x) \to 0$ as $n \to \infty$.*
- *We write $x_n \xrightarrow{d^{\triangleleft}} x$ as $n \to \infty$ if and only if $d(x,x_n) \to 0$ as $n \to \infty$.*
- *We write $x_n \xrightarrow{d} x$ as $n \to \infty$ if and only if*

$$\lim_{n\to\infty} d(x_n,x) = \lim_{n\to\infty} d(x,x_n) = 0.$$

The uniqueness of limits will be discussed below.

Proposition 6. *Let (X,d) be a pre-metric space, and let $\{x_n\}_{n=1}^{\infty}$ be a sequence in X.*

(i) Suppose that d satisfies the $\bowtie$-triangle inequality or $\diamond$-triangle inequality. If $x_n \xrightarrow{d^{\triangleleft}} x$ and $x_n \xrightarrow{d^{\triangleright}} y$, then $x = y$.

(ii) Suppose that d satisfies the $\triangleleft$-triangle inequality. If $x_n \xrightarrow{d^{\triangleright}} x$ and $x_n \xrightarrow{d^{\triangleright}} y$, then $x = y$. In other words, the $d^{\triangleright}$-limit is unique.

(iii) Suppose that d satisfies the $\triangleright$-triangle inequality. If $x_n \xrightarrow{d^{\triangleleft}} x$ and $x_n \xrightarrow{d^{\triangleleft}} y$, then $x = y$. In other words, the $d^{\triangleleft}$-limit is unique.

Proof. To prove part (i), we first assume that d satisfies the $\bowtie$-triangle inequality. Then

$$d(x,y) \le d(x,x_n) + d(x_n,y) \to 0 + 0 = 0,$$

which says that $x = y$. Now suppose that d satisfies the $\diamond$-triangle inequality. Then

$$d(y,x) \le d(x_n,y) + d(x,x_n) \to 0 + 0 = 0$$

which also says that $x = y$. The other cases can be similarly obtained. This completes the proof. □

Let (X,τ) be a topological space. The sequence $\{x_n\}_{n=1}^{\infty}$ in X converges to $x \in X$ with respect to the topology is denoted by $x_n \xrightarrow{\tau} x$ as $n \to \infty$.

Proposition 7. *Let (X,d) be a pre-metric space. Suppose that d satisfies the $\bowtie$-triangle inequality or the $\triangleright$-triangle inequality or the $\triangleleft$-triangle inequality. Assume that $d(x,x) = 0$ for all $x \in X$. Then the following statements hold true.*

(i) Let $\tau^{\triangleright}$ be the topology defined by (1) in Theorem 1, and let $\{x_n\}_{n=1}^{\infty}$ be a sequence in X. Then $x_n \xrightarrow{\tau^{\triangleright}} x$ as $n \to \infty$ if and only if $x_n \xrightarrow{d^{\triangleright}} x$ as $n \to \infty$.

(ii) Let $\tau^{\triangleleft}$ be the topology defined by (2) in Theorem 1, and let $\{x_n\}_{n=1}^{\infty}$ be a sequence in X. Then $x_n \xrightarrow{\tau^{\triangleleft}} x$ as $n \to \infty$ if and only if $x_n \xrightarrow{d^{\triangleleft}} x$ as $n \to \infty$.

Proof. Under the assumptions, Theorem 1 says that we can induce two topologies $\tau^{\triangleright}$ and $\tau^{\triangleleft}$. It suffices to prove part (i). Suppose that $x_n \xrightarrow{\tau^{\triangleright}} x$ as $n \to \infty$. Given any $\epsilon > 0$, there exits $n_\epsilon \in \mathbb{N}$ such that $x_n \in B^{\triangleright}(x;\epsilon)$ for all $n \ge n_\epsilon$, i.e., $d(x_n,x) < \epsilon$ for all $n \ge n_\epsilon$. This says that $d(x_n,x) \to 0$ as $n \to \infty$. Conversely, if $d(x_n,x) \to 0$ as $n \to \infty$, then, given any $\epsilon > 0$, there exists $n_\epsilon \in \mathbb{N}$ such that $d(x_n,x) < \epsilon$ for all $n \ge n_\epsilon$, which says that $x_n \in B^{\triangleright}(x;\epsilon)$ for all $n \ge n_\epsilon$. This shows that $x_n \xrightarrow{\tau^{\triangleright}} x$ as $n \to \infty$, and the proof is complete. □

Let (X, d) be a pre-metric space. We consider the following sets

$$\bar{B}^{\triangleleft}(x;r) = \{y \in X : d(x,y) \leq r\} \text{ and } \bar{B}^{\triangleright}(x;r) = \{y \in X : d(y,x) \leq r\}.$$

If the symmetric condition is satisfied, then we simply write $\bar{B}(x;r)$. We are going to consider the closeness of $\bar{B}^{\triangleleft}(x;r)$ and $\bar{B}^{\triangleright}(x;r)$. Let us recall that, given a topological space (X, τ), we say that a subset F of X is τ-closed if and only if $\tau\text{-cl}(F) = F$, where $\tau\text{-cl}(F)$ denotes the τ-closure of F.

Proposition 8. *Let (X, d) be a pre-metric space. Suppose that d satisfies $d(x,x) = 0$ for all $x \in X$ and the $\bowtie$-triangle inequality. We have the following results.*

- *$\bar{B}^{\triangleleft}(x;r)$ is $\tau^{\triangleright}$-closed. In other words, we have $\tau^{\triangleright}\text{-cl}(\bar{B}^{\triangleleft}(x;r)) = \bar{B}^{\triangleleft}(x;r)$.*
- *$\bar{B}^{\triangleright}(x;r)$ is $\tau^{\triangleleft}$-closed. In other words, we have $\tau^{\triangleleft}\text{-cl}(\bar{B}^{\triangleright}(x;r)) = \bar{B}^{\triangleright}(x;r)$.*

Proof. Under the assumptions, Theorem 1 says that we can induce two topologies $\tau^{\triangleright}$ and $\tau^{\triangleleft}$ satisfying the first axiom of countability. To prove the first statement, for $y \in \tau^{\triangleright}\text{-cl}(\bar{B}^{\triangleleft}(x;r))$, since $(X, \tau^{\triangleright})$ satisfies the first axiom of countability, there exists a sequence $\{y_n\}_{n=1}^{\infty}$ in $\bar{B}^{\triangleleft}(x;r)$ such that $y_n \xrightarrow{\tau^{\triangleright}} y$ as $n \to \infty$. We also have $d(x, y_n) \leq r$ for all n. By part (i) of Proposition 7, we have $d(y_n, y) \to 0$ as $n \to \infty$ for all. The $\bowtie$-triangle inequality says that

$$d(x,y) \leq d(x,y_n) + d(y_n,y) \leq r + d(y_n,y) \to r \text{ as } n \to \infty,$$

which shows $y \in \bar{B}^{\triangleleft}(x;r)$. Therefore we obtain $\tau^{\triangleright}\text{-cl}(\bar{B}^{\triangleleft}(x;r)) = \bar{B}^{\triangleleft}(x;r)$. The second statement can be similarly obtained. This completes the proof. □

Proposition 9. *Let (X, d) be a pre-metric space. Suppose that the following conditions are satisfied.*

- *d satisfies the $\triangleright$-triangle inequality and the $\triangleleft$-triangle inequality simultaneously.*
- *$d(x,x) = 0$ for all $x \in X$.*

Then d satisfies the symmetric condition; that is, (X, d) is a metric space.

Proof. Using part (i) of Proposition 2 and part (v) of Proposition 4, we see that $\tau^{\triangleleft}$ defined by (1) in Theorem 1 is a topology such that $\mathcal{B}^{\triangleleft}$ is a base for $\tau^{\triangleleft}$. Part (iii) of Proposition 2 says that there exists a countable local base at each $x \in X$ for $\tau^{\triangleleft}$, which also says that $\tau^{\triangleleft}$ satisfies the first axiom of countability. For $y \in \tau^{\triangleleft}\text{-cl}(\bar{B}^{\triangleleft}(x;r))$, the first axiom of countability says that there exists a sequence $\{y_n\}_{n=1}^{\infty}$ in $\bar{B}^{\triangleleft}(x;r)$ such that $y_n \xrightarrow{\tau^{\triangleleft}} y$ as $n \to \infty$. We also have $d(x, y_n) \leq r$ for all n. By part (ii) of Proposition 7, we have $d(y, y_n) \to 0$ as $n \to \infty$. The $\triangleright$-triangle inequality says that

$$d(x,y) \leq d(x,y_n) + d(y,y_n) \leq r + d(y,y_n) \to r \text{ as } n \to \infty,$$

which shows $y \in \bar{B}^{\triangleleft}(x;r)$, i.e., $\tau^{\triangleleft}\text{-cl}(\bar{B}^{\triangleleft}(x;r)) = \bar{B}^{\triangleleft}(x;r)$. On the other hand, we also have

$$d(y,x) \leq d(y,y_n) + d(x,y_n) \leq d(y,y_n) + r \to r \text{ as } n \to \infty,$$

which shows $y \in \bar{B}^{\triangleright}(x;r)$. Therefore we obtain the inclusion $\tau^{\triangleleft}\text{-cl}(\bar{B}^{\triangleleft}(x;r)) \subseteq \bar{B}^{\triangleright}(x;r)$, which also says that $\bar{B}^{\triangleleft}(x;r) \subseteq \bar{B}^{\triangleright}(x;r)$. Now, given any $x, y \in X$ with $y \neq x$, we have $d(x,y) > 0$. Let $r = d(x,y)$. Then $y \in \bar{B}^{\triangleleft}(x;r)$. This also says that $y \in \bar{B}^{\triangleright}(x;r)$, i.e.,

$$d(y,x) \leq r = d(x,y).$$

By interchanging the roles of x and y, we can similarly obtain $d(x,y) \leq d(y,x)$. This completes the proof. □

Remark 3. *Let (X, d) be a pre-metric space. Suppose that the conditions presented in Proposition 9 are satisfied. Then (X, d) turns into a metric space. It is well-known that the metric space (X, d) can induce a Hausdorff topological space. More precisely, using the notations in this paper, we see that $\tau^{\triangleleft} = \tau^{\triangleright}$ that is simply written as τ. In other words, (X, τ) is a Hausdorff space such that $\mathcal{B}$ is a base for the topology τ, where $\mathcal{B} = \mathcal{B}^{\triangleleft} = \mathcal{B}^{\triangleright}$. The Hausdorff space (X, τ) also satisfies the first axiom of countability. Moreover, $B(x; r)$ is a τ-open set and $\bar{B}(x; r)$ is a τ-closed set, where $B(x; r) = B^{\triangleleft}(x; r) = B^{\triangleright}(x; r)$ and $\bar{B}(x; r) = \bar{B}^{\triangleleft}(x; r) = \bar{B}^{\triangleright}(x; r)$.*

In a future study, we shall avoid to consider the conditions presented in Proposition 9. Otherwise, the study will become trivial, based on the results of conventional metric space.

Conflicts of Interest: The author declares no conflict of interest.

References

1. Wilson, W.A. On Semi-metric spaces. *Am. J. Math.* **1931**, *53*, 361–373. [CrossRef]
2. Alegre, C.; Marín, J. Modified *w*-distances on quasi-metric spaces and a fixed point theorem on complete quasi-metric spaces. *Topol. Appl.* **2016**, *203*, 32–41. [CrossRef]
3. Ali-Akbari, M.; Pourmahdian, M. Completeness of hyperspaces of compact subsets of quasi-metric spaces. *Acta Math. Hung.* **2010**, *127*, 260–272. [CrossRef]
4. Cao, J.; Rodríguez-López, J. On hyperspace topologies via distance functionals in quasi-metric spaces. *Acta Math. Hung.* **2006**, *112*, 249–268. [CrossRef]
5. Cobzas, S. Completeness in quasi-metric spaces and Ekeland Variational Principle. *Topol. Appl.* **2011**, *158*, 1073–1084. [CrossRef]
6. Collins Agyingi, A.; Haihambo, P.; Künzi, H.-P. A. Endpoints in T_0-quasi-metric spaces. *Topol. Appl.* **2014**, *168*, 82–93. [CrossRef]
7. Doitchinov, D. On completeness in quasi-metric spaces. *Topol. Appl.* **1988**, *30*, 127–148. [CrossRef]
8. Künzi, H.-P.A. A construction of the B-completion of a T_0-quasi-metric space. *Topol. Appl.* **2014**, *170*, 25–39. [CrossRef]
9. Künzi, H.-P.A.; Kivuvu, C.M. The B-completion of a T_0-quasi-metric space. *Topol. Appl.* **2009**, *156*, 2070–2081. [CrossRef]
10. Künzi, H.-P.A.; Yildiz, F. Convexity structures in T_0-quasi-metric spaces. *Topol. Appl.* **2016**, *200*, 2–18. [CrossRef]
11. Matthews, S.G. Partial metric topology. *Ann. N. Y. Acad. Sci.* **1994**, *728*, 183–197. [CrossRef]
12. Romaguera, S.; Antonino, J.A. On convergence complete strong quasi-metrics. *Acta Math. Hung.* **1994**, *64*, 65–73. [CrossRef]
13. Triebel, H. A new approach to function spaces on quasi-metric spaces. *Rev. Mat. Complut.* **2005**, *18*, 7–48. [CrossRef]
14. Ume, J.S. A minimization theorem in quasi-metric spaces and its applications. *Int. J. Math. Math. Sci.* **2002**, *31*, 443–447. [CrossRef]
15. Vitolo, P. A representation theorem for quasi-metric spaces. *Topol. Appl.* **1995**, *65*, 101–104. [CrossRef]
16. Wilson, W.A. On Quasi-metric spaces. *Am. J. Math.* **1931**, *53*, 675–684. [CrossRef]

axioms

MDPI

Article

Yukawa Potential, Panharmonic Measure and Brownian Motion

Antti Rasila [1,*] and Tommi Sottinen [2]

[1] Department of Mathematics and Systems Analysis, School of Science, Aalto University, P.O. Box 1100, FIN-00076 Aalto, Finland

[2] Department of Mathematics and Statistics, Faculty of Technology, University of Vaasa, P.O. Box 700, FIN-65101 Vaasa and Finland; tommi.sottinen@iki.fi

* Correspondence: antti.rasila@iki.fi; Tel.: +358-50-593-6016

Received: 9 April 2018; Accepted: 25 April 2018; Published: 1 May 2018

Abstract: This paper continues our earlier investigation, where a walk-on-spheres (WOS) algorithm for Monte Carlo simulation of the solutions of the Yukawa and the Helmholtz partial differential equations (PDEs) was developed by using the Duffin correspondence. In this paper, we investigate the foundations behind the algorithm for the case of the Yukawa PDE. We study the panharmonic measure, which is a generalization of the harmonic measure for the Yukawa PDE. We show that there are natural stochastic definitions for the panharmonic measure in terms of the Brownian motion and that the harmonic and the panharmonic measures are all mutually equivalent. Furthermore, we calculate their Radon–Nikodym derivatives explicitly for some balls, which is a key result behind the WOS algorithm.

Keywords: potential theory; Brownian motion; Duffin correspondence; harmonic measure; Bessel functions; Monte Carlo simulation; panharmonic measure; walk-on-spheres algorithm; Yukawa equation

PACS: 60J45; 31C45

---

1. Introduction and Preliminaries

The harmonic measure is a fundamental tool in geometric function theory, and it has interesting applications in the study of bounded analytic functions, quasiconformal mappings and potential theory. For example, the harmonic measure has proven very useful in the study of quasidisks and related topics (see, e.g., [1–3]). Results involving the harmonic measure have been given by numerous authors since the 1930s (see [4] and references therein). In this paper, we consider the panharmonic measure, which is a natural counterpart of the classical harmonic measure, whereby the harmonic functions related are replaced with the smooth solutions to the *Yukawa equation:*

$$\Delta u(x) = \mu^2 u(x), \quad \mu^2 \geq 0. \tag{1}$$

Equation (1) first arose from the work of the Japanese physicist Hideki Yukawa in particle physics. Here, $u\colon D \to \mathbb{R}$ is a two-times differentiable function and $D \subset \mathbb{R}^n$, $n \geq 2$ is a domain. The Yukawa equation was first studied in order to describe the nuclear potential of a point charge. This model led to the concept of the Yukawa potential (also called a screened Coulomb potential), which satisfies an equation of the type given by Equation (1). The Yukawa equation also arises from certain problems related to optics (see [5]). Clearly, when $\mu = 0$, we have the Laplace equation, and, indeed, the results given in this paper reduce to the classical results.

Using the terminology of Duffin [6,7], we call a function $u\colon D \to \mathbb{R}$ *panharmonic*, or *μ-panharmonic*, in a domain D if its second derivatives are continuous and if it satisfies the Yukawa equation (Equation (1)) for all $x \in D$. The function u is called panharmonic at $x_0 \in D$ if there is a neighborhood of x_0 where u is panharmonic.

In Definition 1 of the panharmonic measure below, and in all that follows, we always assume that $n \geq 2$, although some results are also true in the dimension $n = 1$. For Definition 1, we need the notions of smallness and the regularity of a domain. These are best given by using the stochastic characterization via the Brownian motion. We refer to any of the classical textbooks [8–10] for further details.

We recall that the n-dimensional Brownian motion $W = (W(t); t \geq 0)$ starting from the point $x \in \mathbb{R}^n$ is the time-homogeneous Markov process with the Markov semigroup

$$P(t)f(x) = \mathbb{E}^x\left[f(W(t))\right]$$

given by

$$P(t) = e^{t\frac{1}{2}\Delta};$$

that is, $\frac{1}{2}\Delta$ is the generator of the Markov semigroup of the Brownian motion.

A domain $D \subset \mathbb{R}^n$ is *regular* if the Brownian motion does not dwell on its boundary; more precisely, D is (Wiener) regular if

$$\mathbb{P}^x\left[\tau_D = 0\right] = 1, \quad \text{for all } x \in \partial D,$$

where $\mathbb{P}^x$ is the probability measure under which $\mathbb{P}^x[W(0) = x] = 1$, and

$$\tau_D = \inf\{t > 0; W(t) \in D^c\}$$

is the first hitting time of the Brownian motion in the set D^c. We call a regular domain D (Wiener) *small* if a Brownian motion starting inside D eventually will leave the domain; that is, D is small if

$$\mathbb{P}^x\left[\tau_D < \infty\right] = 1, \quad \text{for all } x \in D.$$

For example, all bounded domains are small. All half-spaces are also small.

The panharmonic, or μ-panharmonic, measure is a generalization of the harmonic measure:

Definition 1. *Let $D \subset \mathbb{R}^n$ be a small regular domain, and let $\mu^2 \geq 0$. The μ-*panharmonic measure *on a boundary ∂D with a pole at $x \in D$ is the measure $H_\mu^x(D;\cdot)$ such that* any bounded *μ-panharmonic function u on $\bar{D}$ admits the representation*

$$u(x) = \int_{y \in \partial D} u(y)\, H_\mu^x(D; \mathrm{d}y). \tag{2}$$

The existence and uniqueness of a panharmonic measure is established by Theorem 1 and Corollary 2 later. Indeed, by Theorem 1, all *bounded* solutions to the Dirichlet problem $\Delta u - \mu^2 u = 0$ on a *small* regular domain with continuous and bounded boundary data are given by the panharmonic measure as in Equation (2). By Corollary 2, if $\mu^2 > 0$, then the assumption that the domain is small can be removed; that is, all bounded solutions on a regular domain are of the form given by Equation (2) if the boundary data is bounded and continuous. Of course, it is well known that there are unbounded solutions to the Laplace equation that do not admit the harmonic measure representation. The same is true for the Yukawa equation. We refer to Evans [11] for more details on the solutions of the Laplace equation.

We note that if we replace the "killing parameter" μ^2 in the Yukawa equation (Equation (1)) with a "creation parameter" $\lambda < 0$, we obtain another important partial differential equation (PDE), the

Helmholtz equation. In principle, the stochastic approaches taken in this paper can be applied to the solutions of the Helmholtz equation if the domain D is small enough compared to the parameter λ. For details, we refer to Chung and Zhao [8]. If we replace μ^2 by a (positive) function, we obtain the *Schrödinger equation*. Again, the stochastic approaches taken in this paper can be applied, in principle, to the Schrödinger equation, but the results may not be mathematically very tractable. Again, we refer to Chung and Zhao [8] for details. The rest of the paper is organized as follows: In Section 2, we show three different connections between the panharmonic measures and the Brownian motion. The first two (Theorem 1 and Corollary 1) are essentially well known. The third (Corollary 2) is new. In Section 3, we show that the panharmonic measures and the harmonic measures are all mutually equivalent (Theorem 3) and provide some corollaries; namely, we provide a domination principle for the Dirichlet problem related to the Yukawa equation (Corollary 3) and analogs of theorems of Riesz–Riesz, Makarov and Dahlberg for the panharmonic measures (Corollary 4). In Section 4, we consider the panharmonic measures on balls and prove an analogue of the Gauss mean value theorem, or the average property, for the panharmonic functions (Theorem 4), and as a corollary, we obtain the Liouville theorem for panharmonic functions (Corollary 5). Finally, in Section 5, we discuss extensions to the Schrödinger and the Helmholtz PDEs and the walk-on-spheres (WOS) simulation of PDEs.

2. Yukawa Equation and Brownian Motion

We first recall the celebrated connection between the harmonic measure and the Brownian motion first noticed by Kakutani [12] in the 1940s; the harmonic measure is the *hitting measure*:

$$H^x(D;\mathrm{d}y) = \mathbb{P}^x\left[W(\tau_D) \in \mathrm{d}y, \tau_D < \infty\right]. \tag{3}$$

Theorem 1 below is a variant of the Kakutani connection (Equation (3)). A key component of the variant is the following disintegration of the harmonic measure at the time the associated Brownian motion hits the boundary ∂D:

Lemma 1. *Let $D \subset \mathbb{R}^n$ be a regular domain, and let $x \in D$. Then*

$$H^x(D;\mathrm{d}y) = \int_{t=0}^{\infty} h^x(D;\mathrm{d}y,t)\,\mathrm{d}t,$$

where

$$h^x(D;\mathrm{d}y,t) = \mathbb{P}^x\left[W(\tau_D) \in \mathrm{d}y \mid \tau_D = t\right] \frac{\mathrm{d}\mathbb{P}^x}{\mathrm{d}t}\left[\tau_D \le t\right] \tag{4}$$

is the harmonic kernel.

Proof. First, we show the existence of the regular conditional distribution:

$$\mathfrak{p}^x(\mathrm{d}y \,|\, t) = \mathbb{P}^x\left[W(\tau_D) \in \mathrm{d}y \mid \tau_D = t\right]. \tag{5}$$

For this, we note that the random vector $(W(\tau_D), \tau_D)$ can be considered as a function from a space of continuous functions that are the Brownian trajectories equipped with the metric

$$d(f,g) = \sum_{T=1}^{\infty} 2^{-T} \left\| f\mathbf{1}_{[T-1,T)} - g\mathbf{1}_{[T-1,T)} \right\|_{\infty}.$$

For Brownian trajectories, the metric d is almost surely finite because of the independent increments of the Brownian motion and the Borel–Cantelli lemma. Additionally, with the metric d, the space of Brownian paths is a Polish space. Now, by Theorem A1.2 of [13], Polish spaces are Borel spaces. Consequently, for any fixed $x \in D$, by Theorems 6.3 and 6.4 of [13], the probability kernel (Equation (5)) exists and is measurable with respect to t. Consequently, the harmonic kernel is measurable with respect to t.

Second, we show that the distribution of the hitting time τ_D is absolutely continuous with respect to the Lebesgue measure. Let $\varepsilon > 0$ be small enough so that $B = B(x,\varepsilon) \subset D$. Then $\tau_D = \tau_B + (\tau_D - \tau_B)$. Now, the distribution of τ_B is absolutely continuous; see, for example, the section of Bessel processes in Borodin and Salminen [14]. Additionally, because of the rotation symmetry of the Brownian motion, τ_B and $\tau_D - \tau_B$ are independent. Hence, by disintegration and independence, we obtain that

$$\begin{aligned}
\mathbb{P}^x[\tau_D \in \mathrm{d}t] &= \mathbb{P}^x[\tau_B + (\tau_D - \tau_B) \in \mathrm{d}t] \\
&= \int_{s=0}^{\infty} \mathbb{P}^x[t + (\tau_D - \tau_B) \in \mathrm{d}s \,|\, \tau_B = t]\, \mathbb{P}^x[\tau_B \in \mathrm{d}t] \\
&= \int_{s=0}^{\infty} \mathbb{P}^x[t + (\tau_D - \tau_B) \in \mathrm{d}s]\, \mathbb{P}^x[\tau_B \in \mathrm{d}t] \\
&= \varphi^x(t)\, \mathbb{P}^x[\tau_B \in \mathrm{d}t].
\end{aligned}$$

Thus, the distribution of τ_D is absolutely continuous when the distribution of τ_B is absolutely continuous.

Third, we show that the formula given by Equation (4) holds. By disintegrating and conditioning, and by using the continuity of the distribution of τ_D, we obtain that

$$\begin{aligned}
&\mathbb{P}^x\left[W(\tau_D) \in \mathrm{d}y, \tau_D < \infty\right] \\
&= \int_{t=0}^{\infty} \mathbb{P}^x\left[W(\tau_D) \in \mathrm{d}y, \tau_D \in \mathrm{d}t\right] \\
&= \int_{t=0}^{\infty} \mathbb{P}^x\left[W(\tau_D) \in \mathrm{d}y \mid \tau_D = t\right] \mathbb{P}^x\left[\tau_D \in \mathrm{d}t\right] \\
&= \int_{t=0}^{\infty} \mathbb{P}^x\left[W(\tau_D) \in \mathrm{d}y \mid \tau_D = t\right] \frac{\mathrm{d}\mathbb{P}^x}{\mathrm{d}t}\left[\tau_D \le t\right] \mathrm{d}t.
\end{aligned}$$

The claim follows now from the Kakutani connection (Equation (3)). □

The following Theorem 1 is a version of the Kakutani theorem [12] for the Yukawa equation. In some sense, it is a special case of the Kakutani connection to the Schrödinger equation studied extensively by Chung and Zhao [8]. However, it seems that this version with an unbounded and non-small domain D does not appear in any classical texts.

Theorem 1. *Let $D \subset \mathbb{R}^n$ be a regular domain, and let $f : \partial D \to \mathbb{R}$ be bounded and continuous.*

(i) *Then*

$$u(x) = \mathbb{E}^x\left[e^{-\frac{\mu^2}{2}\tau_D} f(W(\tau_D))\,;\, \tau_D < \infty\right] \tag{6}$$

is a solution to the Yukawa–Dirichlet problem:

$$\begin{cases} \Delta u = \mu^2 u & \text{on} \quad D, \\ u = f & \text{on} \quad \partial D. \end{cases}$$

(ii) *Moreover, if u is bounded and D is small, then Equation (6) is* the only *solution to the Yukawa–Dirichlet problem.*

(iii) *As a consequence, the panharmonic measure admits the representation*

$$H_\mu^x(D; \mathrm{d}y) = \int_{t=0}^{\infty} e^{-\frac{\mu^2}{2}t}\, h^x(D; \mathrm{d}y, t)\, \mathrm{d}t, \tag{7}$$

where $h^x(D;\cdot,\cdot)$ is the harmonic kernel defined in Equation (4).

Proof. The first and the second claim of Theorem 1 follow from the classical Kakutani theorem (cf., e.g., [15] Sections 4.4. and 4.6). Indeed, we note that the difficulties involving the Schrödinger equation in [15], Section 4.6 vanish, because

$$\mathbb{E}^x\left[e^{-\frac{\mu^2}{2}\tau_D}\right] \leq 1.$$

To show the third claim, we condition on $\{\tau_D = t\}$ and use the law of total probability:

$$\begin{aligned} u(x) &= \mathbb{E}^x\left[e^{-\frac{\mu^2}{2}\tau_D} f(W(\tau_D))\,;\,\tau_D < \infty\right] \\ &= \int_{y\in\partial D} f(y) \int_{t=0}^{\infty} e^{-\frac{\mu^2}{2}t}\, h^x(D;\mathrm{d}y,t)\,\mathrm{d}t \\ &= \int_{y\in\partial D} f(y)\, H^x_\mu(D;\mathrm{d}y). \end{aligned}$$

□

Remark 1. *Unfortunately, even for very simple D, the harmonic kernel (Equation (4)) is quite difficult to find. The same is true for the regular conditional distribution (Equation (5)). For smooth boundaries ∂D, one can try the following approach: If ∂D is smooth, then the harmonic kernel $h^x(D;\mathrm{d}y,t)$ is absolutely continuous with respect to the Lebesgue measure $\mathrm{d}y$. Indeed, define $p:\mathbb{R}_+\times\mathbb{R}^n\to\mathbb{R}_+$ by*

$$p(t,x) = \frac{1}{(2\pi t)^{n/2}} \exp\left(-\frac{\|x\|^2}{2t}\right). \tag{8}$$

Then p is the Brownian transition kernel*:*

$$p(t,x-y)\,\mathrm{d}y = \mathbb{P}^x\left[W(t)\in\mathrm{d}y\right],$$

and, as a result of [16], Theorem 1, the harmonic kernel can be written as

$$h^x(D;\mathrm{d}y,t) = \frac{1}{2}\frac{\partial p}{\partial \mathrm{n}_y}(D;t,x-y)\,\mathrm{d}y,$$

where n_y is the inward normal at $y\in\partial D$ and $p(D;\cdot,\cdot)$ is the transition density of a Brownian motion that is killed when it hits the boundary ∂D, which can be written as

$$p(D;t,x-y) = p(t,x-y) - \mathbb{E}^x\left[p(t-\tau_D, W(\tau_D)-y)\,;\,\tau_D < t\right] \tag{9}$$

as a result of [10], Equation (3), on page 34.

Consequently, for C^3 boundaries, the harmonic measure admits a Poisson kernel *representation, and therefore, as a result of the representation given by Equation (7), the panharmonic measure also admits a Poisson kernel representation:*

$$\begin{aligned} H^x_\mu(D;\mathrm{d}y) &= \int_{t=0}^{\infty} e^{-\frac{\mu^2}{2}t} h^x(D;\mathrm{d}y,t)\,\mathrm{d}t \\ &= \int_{t=0}^{\infty} e^{-\frac{\mu^2}{2}t}\frac{1}{2}\frac{\partial p}{\partial \mathrm{n}_y}(D;t,x-y)\,\mathrm{d}y\,\mathrm{d}t \\ &= \left[\frac{1}{2}\int_{t=0}^{\infty} e^{-\frac{\mu^2}{2}t}\frac{\partial p}{\partial \mathrm{n}_y}(D;t,x-y)\,\mathrm{d}t\right]\mathrm{d}y. \end{aligned}$$

Theorem 1 gives an interpretation of the panharmonic measure in terms of exponentially discounted Brownian motion. We give a second interpretation in terms of exponentially killed Brownian motion. Indeed, exponential discounting is closely related to exponential killing. The *exponentially killed Brownian motion* W_μ is

$$W_\mu(t) = W(t)\mathbf{1}_{\{Y_\mu > t\}} + \dagger\mathbf{1}_{\{Y_\mu \le t\}},$$

where $\dagger$ is a *coffin state* (by convention $f(\dagger) = 0$ for all functions f) and Y_μ is an independent exponential random variable with mean $2/\mu^2$; that is, $\mathbb{P}\left[Y_\mu > t\right] = e^{-\frac{\mu^2}{2}t}$. Let

$$\tau_D^\mu = \inf\left\{t > 0\,;\, W_\mu(t) \in D^c\right\}.$$

Then we have the following representation of the panharmonic measure:

Corollary 1. *Let $D \subset \mathbb{R}^n$ be a regular domain. Then the panharmonic measure admits the representation*

$$H_\mu^x(D;\mathrm{d}y) = \mathbb{P}^x\left[W_\mu(\tau_D^\mu) \in \mathrm{d}y\,;\, \tau_D^\mu < \infty\right]. \tag{10}$$

Proof. Let $f : \partial D \to \mathbb{R}$ be bounded. Then, by Theorem 1 and the independence of W and Y_μ,

$$\begin{aligned}
&\int_{y\in\partial D} f(y)\, H_\mu^x(D;\mathrm{d}y) \\
&= \mathbb{E}^x\left[e^{-\frac{\mu^2}{2}\tau_D} f\left(W(\tau_D)\right); \tau_D < \infty\right] \\
&= \int_{y\in\partial D} f(y) \int_{t=0}^{\infty} e^{-\frac{\mu^2}{2}t}\, \mathbb{P}^x\left[W(t) \in \mathrm{d}y, \tau_D \in \mathrm{d}t\right] \\
&= \int_{y\in\partial D} f(y) \int_{t=0}^{\infty} \mathbb{P}^x\left[Y_\mu > t\right] \mathbb{P}^x\left[W(t) \in \mathrm{d}y, \tau_D \in \mathrm{d}t\right] \\
&= \int_{y\in\partial D} f(y) \int_{t=0}^{\infty} \mathbb{P}^x\left[Y_\mu > t, W(t) \in \mathrm{d}y, \tau_D \in \mathrm{d}t\right] \\
&= \int_{y\in\partial D} f(y) \int_{t=0}^{\infty} \mathbb{P}^x\left[W_\mu(t) \in \mathrm{d}y, \tau_D^\mu \in \mathrm{d}t\right] \\
&= \mathbb{E}^x\left[f\left(W_\mu(\tau_D^\mu)\right); \tau_D^\mu < \infty\right].
\end{aligned}$$

Because f was arbitrary, the claim follows. □

The two representations, Theorem 1 and Corollary 1, for the panharmonic measures are, at least in spirit, classical. Now we give a third representation for the panharmonic measure in terms of an *escaping Brownian motion*. This representation is apparently new in spirit. The representation is due to the following *Duffin correspondence* [6]: Let $D \subset \mathbb{R}^n$ be a regular domain, and let $u : D \to \mathbb{R}$. Let $I \subset \mathbb{R}$ be any open interval that contains 0. Set $\bar{D} = D \times I$ and define $\bar{u} : \bar{D} \to \mathbb{R}$ by

$$\bar{u}(\bar{x}) = \bar{u}(x, \tilde{x}) = u(x)\cos(\mu\tilde{x}). \tag{11}$$

Theorem 2. *The function $\bar{u}$ defined by Equation (11) is harmonic on $\bar{D}$ if and only if u is μ-panharmonic on D.*

Proof. We first show that D is regular if and only if $\bar{D}$ is regular. Let $\bar{W} = (W, \tilde{W})$ be $(n+1)$-dimensional Brownian motion. Denote

$$\begin{aligned} \tau &= \inf\{t > 0\,;\, W(t) \in D^c\}, \\ \tilde{\tau} &= \inf\{t > 0\,;\, \tilde{W}(t) \in I^c\}, \\ \bar{\tau} &= \inf\{t > 0\,;\, \bar{W}(t) \in \bar{D}^c\}. \end{aligned}$$

We note that for $\{\tilde{\tau} = \tilde{x}\}$ to happen, $\tilde{x}$ must be an endpoint of the interval I. Then, by the independence of W and $\tilde{W}$,

$$\begin{aligned} \mathbb{P}^{x,\tilde{x}}[\bar{\tau} = 0] &= \mathbb{P}^{x,\tilde{x}}[\tau = 0, \tilde{\tau} = 0] \\ &= \mathbb{P}^{x}[\tau = 0]\mathbb{P}^{\tilde{x}}[\tilde{\tau} = 0] \\ &= \mathbb{P}^{x}[\tau = 0], \end{aligned}$$

because I is clearly regular. This shows that $\bar{D}$ is regular if and only if D is regular.

We then show that u satisfies the Laplace equation if and only if $\bar{u}$ satisfies the Yukawa equation; this is straightforward calculus:

$$\begin{aligned} \Delta_{\bar{x}}\bar{u}(\bar{x}) &= \Delta_{x,\tilde{x}}\left[u(x)\cos(\mu\tilde{x})\right] \\ &= \cos(\mu\tilde{x})\Delta_x u(x) + u(x)\frac{\mathrm{d}^2}{\mathrm{d}\tilde{x}^2}\cos(\mu\tilde{x}) \\ &= \cos(\mu\tilde{x})\Delta_x u(x) - \mu^2\cos(\mu\tilde{x})u(x) \\ &= \cos(\mu\tilde{x})\left(\Delta_x u(x) - \mu^2 u(x)\right) \\ &= 0 \end{aligned}$$

if and only if $\Delta_x u(x) = \mu^2 u(x)$. □

Let $\tilde{W}$ be a 1-dimensional standard Brownian motion that is independent of W. Then $\bar{W} = (W, \tilde{W})$ is a $(n+1)$-dimensional standard Brownian motion.

Now the idea of how to use the Duffin correspondence is clear. We start the Brownian particle $\bar{W}$ and count the boundary data on the side of the cylinder $\bar{D} = D \times I$, if the Brownian motion does not escape the cylinder from the bottom or from the top. In that case we count zero in the boundary; whence the name *escaping Brownian motion.*

Corollary 2. *Let $D \subset \mathbb{R}^n$ be a regular domain. Then the panharmonic measure admits the representation*

$$\begin{aligned} & H^x_\mu(D; \mathrm{d}y) \\ &= \mathbb{E}^{x,0}\left[\cos\left(\mu\tilde{W}(\tau_D)\right); W(\tau_D) \in \mathrm{d}y, \sup_{t \le \tau_D}|\tilde{W}(t)| < \frac{\pi}{2\mu}\right] \\ &= \int_{\tilde{y}=-\frac{\pi}{2\mu}}^{\frac{\pi}{2\mu}} \cos\left(\mu\tilde{y}\right) H^{x,0}\left(D \times \left(-\frac{\pi}{2\mu}, \frac{\pi}{2\mu}\right); \mathrm{d}y \otimes \mathrm{d}\tilde{y}\right). \end{aligned} \tag{12}$$

Here we have chosen $I = (-\frac{\pi}{2\mu}, \frac{\pi}{2\mu})$ in the Duffin correspondence.

Consequently, all bounded solutions to the Yukawa–Dirichlet problem on a regular domain with $\mu^2 > 0$ and continuous and bounded boundary data are given by the panharmonic measure.

Proof. The claim follows by combining the Kakutani connection (Equation (3)) with the Duffin correspondence (Equation (11)) by noticing that it is enough to integrate over $\partial D \times (-\pi/(2\mu), \pi/(2\mu))$, as $\cos(\mu\tilde{y}) = 0$ on the boundary $\partial(-\pi/(2\mu), \pi/(2\mu))$.

Finally, we note that for a regular domain D, the domain $\bar{D}$ is regular and small. □

Remark 2. *Equation (12) is exceptionally well suited for calculations of the panharmonic measures on upper half-spaces* $\mathbb{H}^n_+ = \{x \in \mathbb{R}^n; x_n > 0\}$. *Indeed, Duffin ([6], Theorem 5) used it to calculate the Poisson kernel representation for panharmonic measures in the dimension* $n = 2$. *Similar calculations can also be carried out for the general case,* $n \geq 2$.

3. Equivalence of Harmonic and Panharmonic Measures

The probabilistic interpretation provided by Corollary 1 implies that the harmonic measure and the panharmonic measures are equivalent. Indeed, the harmonic measure counts the Brownian particles on the boundary, and the panharmonic measures count the killed Brownian particles on the boundary. However the killing happens with independent exponential random variables. Thus, if the Brownian motion can reach the boundary with positive probability, so can the killed Brownian motion, and vice versa. Additionally, it does not matter, as far as the equivalence is concerned, what the starting point is of the Brownian motion, killed or not.

Theorem 3 below makes the heuristics above precise. As corollaries of Theorem 3, we obtain a domination principle for the Dirichlet problem related to the Yukawa equation (Corollary 3) and analogs of theorems of Riesz–Riesz, Makarov and Dahlberg for the panharmonic measures (Corollary 4).

The same arguments that give the existence of the regular conditional law (Equation (5)) in the proof of Lemma 1 also give the existence and regular measurability of the following conditional Radon–Nikodym derivative:

$$Z^x_\mu(D;y) = \mathbb{E}^x\left[e^{-\frac{\mu^2}{2}\tau_D}\,\Big|\, W(\tau_D) = y\right]. \tag{13}$$

Theorem 3. *Let* D *be a regular domain. Then all the panharmonic measures* $H^x_\mu(D;\cdot)$, $\mu \geq 0, x \in D$ *are mutually equivalent. The Radon–Nikodym derivative of* $H^x_\mu(D;\cdot)$ *with respect to* $H^x(D;\cdot)$ *is the function* $Z^x_\mu(D;\cdot)$ *given by Equation (13). Moreover* $Z^x_\mu(D;y)$ *is strictly decreasing in* μ, *and* $0 < Z^x_\mu(D;y) \leq 1$.

Remark 3. *By Corollary 1, the Radon–Nikodym derivative* $Z^x_\mu(D;\cdot)$ *in Equation (13) can be interpreted as the probability that a Brownian motion killed with intensity* $\mu^2/2$, *and that would exit the domain* D *at* $y \in \partial D$, *survives to the boundary* ∂D:

$$Z^x_\mu(D;y) = \mathbb{P}^x\left[Y_\mu > \tau_D \,|\, W(\tau_D) = y\right], \tag{14}$$

where Y_μ *is an exponentially distributed random variable with mean* $2/\mu^2$ *that is independent of the Brownian motion* W.

Proof of Theorem 3. Let $x, y \in D$, and let $D_0 \subset D$ be a subdomain of D such that $x \in D_0$ and $y \in \partial D_0$. Then, by the Markov property of the Brownian motion and the Kakutani connection (Equation (3)), we have

$$H^x(D;A) = \int_{y\in\partial D_0} H^y(D;A)H^x(D_0;\mathrm{d}y)$$

for all measurable $A \subset \partial D$. This shows that the harmonic measures $H^x(D;\cdot)$, $x \in D$ are mutually equivalent.

To see that $Z^x_\mu(D;\cdot)$ is the Radon–Nikodym derivative, we note that, by the representation given by Equation (7) and the Kakutani connection (Equation (3)),

$$\begin{aligned}
H^x_\mu(D;\mathrm{d}y) &= \int_{t=0}^{\infty} e^{-\frac{\mu^2}{2}t}\, h^x(D;\mathrm{d}y,t)\,\mathrm{d}t \\
&= \int_{t=0}^{\infty} e^{-\frac{\mu^2}{2}t}\, \mathbb{P}^x\left[W(\tau_D)\in \mathrm{d}y, \tau_D \in \mathrm{d}t\right] \\
&= \int_{y\in\partial D} \mathbb{E}^x\left[e^{-\frac{\mu^2}{2}\tau_D} \,\middle|\, W(\tau_D)=y\right] \mathbb{P}^x\left[W(\tau_D)\in\mathrm{d}y\right] \\
&= \int_{y\in\partial D} Z^x_\mu(D;y)\, H^x(D;\mathrm{d}y).
\end{aligned}$$

Finally, the fact that $0 < Z^x_\mu(D;\cdot) \leq 1$ is clear from the representation given by Equation (13). The fact that $Z^z_\mu(D;\cdot)$ is strictly decreasing follows immediately from the representation given by Equation (14). □

From Theorem 3, we obtain immediately the following *domination principle* for the Dirichlet problem related to panharmonic functions:

Corollary 3. *Let D be a regular domain, and let $u_\mu \geq 0$ be μ-panharmonic and $u_\nu \geq 0$ be ν-panharmonic, respectively, on D with $\mu \leq \nu$. Then, $u_\nu \leq u_\mu$ on ∂D implies $u_\nu \leq u_\mu$ on D.*

Because domains with a rectifiable boundary are regular, we obtain immediately from Theorem 3 the following analogs of the theorems of F. Riesz and M. Riesz, Makarov and Dalhberg (see [17–19], respectively).

Corollary 4. *Let $\mathcal{H}^s(D;\cdot)$ be the s-dimensional Hausdorff measure on ∂D.*

(i) *Let $D \subset \mathbb{R}^2$ be a simply connected planar domain bounded by a rectifiable curve. Then $H^x_\mu(D;\cdot)$ and $\mathcal{H}^1(D;\cdot)$ are equivalent for all $\mu \geq 0$ and $x \in D$.*
(ii) *Let $D \subset \mathbb{R}^2$ be a simply connected planar domain. If $E \subset \partial D$ and $\mathcal{H}^s(D;E) = 0$ for some $s < 1$, then $H^x_\mu(D;E) = 0$ for all $\mu \geq 0$ and $x \in D$. Moreover, $H^x_\mu(D;\cdot)$ and $\mathcal{H}^t(D;\cdot)$ are singular for all $\mu \geq 0$ and $x \in D$ if $t > 1$.*
(iii) *Let $D \subset \mathbb{R}^n$ is a bounded Lipschitz domain. Then $H^x_\mu(D;\cdot)$ and $\mathcal{H}^{n-1}(D;\cdot)$ are equivalent for all $\mu \geq 0$ and $x \in D$.*

4. The Average Property for Panharmonic Measures and Functions

By using the representation given by Equation (7), one can calculate the panharmonic measures if one can calculate the corresponding harmonic kernels; or, equivalently, one can calculate the panharmonic measures if one can calculate the corresponding harmonic measures and the Radon–Nikodym derivatives given by Equation (13).

The harmonic kernels for balls are calculated in [16]. We do not present the general formula here. Instead, we confine ourselves to the case in which the center of the ball and the pole of the panharmonic measure coincide, and give the Gauss mean value theorem, or the average property, for panharmonic measures. As a corollary, we have the Liouville theorem for the panharmonic measures.

Let $D \subset \mathbb{R}^n$ be a regular domain. For the harmonic measure, the Gauss mean value theorem states that a function $u : D \to \mathbb{R}$ is harmonic if and only if for all balls $B_n(x,r) \subset D$ we have the *average property*:

$$u(x) = \int_{y\in\partial B_n(x,r)} u(y)\,\sigma_n(r;\mathrm{d}y),$$

where

$$\sigma_n(r; \mathrm{d}y) = \frac{\Gamma(n/2)}{2\pi^{n/2}} r^{1-n}\, \mathrm{d}y$$

is the uniform probability measure on the sphere $\partial B_n(x,r)$.

For the panharmonic measures, the situation is similar to the harmonic measure: the only difference is that the uniform probability measure has to be replaced by a uniform sub-probability measure that depends on the killing parameter μ and the radius of the ball r. Indeed, we denote

$$\psi_n(\mu) = \frac{\mu^\nu}{2^\nu \Gamma(\nu+1) I_\nu(\mu)}, \quad \mu > 0, \tag{15}$$

where $\nu = (n-2)/2$, and

$$I_\nu(x) = \sum_{m=0}^{\infty} \frac{1}{m!\Gamma(m+\nu+1)} \left(\frac{x}{2}\right)^{2m+\nu}$$

is the modified Bessel function of the first kind of order ν.

Theorem 4. *Let $D \subset \mathbb{R}^n$ be a regular domain, and let $\mu > 0$. A function $u: D \to \mathbb{R}$ is μ-panharmonic if and only if it has the average property*

$$u(x) = \psi_n(\mu r) \int_{y \in \partial B_n(x,r)} u(y)\, \sigma_n(r; \mathrm{d}y).$$

for all open balls $B_n(x,r) \subset D$. Equivalently,

$$H_\mu^x(B_n(x,r); \mathrm{d}y) = \psi_n(\mu r)\, \sigma_n(r; \mathrm{d}y).$$

Remark 4. *Theorem 4 states that $\psi_n(\mu r)$ is the Radon–Nikodym derivative:*

$$\psi_n(\mu r) = Z_\mu^x(B_n(x,r); y) = \mathbb{E}^x\left[e^{-\frac{\mu^2}{2}\tau_{B_n(x,r)}} \Big| W\left(\tau_{B_n(x,r)}\right) = y \right].$$

Proof of Theorem 4. We note that we may assume $x = 0$.

Denote by τ_r^n the first hitting time of the Brownian motion W on the boundary $\partial B_n(0,r)$; that is, τ_r^n is identical in law to the first hitting time of the Bessel process with index $\nu = (n-2)/2$ reaching the level r when it starts from zero.

From the rotation symmetry of the Brownian motion, it follows that the hitting place is uniformly distributed on $\partial B_n(0,r)$ for all hitting times t. Consequently, by Theorem 1 and the independence of the hitting time τ_r^n and place $W(\tau_r^n)$,

$$\begin{aligned} H_\mu^0(B_n(0,r); \mathrm{d}y) &= \mathbb{E}^0\left[e^{-\frac{\mu^2}{2}\tau_r^n}; W(\tau_r^n) \in \mathrm{d}y \right] \\ &= \mathbb{E}^0\left[e^{-\frac{\mu^2}{2}\tau_r^n} \right] \mathbb{P}^0\left[W(\tau_r^n) \in \mathrm{d}y \right] \\ &= \mathbb{E}^0\left[e^{-\frac{\mu^2}{2}\tau_r^n} \right] \sigma_n(r; \mathrm{d}y). \end{aligned}$$

The hitting-time distributions for the Bessel process are well known. By, for example, Wendel ([20], Theorem 4),

$$\mathbb{E}^0\left[e^{-\frac{\mu^2}{2}\tau_r^n} \right] = \frac{(\mu r)^\nu}{2^\nu \Gamma(\nu+1) I_\nu(\mu r)}.$$

The claim follows from this. □

Remark 5. *The Radon–Nikodym derivative, or the "killing constant", $\psi_n(\mu)$, is rather complicated. However, some of its properties are easy to see (cf. Figure 1):*

(i) $\psi_n(\mu)$ *is continuous in* μ,
(ii) $\psi_n(\mu)$ *is strictly decreasing in* μ,
(iii) $\psi_n(\mu) \to 0$ *as* $\mu \to \infty$,
(iv) $\psi_n(\mu) \to 1$ *as* $\mu \to 0$,
(v) $\psi_n(\mu)$ *is increasing in* n.

The items (i)–(iv) are clear, because $\psi_n(\mu)$ is the probability that an exponentially killed Brownian motion starting from the origin and with killing intensity $\mu^2/2$ is not killed before it hits the boundary of the unit ball. A non-probabilistic argument for (i)–(iv) is to note that

$$\psi_n(\mu r) = \mathbb{E}^0\left[e^{-\frac{\mu^2}{2}\tau_r^n}\right]$$

and to use the monotone convergence. The item (v) is somewhat surprising: the higher the dimension n, the more likely it is for the killed Brownian motion to survive to the boundary of the unit ball. A possible intuitive explanation is that the higher the dimension, the more transitive the unit ball is, combined with the remarkable result by Ciesielski and Taylor [21] that the probability distribution for the total time spent in a ball by $(n+2)$-dimensional Brownian motion is the same as the probability distribution of the hitting time of n-dimensional Brownian motion on the boundary of the ball.

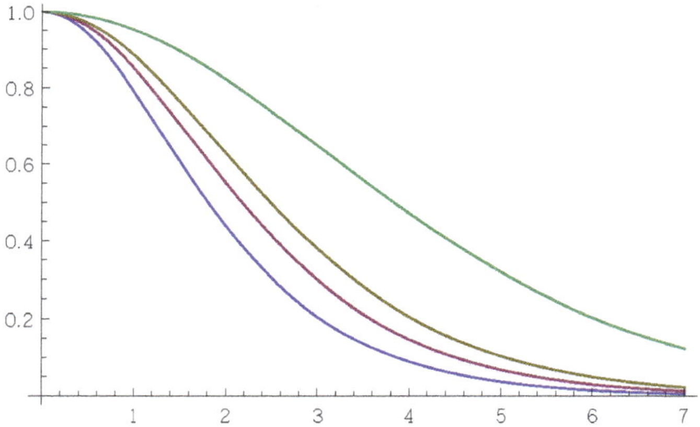

Figure 1. Function ψ_n with (from bottom to top) $n = 2, 3, 4, 10$.

Corollary 5. *Let $\mu > 0$, and let u be μ-panharmonic on the entire space $\mathbb{R}^n$. If u is bounded, then $u \equiv 0$.*

Proof. By Theorem 4,

$$\begin{aligned}
&|u(x) - u(0)| \\
=\ & \left|\psi_n(\mu r)\int_{\partial B_n(x,r)} u(y)\,\sigma_n(r;\mathrm{d}y) - \psi_n(\mu r)\int_{\partial B_n(0,r)} u(y)\,\sigma_n(r;\mathrm{d}y)\right| \\
\le\ & \left|\psi_n(\mu r)\int_{\partial B_n(x,r)} u(y)\,\sigma_n(r;\mathrm{d}y)\right| + \left|\psi_n(\mu r)\int_{\partial B_n(0,r)} u(y)\,\sigma_n(r;\mathrm{d}y)\right| \\
\le\ & 2\psi_n(\mu r)\|u\|_\infty,
\end{aligned}$$

which tends to 0 as $r \to \infty$ by property (iii) of Remark 5. This shows that u is constant. However, for a constant u, the Yukawa equation (Equation (1)) yields $0 = \mu^2 u$, which implies that $u \equiv 0$. □

5. Discussion on Extensions and Simulation

The Yukawa equation (Equation (1)) is a special case of the Schrödinger equation:

$$\Delta u(x) = q(x)u(x). \tag{16}$$

The Schrödinger equation and its connection to the Brownian motion has been studied, for example, by Chung and Zhao [8]. Our investigation here can be seen as a special case. For example, analogs of Theorem 1 and Corollary 1 are known for the Schrödinger equation. However, analogs of the Duffin correspondence (Equation (11)) and Corollary 2 are not known even to exist. Moreover, the results given here cannot easily be calculated for the Schrödinger equation. The problem is that the prospective Radon–Nikodym derivate of the measure associated with the solutions of the Schrödinger equation with respect to the harmonic measure takes the form

$$Z_q^x(D;y) = \mathbb{E}^x\left[e_q(\tau_D)\middle|W(\tau_D) = y\right], \tag{17}$$

where

$$e_q(t) = e^{-\frac{1}{2}\int_0^t q(W(s))\,\mathrm{d}s}$$

is the *Feynman–Kac functional*. Thus, we see that in order to calculate the Radon–Nikodym derivative, we need to know the joint density of the Feynman–Kac functional and the Brownian motion when the Brownian motion hits the boundary ∂D. If q is constant, that is, we have either the Yukawa equation or the Helmholtz equation, then it is enough to know the joint distribution of the hitting time and place of the Brownian motion on the boundary ∂D. These distributions are well studied (see, e.g., [14,16,21–23]), but few joint distributions involving the Feynman–Kac functionals are known.

In addition to the Yukawa equation, the other important special case of the Schrödinger equation (Equation (16)) is the Helmholtz equation:

$$\Delta u(x) = -\lambda u(x), \quad \lambda \geq 0. \tag{18}$$

It is possible to also provide a Duffin correspondence for the Helmholtz equation. Indeed, for example, setting

$$\tilde{u}(\tilde{x}) = \tilde{u}(x,\tilde{x}) = u(x)\cosh(\lambda\tilde{x})$$

provides a correspondence (see [24] for details). Thus, our results extend in a straightforward manner to the Helmholtz equation (Equation (18)) for domains that are small enough with respect to the creation parameter λ that the associated Feynman–Kac functional is finite:

$$\mathbb{E}^x\left[e^{\frac{\lambda}{2}\tau_D}\right] < \infty. \tag{19}$$

Finally, we note that Theorem 1, Corollary 1 and Corollary 2 give three different ways to simulate the panharmonic measures. Indeed, in [24,25], the classical WOS algorithm given by Muller [26] was extended for the Yukawa PDE, and also for the Helmholtz PDE, by using the results mentioned above.

Author Contributions: Antti Rasila was responsible for the general idea of connecting the Yukawa equation to the harmonic measure and the Brownian motion through the Duffin correspondence, connections to classical analysis and potential theory, and computer experiments with Mathematica. Tommi Sottinen was responsible for the proofs and methodologies making use of stochastic analysis. Much of the paper was shaped in discussions between the authors.

Acknowledgments: T. Sottinen was partially funded by the Finnish Cultural Foundation (National Foundations' Professor Pool).

Conflicts of Interest: The authors declare no conflict of interest.

Reference

1. Bishop, C.J.; Jones, P.W. Harmonic measure and arclength. *Ann. Math.* **1990**, *132*, 511–547. [CrossRef]
2. Gehring, F.W.; Hag, K. *The Ubiquitous Quasidisk*; No. 184.; American Mathematical Society: Providence, RI, USA, 2012.
3. Krzyż, J.G. Quasicircles and harmonic measure. *Ann. Acad. Sci. Fenn.* **1987**, *12*, 19–24. [CrossRef]
4. Garnett, J.B.; Marshall, D.E. *Harmonic Measure*; Cambridge University Press: Cambridge, UK, 2005.
5. Harrach, B. On uniqueness in diffuse optical tomography. *Inverse Probl.* **2009**, *25*, 1–14. [CrossRef]
6. Duffin, R.J. Yukawan potential theory. *J. Math. Anal. Appl.* **1971**, *35*, 105–130. [CrossRef]
7. Duffin, R.J. Hilbert transforms in Yukawan potential theory. *Proc. Nat. Acad. Sci. USA* **1972**, *69*, 3677–3679. [CrossRef] [PubMed]
8. Chung, K.L.; Zhao, Z. *From Brownian Motion to Schrödinger's Equation*, 2nd ed.; Springer: Berlin/ Heidelberg, Germany, 2001.
9. Doob, J.L. *Classical Potential Theory and Its Probabilistic Counterpart;* Grundlehren der Mathematischen Wissenschaften 262; Springer: New York, NY, USA, 1984.
10. Port, S.C.; Stone, C.J. *Brownian Motion and Classical Potential Theory*; Academic Press: New York, NY, USA, 1978.
11. Evans, L.C. *Partial Differential Equations (Graduate Studies in Mathematics)*, 2nd ed.; American Mathematical Society: Providence, RI, USA, 2010.
12. Kakutani, S. On Brownian motion in n-space. *Proc. Imp. Acad.* **1944**, *20*, 648–652. [CrossRef]
13. Kallenberg, O. *Foundations of Modern Probability*, 2nd ed.; Springer: New York, NY, USA, 2002.
14. Borodin, A.; Salminen, P. *Handbook of Brownian Motion—Facts and Formulae*, 2nd ed.; Birkhäuser: Basel, Switzerland, 2002.
15. Durrett, R. *Stochastic Calculus: A Practical Introduction;* CRC Press: Boca Raton, FL, USA, 1996.
16. Hsu, P. Brownian exit distribution of a ball. In *Seminar on Stochastic Processes 1985*; Birkhäuser: Boston, MA, USA, 1986.
17. Dahlberg, B. Estimates of harmonic measure. *Arch. Rat. Mech. Anal.* **1977**, *65*, 275–288. [CrossRef]
18. Makarov, N.G. On the Distortion of Boundary Sets Under Conformal Maps. *Proc. Lond. Math. Soc.* **1985**, *52*, 369–384. [CrossRef]
19. Riesz, F.; Riesz, M. Über die Randwerte einer analytischen Funktion. In *Quatrième Congrès des Mathématiciens Scandinaves*; Stockholm, Sweden, 1916; pp. 27–44.
20. Wendel, J.G. Hitting Spheres with Brownian Motion. *Ann. Probab.* **1980**, *8*, 164–169. [CrossRef]
21. Ciesielski, Z.; Taylor, S.J. First passage times and sojourn times for Brownian motion in space and the exact Hausdorff measure of the sample path. *Trans. Am. Math. Soc.* **1962**, *103*, 434–450. [CrossRef]
22. Kent, T. Eigenvalue expansion for diffusion hitting times. *Z. Wahrscheinlichkeitstheorie Ver. Gebiete* **1980**, *52*, 309–319. [CrossRef]
23. Lévy, P. La mesure de Hausdorff de la courbe du mouvement brownien. *Giorn. Ist. Ital. Attuari* **1953**, *16*, 1–37.
24. Yang, X.; Rasila, A.; Sottinen, T. Walk On Spheres Algorithm for Helmholtz and Yukawa Equations via Duffin Correspondence. *Methodol. Comput. Appl. Probab.* **2017**, *19*, 589–602. [CrossRef]
25. Yang, X.; Rasila, A.; Sottinen, T. Efficient simulation of Schrödinger equation with piecewise constant positive potential. *arXiv* **2015**, arXiv:1512.01306.
26. Muller, M.E. Some continuous Monte Carlo methods for the Dirichlet problem. *Ann. Math. Stat.* **1956**, *27*, 569–589. [CrossRef]

axioms

MDPI

Article

Subordination Properties for Multivalent Functions Associated with a Generalized Fractional Differintegral Operator

Hanaa M. Zayed [1,*], Mohamed Kamal Aouf [2] and Adela O. Mostafa [2]

[1] Department of Mathematics, Faculty of Science, Menofia University, Shebin Elkom 32511, Egypt
[2] Department of Mathematics, Faculty of Science, Mansoura University, Mansoura 35516, Egypt; mkaouf@mans.edu.eg (M.K.A.); adelaeg254@yahoo.com (A.O.M.)
* Correspondence: hanaa_zayed42@yahoo.com

Received: 19 January 2018; Accepted: 17 April 2018; Published: 24 April 2018

Abstract: Using of the principle of subordination, we investigate some subordination and convolution properties for classes of multivalent functions under certain assumptions on the parameters involved, which are defined by a generalized fractional differintegral operator under certain assumptions on the parameters involved.

Keywords: differential subordination; p-valent functions; generalized fractional differintegral operator

JEL Classification: 30C45; 30C50

1. Introduction and Definitions

Denote by $\mathcal{A}(p)$ the class of analytic and p-valent functions of the form:

$$f(z) = z^p + \sum_{n=1}^{\infty} a_{p+n} z^{p+n} \ (p \in \mathbb{N} = \{1, 2, ...\}; z \in \mathbb{U} = \{z \in \mathbb{C} : |z| < 1\}). \tag{1}$$

For functions f, g analytic in $\mathbb{U}$, f is subordinate to g, written $f(z) \prec g(z)$ if there exists a function w, analytic in $\mathbb{U}$ with $w(0) = 0$ and $|w(z)| < 1$, such that $f(z) = g(w(z))$, $z \in \mathbb{U}$. If g is univalent in $\mathbb{U}$, then (see [1,2]):

$$f(z) \prec g(z) \Leftrightarrow f(0) = g(0) \text{ and } f(\mathbb{U}) \subset g(\mathbb{U}).$$

If $\varphi(z)$ is analytic in $\mathbb{U}$ and satisfies:

$$H(\varphi(z), z\varphi'(z)) \prec h(z), \tag{2}$$

then φ is a solution of (2). The univalent function q is called dominant, if $\varphi(z) \prec q(z)$ for all φ. A dominant $\widetilde{q}$ is called the best dominant, if $\widetilde{q}(z) \prec q(z)$ for all dominants q.

Let ${}_2F_1(a, b; c; z)$ $(c \neq 0, -1, -2, \dots)$ be the well-known (Gaussian) hypergeometric function defined by:

$${}_2F_1(a, b; c; z) := \sum_{n=0}^{\infty} \frac{(a)_n (b)_n}{(c)_n (1)_n} z^n, \ z \in \mathbb{U},$$

where:

$$(\lambda)_n := \frac{\Gamma(\lambda + n)}{\Gamma(\lambda)}.$$

We will recall some definitions that will be used in our paper.

Definition 1. *For $f(z) \in \mathcal{A}(p)$, the fractional integral and fractional derivative operators of order λ are defined by Owa [3] (see also [4]) as:*

$$D_z^{-\lambda} f(z) := \frac{1}{\Gamma(\lambda)} \int_0^z \frac{f(\zeta)}{(z-\zeta)^{1-\lambda}}\, d\zeta \quad (\lambda > 0),$$

$$D_z^{\lambda} f(z) := \frac{1}{\Gamma(1-\lambda)} \frac{d}{dz} \int_0^z \frac{f(\zeta)}{(z-\zeta)^{\lambda}}\, d\zeta \quad (0 \leq \lambda < 1),$$

where f is an analytic function in a simply-connected region of the complex z-plane containing the origin, and the multiplicity of $(z-\zeta)^{\lambda-1}$ $((z-\zeta)^{-\lambda})$ is removed by requiring $\log(z-\zeta)$ to be real when $z-\zeta > 0$.

Definition 2. *For $f(z) \in \mathcal{A}(p)$ and in terms of ${}_2F_1$, the generalized fractional integral and generalized fractional derivative operators defined by Srivastava et al. [5] (see also [6]) as:*

$$I_{0,z}^{\lambda,\mu,\eta} f(z) := \frac{z^{-\lambda-\mu}}{\Gamma(\lambda)} \int_0^z (z-\zeta)^{\lambda-1} f(\zeta)\, {}_2F_1\left(\mu+\lambda, -\eta; \lambda; 1-\frac{\zeta}{z}\right) d\zeta \ (\lambda > 0,\ \mu,\eta \in \mathbb{R}),$$

$$J_{0,z}^{\lambda,\mu,\eta} f(z) := \begin{cases} \frac{d}{dz}\left\{ \frac{z^{\lambda-\mu} \int_0^z (z-\zeta)^{-\lambda} f(\zeta) {}_2F_1\left(\mu-\lambda, 1-\eta; 1-\lambda; 1-\frac{\zeta}{z}\right) d\zeta}{\Gamma(1-\lambda)} \right\} (0 \leq \lambda < 1), \\ \\ \frac{d^n}{dz^n} J_{0,z}^{\lambda-n,\mu,\eta} f(z) \ (n \leq \lambda < n+1;\ n \in \mathbb{N}), \end{cases}$$

where $f(z)$ is an analytic function in a simply-connected region of the complex z-plane containing the origin with the order $f(z) = O(|z|^{\varepsilon})$, $z \to 0$ when $\varepsilon > \max\{0, \mu-\eta\} - 1$, and the multiplicity of $(z-\zeta)^{\lambda-1}$ $((z-\zeta)^{-\lambda})$ is removed by requiring $\log(z-\zeta)$ to be real when $z-\zeta > 0$.

We note that:

$$I_{0,z}^{\lambda,-\lambda,\eta} f(z) = D_z^{-\lambda} f(z) \ (\lambda > 0) \text{ and } J_{0,z}^{\lambda,\lambda,\eta} f(z) = D_z^{\lambda} f(z) \ (0 \leq \lambda < 1),$$

where $D_z^{-\lambda} f(z)$ and $D_z^{\lambda} f(z)$ are the fractional integral and fractional derivative operators studied by Owa [3].

Goyal and Prajapat [7] (see also [8]) defined the operator:

$$S_{0,z}^{\lambda,\mu,\eta,p} f(z) = \begin{cases} \frac{\Gamma(p+1-\mu)\Gamma(p+1-\lambda+\eta)}{\Gamma(p+1)\Gamma(p+1-\mu+\eta)} z^{\mu} J_{0,z}^{\lambda,\mu,\eta} f(z) \ (0 \leq \lambda < \eta+p+1;\ z \in \mathbb{U}), \\ \\ \frac{\Gamma(p+1-\mu)\Gamma(p+1-\lambda+\eta)}{\Gamma(p+1)\Gamma(p+1-\mu+\eta)} z^{\mu} I_{0,z}^{-\lambda,\mu,\eta} f(z) \ (-\infty < \lambda < 0;\ z \in \mathbb{U}). \end{cases}$$

For $f(z) \in \mathcal{A}(p)$, we have:

$$\begin{aligned} S_{0,z}^{\lambda,\mu,\eta,p} f(z) &= z^p {}_3F_2(1, 1+p, 1+p+\eta-\mu; 1+p-\mu, 1+p+\eta-\lambda; z) * f(z) \\ &= z^p + \sum_{n=1}^{\infty} \frac{(p+1)_n (p+1-\mu+\eta)_n}{(p+1-\mu)_n (p+1-\lambda+\eta)_n} a_{p+n} z^{p+n} \\ (p &\in \mathbb{N};\ \mu,\eta \in \mathbb{R};\ \mu < p+1;\ -\infty < \lambda < \eta+p+1), \end{aligned}$$

where "$*$" stands for convolution of two power series, and ${}_qF_s$ $(q \leq s+1;\ q,s \in \mathbb{N}_0 = \mathbb{N} \cup \{0\})$ is the well-known generalized hypergeometric function.

Let:

$$\begin{aligned} G^{\lambda}_{p,\eta,\mu}(z) &= z^p + \sum_{n=1}^{\infty} \frac{(p+1)_n(p+1-\mu+\eta)_n}{(p+1-\mu)_n(p+1-\lambda+\eta)_n} z^{p+n} \\ (p &\in \mathbb{N};\ \mu,\eta \in \mathbb{R};\ \mu < p+1;\ -\infty < \lambda < \eta+p+1). \end{aligned}$$

and:

$$G^{\lambda}_{p,\eta,\mu}(z) * \left[G^{\lambda}_{p,\eta,\mu}(z)\right]^{-1} = \frac{z^p}{(1-z)^{\delta+p}}\ (\delta > -p;\ z \in \mathbb{U}).$$

Tang et al. [9] (see also [10–15]) defined the operator $H^{\lambda,\delta}_{p,\eta,\mu} : \mathcal{A}(p) \to \mathcal{A}(p)$, where:

$$\begin{aligned} H^{\lambda,\delta}_{p,\eta,\mu} f(z) &= z^p + \sum_{n=1}^{\infty} \frac{(\delta+p)_n(p+1-\mu)_n(p+1-\lambda+\eta)_n}{(1)_n(p+1)_n(p+1-\mu+\eta)_n} a_{p+n} z^{p+n} \\ (p &\in \mathbb{N},\ \delta > -p,\ \mu,\eta \in \mathbb{R},\ \mu < p+1,\ -\infty < \lambda < \eta+p+1). \end{aligned}$$

It is easy to verify that:

$$z\left(H^{\lambda,\delta}_{p,\eta,\mu} f(z)\right)' = (\delta+p) H^{\lambda,\delta+1}_{p,\eta,\mu} f(z) - \delta H^{\lambda,\delta}_{p,\eta,\mu} f(z), \tag{3}$$

and:

$$z\left(H^{\lambda+1,\delta}_{p,\eta,\mu} f(z)\right)' = (p+\eta-\lambda) H^{\lambda,\delta}_{p,\eta,\mu} f(z) - (\eta-\lambda) H^{\lambda+1,\delta}_{p,\eta,\mu} f(z). \tag{4}$$

By using the operator $H^{\lambda,\delta}_{p,\eta,\mu}$, we introduce the following class.

Definition 3. *For $A,\ B\ (-1 \le B < A \le 1)$, $f \in \mathcal{A}(p)$ is in the class $\mathcal{T}^{\lambda,\delta}_{p,\eta,\mu}(A,B)$ if*

$$\frac{(H^{\lambda,\delta}_{p,\eta,\mu} f(z))'}{pz^{p-1}} \prec \frac{1+Az}{1+Bz}\ (z \in \mathbb{U};\ p \in \mathbb{N}),$$

which is equivalent to:

$$\left| \frac{\dfrac{(H^{\lambda,\delta}_{p,\eta,\mu} f(z))'}{pz^{p-1}} - 1}{B\dfrac{(H^{\lambda,\delta}_{p,\eta,\mu} f(z))'}{pz^{p-1}} - A} \right| < 1\ (z \in \mathbb{U}).$$

For convenience, we write $\mathcal{T}^{\lambda,\delta}_{p,\eta,\mu}\left(1 - \frac{2\xi}{p}, -1\right) = \mathcal{T}^{\lambda,\delta}_{p,\eta,\mu}(\xi)\ (0 \le \xi < p)$, which satisfies the inequality:

$$\Re\left\{\frac{(H^{\lambda,\delta}_{p,\eta,\mu} f(z))'}{z^{p-1}}\right\} > \xi\ (0 \le \xi < p).$$

In this paper, we investigate some subordination and convolution properties for classes of multivalent functions, which are defined by a generalized fractional differintegral operator. The theory of subordination received great attention, particularly in many subclasses of univalent and multivalent functions (see, for example, [13,15–17]).

2. Preliminaries

To prove our main results, we shall need the following lemmas.

Lemma 1. *[18]. Let h be an analytic and convex (univalent) function in $\mathbb{U}$ with $h(0) = 1$. Additionally, let ϕ given by:*

$$\phi(z) = 1 + c_n z^n + c_{n+1} z^{n+1} + \ldots \tag{5}$$

be analytic in $\mathbb{U}$. *If:*

$$\phi(z) + \frac{z\phi'(z)}{\sigma} \prec h(z) \ (\Re(\sigma) \geqslant 0;\ \sigma \neq 0), \tag{6}$$

then:

$$\phi(z) \prec \psi(z) = \frac{\sigma}{n} z^{-\frac{\sigma}{n}} \int_0^z t^{\frac{\sigma}{n}-1} h(t) dt \prec h(z), \tag{7}$$

and ψ *is the best dominant of* (6).

Denote by $P(\varsigma)$ the class of functions Φ given by:

$$\Phi(z) = 1 + c_1 z + c_1 z^2 + ..., \tag{8}$$

which are analytic in $\mathbb{U}$ and satisfy the following inequality:

$$\Re\{\Phi(z)\} > \varsigma \ (0 \leq \varsigma < 1).$$

Using the well-known growth theorem for the Carathéodory functions (cf., e.g., [19]), we may easily deduce the following result:

Lemma 2. *[19]. If* $\Phi \in P(\varsigma)$. *Then*

$$\Re\{\Phi(\varsigma)\} \geqslant 2\varsigma - 1 + \frac{2(1-\varsigma)}{1+|z|} \ (0 \leq \varsigma < 1).$$

Lemma 3. *[20]. For* $0 \leq \varsigma_1, \varsigma_2 < 1$,

$$P(\varsigma_1) * P(\varsigma_2) \subset P(\varsigma_3) \ (\varsigma_3 = 1 - 2(1-\varsigma_1)(1-\varsigma_2)).$$

The result is the best possible.

Lemma 4. *[21]. Let* φ *be such that* $\varphi(0) = 1$ *and* $\varphi(z) \neq 0$ *and* $A, B \in \mathbb{C}$, *with* $A \neq B$, $|B| \leq 1, \nu \in \mathbb{C}^*$.

(i) If $\left|\frac{\nu(A-B)}{B} - 1\right| \leq 1$ or $\left|\frac{\nu(A-B)}{B} + 1\right| \leq 1, B \neq 0$ and $\varphi(z)$ satisfies:

$$1 + \frac{z\varphi'(z)}{\nu\varphi(z)} \prec \frac{1+Az}{1+Bz},$$

then:

$$\varphi(z) \prec (1+Bz)^{\nu\left(\frac{A-B}{B}\right)}$$

and this is the best dominant.

(ii) If $B = 0$ and $|\nu A| < \pi$ and if φ satisfies:

$$1 + \frac{z\varphi'(z)}{\nu\varphi(z)} \prec 1 + Az,$$

then:

$$\varphi(z) \prec e^{\nu A z},$$

and this is the best dominant.

Lemma 5. *[2]. Let $\Omega \subset \mathbb{C}$, $b \in \mathbb{C}$, $\Re(b) > 0$ and $\psi : \mathbb{C}^2 \times \mathbb{U} \to \mathbb{C}$ satisfy $\psi(ix, y; z) \notin \Omega$ for all $x, y \leq -\frac{|b-ix|^2}{2\Re(b)}$ and all $z \in \mathbb{U}$. If $p(z) = 1 + p_1 z + p_2 z^2 + ...$, is analytic in $\mathbb{U}$ and if:*

$$\psi\left(p(z), zp'(z); z\right) \in \Omega,$$

then $\Re\{p(z)\} > 0$ in $\mathbb{U}$.

Lemma 6. *[22]. Let $\psi(z)$ be analytic in $\mathbb{U}$ with $\psi(0) = 1$ and $\psi(z) \neq 0$ for all z. If there exist two points $z_1,\ z_2 \in \mathbb{U}$ such that:*

$$-\frac{\pi}{2}\rho_1 = \arg\{\psi(z_1)\} < \arg\{\psi(z)\} < \frac{\pi}{2}\rho_2 = \arg\{\psi(z_2)\}, \tag{9}$$

for some ρ_1 and ρ_2 $(\rho_1,\ \rho_2 > 0)$ and for all z $(|z| < |z_1| = |z_2|)$, then:

$$\frac{z_1\psi'(z_1)}{\psi(z_1)} = -i\left(\frac{\rho_1 + \rho_2}{2}\kappa\right) \text{ and } \frac{z_2\psi'(z_2)}{\psi(z_2)} = i\left(\frac{\rho_1 + \rho_2}{2}\kappa\right), \tag{10}$$

where:

$$\kappa \geq \frac{1 - |a|}{1 + |a|} \text{ and } a = i\tan\left(\frac{\rho_2 - \rho_1}{\rho_2 + \rho_1}\right). \tag{11}$$

3. Properties Involving $H_{p,\eta,\mu}^{\lambda,\delta}$

Unless otherwise mentioned, we assume throughout this paper that $p \in \mathbb{N}$, $\delta > -p$, $\mu, \eta \in \mathbb{R}$, $\mu < p + 1$, $-\infty < \lambda < \eta + p + 1$, $-1 \leq B < A \leq 1$, $\theta > 0$, and the powers are considered principal ones.

Theorem 1. *Let $f \in \mathcal{A}(p)$ satisfy:*

$$(1-\theta)\frac{\left(H_{p,\eta,\mu}^{\lambda,\delta} f(z)\right)'}{pz^{p-1}} + \theta\frac{\left(H_{p,\eta,\mu}^{\lambda,\delta+1} f(z)\right)'}{pz^{p-1}} \prec \frac{1 + Az}{1 + Bz}. \tag{12}$$

Then:

$$\Re\left(\frac{\left(H_{p,\eta,\mu}^{\lambda,\delta} f(z)\right)'}{pz^{p-1}}\right)^{\frac{1}{\tau}} > \left(\frac{\delta + p}{\theta}\int_0^1 u^{\frac{\delta+p}{\theta}-1}\left(\frac{1 - Au}{1 - Bu}\right)du\right)^{\frac{1}{\tau}},\ \tau \geq 1. \tag{13}$$

The estimate in (13) is sharp.

Proof. Let:

$$\phi(z) = \frac{\left(H_{p,\eta,\mu}^{\lambda,\delta} f(z)\right)'}{pz^{p-1}}\ (z \in \mathbb{U}). \tag{14}$$

Then, ϕ is analytic in $\mathbb{U}$. After some computations, we get:

$$(1-\theta)\frac{\left(H_{p,\eta,\mu}^{\lambda,\delta} f(z)\right)'}{pz^{p-1}} + \theta\frac{\left(H_{p,\eta,\mu}^{\lambda,\delta+1} f(z)\right)'}{pz^{p-1}} = \phi(z) + \frac{\theta z\phi'(z)}{\delta + p} \prec \frac{1 + Az}{1 + Bz}.$$

Now, by using Lemma 1, we deduce that:

$$\frac{\left(H_{p,\eta,\mu}^{\lambda,\delta} f(z)\right)'}{pz^{p-1}} \prec \frac{\delta + p}{\theta} z^{-\frac{\delta+p}{\theta}} \int_0^z t^{\frac{\delta+p}{\theta}-1}\left(\frac{1 + At}{1 + Bt}\right)dt, \tag{15}$$

or, equivalently,

$$\frac{\left(H_{p,\eta,\mu}^{\lambda,\delta} f(z)\right)'}{pz^{p-1}} = \frac{\delta+p}{\theta}\int_0^1 u^{\frac{\delta+p}{\theta}-1}\left(\frac{1+Auw(z)}{1+Buw(z)}\right)du,$$

and so:

$$\Re\left(\frac{\left(H_{p,\eta,\mu}^{\lambda,\delta} f(z)\right)'}{pz^{p-1}}\right) > \left(\frac{\delta+p}{\theta}\int_0^1 u^{\frac{\delta+p}{\theta}-1}\left(\frac{1-Au}{1-Bu}\right)du\right). \tag{16}$$

Since:

$$\Re\left(\chi^{\frac{1}{\tau}}\right) \geq \left(\Re(\chi)\right)^{\frac{1}{\tau}} \ (\chi \in \mathbb{C},\ \Re\{\chi\} \geq 0,\ \tau \geq 1). \tag{17}$$

The inequality (13) now follows from (16) and (17). To prove that the result is sharp, let:

$$\frac{\left(H_{p,\eta,\mu}^{\lambda,\delta} f(z)\right)'}{pz^{p-1}} = \frac{\delta+p}{\theta}\int_0^1 u^{\frac{\delta+p}{\theta}-1}\left(\frac{1+Auz}{1+Buz}\right)du. \tag{18}$$

Now, for $f(z)$ defined by (18), we have:

$$(1-\theta)\frac{\left(H_{p,\eta,\mu}^{\lambda,\delta} f(z)\right)'}{pz^{p-1}} + \theta\frac{\left(H_{p,\eta,\mu}^{\lambda,\delta+1} f(z)\right)'}{pz^{p-1}} = \frac{1+Az}{1+Bz} \ (z \in \mathbb{U}),$$

Letting $z \to -1$, we obtain:

$$\frac{\left(H_{p,\eta,\mu}^{\lambda,\delta} f(z)\right)'}{pz^{p-1}} \to \frac{\delta+p}{\theta}\int_0^1 u^{\frac{\delta+p}{\theta}-1}\left(\frac{1-Au}{1-Bu}\right)du,$$

which ends our proof. □

Putting $\theta = 1$ and using Lemma 1 for Equation (15) in Theorem 1, we obtain the following example.

Example 1. *Let the function $f(z) \in \mathcal{A}(p)$. Then, following containment property holds,*

$$\mathcal{T}_{p,\mu,\eta}^{\lambda,\delta+1}(A,B) \subset \mathcal{T}_{p,\mu,\eta}^{\lambda,\delta}(A,B).$$

Using (4) instead of (3) in Theorem 1, one can prove the following theorem.

Theorem 2. *Let $f \in \mathcal{A}(p)$ satisfy*

$$(1-\theta)\frac{\left(H_{p,\eta,\mu}^{\lambda+1,\delta} f(z)\right)'}{pz^{p-1}} + \theta\frac{\left(H_{p,\eta,\mu}^{\lambda,\delta} f(z)\right)'}{pz^{p-1}} \prec \frac{1+Az}{1+Bz}.$$

Then:

$$\Re\left(\frac{\left(H_{p,\eta,\mu}^{\lambda+1,\delta} f(z)\right)'}{pz^{p-1}}\right)^{\frac{1}{\tau}} > \left(\frac{p+\eta-\lambda}{\theta}\int_0^1 u^{\frac{p+\eta-\lambda}{\theta}-1}\left(\frac{1-Au}{1-Bu}\right)du\right)^{\frac{1}{\tau}},\ \tau \geq 1. \tag{19}$$

The result is sharp.

Putting $\theta = 1$ in Theorem 2, we obtain the following example.

Example 2. *Let the function $f(z) \in \mathcal{A}(p)$. Then, following inclusion property holds*

$$\mathcal{T}^{\lambda,\delta}_{p,\mu,\eta}(A,B) \subset \mathcal{T}^{\lambda+1,\delta}_{p,\mu,\eta}(A,B).$$

For a function $f \in \mathcal{A}(p)$, the generalized Bernardi–Libera–Livingston integeral operator $F_{p,\gamma}$ is defined by (see [23]):

$$\begin{aligned} F_{p,\gamma}f(z) &= \frac{\gamma+p}{z^p}\int_0^z t^{\gamma-1}f(t)dt \\ &= \left(z^p + \sum_{k=1}^{\infty}\frac{\gamma+p}{\gamma+p+k}z^{p+k}\right) * f(z)\ (\gamma > -p) \\ &= z^p\, {}_3F_2\left(1,1,\gamma+p;1,\gamma+p+1;z\right) * f(z). \end{aligned} \tag{20}$$

Lemma 7. *If $f \in \mathcal{A}(p)$, prove that:*
(i) $H^{\lambda,\delta}_{p,\eta,\mu}\left(F_{p,\gamma}f\right) = F_{p,\gamma}\left(H^{\lambda,\delta}_{p,\eta,\mu}f\right)$,
(ii)

$$z\left(H^{\lambda,\delta}_{p,\eta,\mu}F_{p,\gamma}f(z)\right)' = (p+\gamma)H^{\lambda,\delta}_{p,\eta,\mu}f(z) - \gamma H^{\lambda,\delta}_{p,\eta,\mu}F_{p,\gamma}f(z). \tag{21}$$

Proof. Since

$$\begin{aligned} H^{\lambda,\delta}_{p,\eta,\mu}\left(F_{p,\gamma}f\right) &= \left[z^p\, {}_3F_2\left(\delta+p,p+1-\mu,p+1-\lambda+\eta;p+1,p+1-\mu+\eta;z\right)\right] * \left(F_{p,\gamma}f\right) \\ &= \left[z^p\, {}_3F_2\left(\delta+p,p+1-\mu,p+1-\lambda+\eta;p+1,p+1-\mu+\eta;z\right)\right] * \\ &\quad \left[z^p\, {}_3F_2\left(1,1,\gamma+p;1,\gamma+p+1;z\right) * f(z)\right], \end{aligned}$$

and:

$$\begin{aligned} F_{p,\gamma}\left(H^{\lambda,\delta}_{p,\eta,\mu}f\right) &= z^p\, {}_3F_2\left(1,1,\gamma+p;1,\gamma+p+1;z\right) * \left(H^{\lambda,\delta}_{p,\eta,\mu}f\right) \\ &= z^p\, {}_3F_2\left(1,1,\gamma+p;1,\gamma+p+1;z\right) * \\ &\quad \left[z^p\, {}_3F_2\left(\delta+p,p+1-\mu,p+1-\lambda+\eta;p+1,p+1-\mu+\eta;z\right) * f(z)\right]. \end{aligned}$$

Now, the first part of this lemma follows. Furthermore,

$$z\left(F_{p,\gamma}f(z)\right)' = (p+\gamma)f(z) - \gamma F_{p,\gamma}f(z). \tag{22}$$

If we replace $f(z)$ by $H^{\lambda,\delta}_{p,\eta,\mu}f(z)$ and using the first part of this lemma, we get (21). □

Theorem 3. *Suppose that $p+\gamma>0$, $f \in \mathcal{T}^{\lambda,\delta}_{p,\eta,\mu}(A,B)$ and $F_{p,\gamma}$ defined by (20). Then:*

$$\Re\left(\frac{\left(H^{\lambda,\delta}_{p,\eta,\mu}F_{p,\gamma}f(z)\right)'}{pz^{p-1}}\right)^{\frac{1}{\tau}} > \left((p+\gamma)\int_0^1 u^{p+\gamma-1}\left(\frac{1-Au}{1-Bu}\right)du\right)^{\frac{1}{\tau}},\ \tau \geq 1. \tag{23}$$

The result is sharp.

Proof. Let:

$$\phi(z) = \frac{\left(H^{\lambda,\delta}_{p,\eta,\mu}F_{p,\gamma}f(z)\right)'}{pz^{p-1}}\ (z \in \mathbb{U}). \tag{24}$$

Then, ϕ is analytic in $\mathbb{U}$. After some calculations, we have:

$$\frac{(H_{p,\eta,\mu}^{\lambda,\delta} f(z))'}{pz^{p-1}} = \phi(z) + \frac{z\phi'(z)}{p+\gamma} \prec \frac{1+Az}{1+Bz}.$$

Employing the same technique that was used in proving Theorem 1, the remaining part of the theorem can be proven. □

Theorem 4. *Let $-1 \le B_i < A_i \le 1$ $(i = 1,2)$. If each of the functions $f_i \in \mathcal{A}(p)$ satisfies:*

$$(1-\theta)\frac{H_{p,\eta,\mu}^{\lambda,\delta} f_i(z)}{z^p} + \theta\frac{H_{p,\eta,\mu}^{\lambda,\delta+1} f_i(z)}{z^p} \prec \frac{1+A_i z}{1+B_i z} \ (i = 1,2), \tag{25}$$

then:

$$(1-\theta)\frac{H_{p,\eta,\mu}^{\lambda,\delta} F(z)}{z^p} + \theta\frac{H_{p,\eta,\mu}^{\lambda,\delta+1} F(z)}{z^p} \prec \frac{1+(1-2\varrho)z}{1-z}, \tag{26}$$

where:

$$F(z) = H_{p,\eta,\mu}^{\lambda,\delta}(f_1 * f_2)(z) \tag{27}$$

and:

$$\varrho = 1 - \frac{4(A_1-B_1)(A_2-B_2)}{(1-B_1)(1-B_2)}\left[1 - \frac{1}{2}\,{}_2F_1\left(1,1;\frac{\delta+p}{\theta}+1;\frac{1}{2}\right)\right]. \tag{28}$$

The result is possible when $B_1 = B_2 = -1$.

Proof. Suppose that $f_i \in \mathcal{A}(p)$ $(i = 1,2)$ satisfy the condition (25). Setting:

$$p_i(z) = (1-\theta)\frac{H_{p,\eta,\mu}^{\lambda,\delta} f_i(z)}{z^p} + \theta\frac{H_{p,\eta,\mu}^{\lambda,\delta+1} f_i(z)}{z^p} \ (i = 1,2), \tag{29}$$

we have:

$$p_i(z) \in P(\varsigma_i) \ \left(\varsigma_i = \frac{1-A_i}{1-B_i},\ i = 1,2\right).$$

Thus, by making use of the identity (3) in (29), we get:

$$H_{p,\eta,\mu}^{\lambda,\delta} f_i(z) = \frac{\delta+p}{\theta} z^{p-\frac{\delta+p}{\theta}} \int_0^z t^{\frac{\delta+p}{\theta}-1} p_i(t)dt \ (i = 1,2), \tag{30}$$

which, in view of F given by (27) and (30), yields:

$$H_{p,\eta,\mu}^{\lambda,\delta} F(z) = \frac{\delta+p}{\theta} z^{p-\frac{\delta+p}{\theta}} \int_0^z t^{\frac{\delta+p}{\theta}-1} F(t)dt, \tag{31}$$

where:

$$F(z) = (1-\theta)\frac{H_{p,\eta,\mu}^{\lambda,\delta} F(z)}{z^p} + \theta\frac{H_{p,\eta,\mu}^{\lambda,\delta+1} F(z)}{z^p} = \frac{\delta+p}{\theta} z^{-\frac{\delta+p}{\theta}} \int_0^z t^{\frac{\delta+p}{\theta}-1} (p_1 * p_2)(t)dt. \tag{32}$$

Since $p_i(z) \in P(\varsigma_i)$ $(i = 1,2)$, it follows from Lemma 3 that:

$$(p_1 * p_2)(z) \in P(\varsigma_3) \ (\varsigma_3 = 1 - 2(1-\varsigma_1)(1-\varsigma_2)). \tag{33}$$

Now, by using (33) in (32) and then appealing to Lemma 2, we have:

$$\begin{aligned}
\Re\{F(z)\} &= \frac{\delta+p}{\theta}\int_0^1 u^{\frac{\delta+p}{\theta}-1}\Re\{(p_1 * p_2)(uz)\}\,du \\
&\geqslant \frac{\delta+p}{\theta}\int_0^1 u^{\frac{\delta+p}{\theta}-1}\left(2\varsigma_3-1+\frac{2(1-\varsigma_3)}{1+u\,|z|}\right)du \\
&> \frac{\delta+p}{\theta}\int_0^1 u^{\frac{\delta+p}{\theta}-1}\left(2\varsigma_3-1+\frac{2(1-\varsigma_3)}{1+u}\right)du \\
&= 1-\frac{4(A_1-B_1)(A_2-B_2)}{(1-B_1)(1-B_2)}\left[1-\tfrac{\delta+p}{\theta}\int_0^1 u^{\frac{\delta+p}{\theta}-1}(1+u)^{-1}du\right] \\
&= 1-\frac{4(A_1-B_1)(A_2-B_2)}{(1-B_1)(1-B_2)}\left[1-\frac{1}{2}\,{}_2F_1\left(1,1;\tfrac{\delta+p}{\theta}+1;\frac{1}{2}\right)\right] = \varrho.
\end{aligned}$$

When $B_1 = B_2 = -1$, we consider the functions $f_i(z) \in \mathcal{A}(p)$ $(i = 1,2)$, which satisfy (25), are defined by:

$$H^{\lambda,\delta}_{p,\eta,\mu}f_i(z) = \tfrac{\delta+p}{\theta}z^{p-\frac{\delta+p}{\theta}}\int_0^z t^{\frac{\delta+p}{\theta}-1}\left(\frac{1+A_it}{1-t}\right)dt\ (i=1,2).$$

Thus, it follows from (32) that:

$$F(z) = \tfrac{\delta+p}{\theta}\int_0^1 u^{\frac{\delta+p}{\theta}-1}\left[1-(1+A_1)(1+A_2)+\frac{(1+A_1)(1+A_2)}{(1-uz)}\right]du$$

$$= 1-(1+A_1)(1+A_2)+(1+A_1)(1+A_2)(1-z)^{-1}\,{}_2F_1\left(1,1;\tfrac{\delta+p}{\theta}+1;\frac{z}{z-1}\right)$$

$$\to 1-(1+A_1)(1+A_2)+\frac{1}{2}(1+A_1)(1+A_2)\,{}_2F_1\left(1,1;\tfrac{\delta+p}{\theta}+1;\frac{1}{2}\right) \text{ as } z\to -1,$$

which evidently ends the proof. □

Theorem 5. *Let $v \in \mathbb{C}^*$, and let A, $B \in \mathbb{C}$ with $A \neq B$ and $|B| \le 1$. Suppose that:*

$$\begin{cases} \left|\frac{v(\delta+p)(A-B)}{B}-1\right| \le 1 \text{ or } \left|\frac{v(\delta+p)(A-B)}{B}+1\right| \le 1 & \text{if } B\neq 0, \\ |v(\delta+p)A| \le \pi & \text{if } B=0. \end{cases}$$

If $f \in \mathcal{A}(p)$ with $H^{\lambda,\delta}_{p,\eta,\mu}f(z) \neq 0$ for all $z \in \mathbb{U}^* = \mathbb{U}\backslash\{0\}$, then:

$$\frac{H^{\lambda,\delta+1}_{p,\eta,\mu}f(z)}{H^{\lambda,\delta}_{p,\eta,\mu}f(z)} \prec \frac{1+Az}{1+Bz},$$

implies:

$$\left(\frac{H^{\lambda,\delta}_{p,\eta,\mu}f(z)}{z^p}\right)^v \prec q(z),$$

where:

$$q(z) = \begin{cases} (1+Bz)^{v(\delta+p)(A-B)/B} & \text{if } B\neq 0, \\ e^{v(\delta+p)Az} & \text{if } B=0, \end{cases}$$

is the best dominant.

Proof. Putting:

$$\Delta(z) = \left(\frac{H^{\lambda,\delta}_{p,\eta,\mu}f(z)}{z^p}\right)^v \quad (z\in\mathbb{U}). \tag{34}$$

Then, Δ is analytic in $\mathbb{U}$, $\Delta(0) = 1$ and $\Delta(z) \neq 0$ for all $z \in \mathbb{U}$. Taking the logarithmic derivatives on both sides of (34) and using (3), we have:

$$1 + \frac{z\Delta'(z)}{v(\delta + p)\Delta(z)} = \frac{H^{\lambda,\delta+1}_{p,\eta,\mu} f(z)}{H^{\lambda,\delta}_{p,\eta,\mu} f(z)} \prec \frac{1 + Az}{1 + Bz}.$$

Now, the assertions of Theorem 5 follow by Lemma 4. □

Theorem 6. *Let* $0 \leq \alpha \leq 1$, $\zeta > 1$. *If* $f(z) \in \mathcal{A}(p)$ *satisfies:*

$$\Re \left((1-\alpha) \frac{\left(H^{\lambda,\delta+2}_{p,\eta,\mu} f(z)\right)'}{\left(H^{\lambda,\delta+1}_{p,\eta,\mu} f(z)\right)'} + \alpha \frac{\left(H^{\lambda,\delta+1}_{p,\eta,\mu} f(z)\right)'}{\left(H^{\lambda,\delta}_{p,\eta,\mu} f(z)\right)'} \right) < \zeta, \tag{35}$$

then:

$$\Re \left(\frac{H^{\lambda,\delta+1}_{p,\eta,\mu} f(z)}{H^{\lambda,\delta}_{p,\eta,\mu} f(z)} \right) < \beta,$$

where $\beta \in (1, \infty)$ *is the positive root of the equation:*

$$2(\delta + p + \alpha)\beta^2 - [2\zeta(\delta + p + 1) - (1-\alpha)]\beta - (1-\alpha) = 0. \tag{36}$$

Proof. Let:

$$\frac{H^{\lambda,\delta+1}_{p,\eta,\mu} f(z)}{H^{\lambda,\delta}_{p,\eta,\mu} f(z)} = \beta + (1-\beta)\varphi(z). \tag{37}$$

Then, φ is analytic in $\mathbb{U}$, $\varphi(0) = 1$ and $\varphi(z) \neq 0$ for all $z \in \mathbb{U}$. Taking the logarithmic derivatives on both sides of (37) and using the identity (3), we have:

$$(\delta + p + 1) \frac{\left(H^{\lambda,\delta+2}_{p,\eta,\mu} f(z)\right)'}{\left(H^{\lambda,\delta+1}_{p,\eta,\mu} f(z)\right)'} - (\delta + p) \frac{\left(H^{\lambda,\delta+1}_{p,\eta,\mu} f(z)\right)'}{\left(H^{\lambda,\delta}_{p,\eta,\mu} f(z)\right)'} = 1 + \frac{(1-\beta) z\varphi'(z)}{\beta + (1-\beta)\varphi(z)},$$

and so:

$$\begin{aligned}(1-\alpha) \frac{\left(H^{\lambda,\delta+2}_{p,\eta,\mu} f(z)\right)'}{\left(H^{\lambda,\delta+1}_{p,\eta,\mu} f(z)\right)'} + \alpha \frac{\left(H^{\lambda,\delta+1}_{p,\eta,\mu} f(z)\right)'}{\left(H^{\lambda,\delta}_{p,\eta,\mu} f(z)\right)'} &= \alpha\beta + \frac{(1-\alpha)(\delta + p)\beta}{\delta + p + 1} \\ &+ \frac{(1-\beta)[\alpha + (1-\alpha)(\delta + p)]}{\delta + p + 1}\varphi(z) + \frac{(1-\alpha)(1-\beta)}{[\beta + (1-\beta)\varphi(z)](\delta + p + 1)} z\varphi'(z).\end{aligned}$$

Let:

$$\begin{aligned}\Psi(r, s; z) &= \alpha\beta + \frac{(1-\alpha)(\delta + p)\beta}{\delta + p + 1} + \frac{(1-\beta)[\alpha + (1-\alpha)(\delta + p)]}{\delta + p + 1} r \\ &+ \frac{(1-\alpha)(1-\beta)}{[\beta + (1-\beta)\varphi(z)](\delta + p + 1)} s,\end{aligned}$$

and:

$$\Omega = \{w \in \mathbb{C} : \Re(w) < \zeta\}.$$

Then, for $x, y \leq -\frac{1+x^2}{2}$, we have:

$$\begin{aligned}\Re\{\Psi(ix, y; z)\} &= \alpha\beta + \frac{(1-\alpha)(\delta+p)\beta}{\delta+p+1} + \frac{(1-\alpha)(1-\beta)\beta y}{\left[\beta^2 + (1-\beta)^2 x^2\right](\delta+p+1)} \\ &\geq \alpha\beta + \frac{(1-\alpha)(\delta+p)\beta}{\delta+p+1} - \frac{(1-\alpha)(1-\beta)}{2\beta(\delta+p+1)} = \zeta,\end{aligned}$$

where β is the positive root of Equation (36). Suppose that:

$$R(\beta) = 2(\delta+p+\alpha)\beta^2 - [2\zeta(\delta+p+1) - (1-\alpha)]\beta - (1-\alpha) = 0.$$

For $\beta = 0$, $R(0) = -(1-\alpha) \leq 0$ and for $\beta = 1$, $R(1) = 2(\delta+p)(1-\zeta) + 2(\alpha-\zeta) \leq 0$. This proves that $\beta \in (1, \infty)$. Thus, for $z \in \mathbb{U}$, $\Psi(ix, y; z) \notin \Omega$, and so, we obtain the required result by an application of Lemma 5. □

Theorem 7. *Suppose that* $0 < \varepsilon_1, \varepsilon_2 \leq 1$. *If:*

$$-\frac{\pi}{2}\varepsilon_1 < \arg\left\{(1-\theta)\frac{\left(H^{\lambda,\delta}_{p,\eta,\mu}f(z)\right)'}{pz^{p-1}} + \theta\frac{\left(H^{\lambda,\delta+1}_{p,\eta,\mu}f(z)\right)'}{pz^{p-1}}\right\} < \frac{\pi}{2}\varepsilon_2, \tag{38}$$

then:

$$-\frac{\pi}{2}\xi_1 < \arg\left(\frac{\left(H^{\lambda,\delta}_{p,\eta,\mu}f(z)\right)'}{pz^{p-1}}\right) < \frac{\pi}{2}\xi_2, \tag{39}$$

where:

$$\varepsilon_1 = \xi_1 + \frac{2}{\pi}\arctan\left(\frac{(\xi_1+\xi_2)\,\theta}{2(\delta+p)}\frac{1-|a|}{1+|a|}\right), \ \varepsilon_2 = \xi_2 + \frac{2}{\pi}\arctan\left(\frac{(\xi_1+\xi_2)\,\theta}{2(\delta+p)}\frac{1-|a|}{1+|a|}\right). \tag{40}$$

Proof. Let:

$$\phi(z) = \frac{\left(H^{\lambda,\delta}_{p,\eta,\mu}f(z)\right)'}{pz^{p-1}} \quad (z \in \mathbb{U}).$$

Then, from Theorem 1, we have:

$$(1-\theta)\frac{\left(H^{\lambda,\delta}_{p,\eta,\mu}f(z)\right)'}{pz^{p-1}} + \theta\frac{\left(H^{\lambda,\delta+1}_{p,\eta,\mu}f(z)\right)'}{pz^{p-1}} = \phi(z) + \frac{\theta z\phi'(z)}{\delta+p}.$$

Let $U(z)$ be the function that maps $\mathbb{U}$ onto the domain:

$$\left\{w \in \mathbb{C} : -\frac{\pi}{2}\varepsilon_1 < \arg(w) < \frac{\pi}{2}\varepsilon_2\right\},$$

with $U(0) = 1$, then:

$$\phi(z) + \frac{\theta z\phi'(z)}{\delta+p} \prec U(z).$$

Assume that z_1, z_2 are two points in $\mathbb{U}$ such that the condition (9) is satisfied, then by Lemma 6, we obtain (10) under the constraint (11). Therefore,

$$\begin{aligned}
\arg\left[(\delta+p)\,\phi(z_1)+\theta z_1\phi'(z_1)\right] &= \arg\phi(z_1)\left[(\delta+p)+\theta\frac{z_1\phi'(z_1)}{\phi(z_1)}\right] \\
&= \arg\phi(z_1)+\arg\left[(\delta+p)+\theta\frac{z_1\phi'(z_1)}{\phi(z_1)}\right] \\
&= -\frac{\pi}{2}\xi_1+\arg\left[(\delta+p)-i\theta\frac{(\xi_1+\xi_2)\,\kappa}{2}\right] \\
&= -\frac{\pi}{2}\xi_1-\arctan\left[\frac{(\xi_1+\xi_2)\,\theta\kappa}{2\,(\delta+p)}\right] \\
&\le -\frac{\pi}{2}\xi_1-\arctan\left[\frac{(\xi_1+\xi_2)\,\theta}{2\,(\delta+p)}\frac{1-|a|}{1+|a|}\right],
\end{aligned}$$

and:

$$\arg\left[(\delta+p)\,\phi(z_2)+\theta z_2\phi'(z_2)\right] \ge \frac{\pi}{2}\xi_2+\arctan\left[\frac{(\xi_1+\xi_2)\,\theta}{2\,(\delta+p)}\frac{1-|a|}{1+|a|}\right].$$

which contradicts the assumption (38). This evidently completes the proof of Theorem 7. □

Acknowledgments: The authors would like to thank all referees for their valuable comments which led to the improvement of this paper.

Author Contributions: All the authors read and approved the final manuscript as a consequence of the authors meetings.

Conflicts of Interest: The authors declare no conflict of interest.

References

1. Bulboacă, T. *Differential Subordinations and Superordinations, Recent Results*; House of Scientific Book Publ.: Cluj-Napoca, Romania, 2005.
2. Miller, S.S.; Mocanu, P.T. *Differential Subordination: Theory and Applications, Series on Monographs and Textbooks in Pure and Applied Mathematics*; Marcel Dekker Inc.: New York, NY, USA, 2000; Volume 225.
3. Owa, S. On the distortion theorems I. *Kyungpook Math. J.* **1978**, *18*, 53–59.
4. Owa, S.; Srivastava, H.M. Univalent and starlike generalized hypergeometric functions. *Can. J. Math.* **1987**, *39*, 1057–1077. [CrossRef]
5. Srivastava, H.M.; Saigo, M.; Owa, S. A class of distortion theorems involving certain operators of fractional calculus. *J. Math. Anal. Appl.* **1988**, *131*, 412–420. [CrossRef]
6. Prajapat, J.K.; Raina, R.K.; Srivastava, H.M. Some inclusion properties for certain subclasses of strongly starlike and strongly convex functions involving a family of fractional integral operators. *Integral Transform. Spec. Funct.* **2007**, *18*, 639–651. [CrossRef]
7. Goyal, S.P.; Prajapat, J.K. A new class of analytic *p*-valent functions with negative coefficients and fractional calculus operators. *Tamsui Oxf. J. Math. Sci.* **2004**, *20*, 175–186.
8. Prajapat, J.K.; Aouf, M.K. Majorization problem for certain class of *p*-valently analytic function defined by generalized fractional differintegral operator. *Comput. Math. Appl.* **2012**, *63*, 42–47. [CrossRef]
9. Tang, H.; Deng, G.; Li, S.; Aouf, M.K. Inclusion results for certain subclasses of spiral-like multivalent functions involving a generalized fractional differintegral operator. *Integral Transform. Spec. Funct.* **2013**, *24*, 873–883. [CrossRef]
10. Aouf, M.K.; Mostafa, A.O.; Zayed, H.M. Some characterizations of integral operators associated with certain classes of *p*-valent functions defined by the Srivastava-Saigo-Owa fractional differintegral operator. *Complex Anal. Oper. Theory* **2016**, *10*, 1267–1275. [CrossRef]
11. Aouf, M.K.; Mostafa, A.O.; Zayed, H.M. Subordination and superordination properties of *p*-valent functions defined by a generalized fractional differintegral operator. *Quaest. Math.* **2016**, *39*, 545–560. [CrossRef]

12. Aouf, M.K.; Mostafa, A.O.; Zayed, H.M. On certain subclasses of multivalent functions defined by a generalized fractional differintegral operator. *Afr. Mat.* **2017**, *28*, 99–107. [CrossRef]
13. Mostafa, A.O.; Aouf, M.K.; Zayed, H.M.; Bulboacă, T. Multivalent functions associated with Srivastava-Saigo -Owa fractional differintegral operator. *RACSAM* **2017**. [CrossRef]
14. Mostafa, A.O.; Aouf, M.K.; Zayed, H.M. Inclusion relations for subclasses of multivalent functions defined by Srivastava–Saigo–Owa fractional differintegral operator. *Afr. Mat.* **2018**, -018-0567-3. [CrossRef]
15. Mostafa, A.O.; Aouf, M.K.; Zayed, H.M. Subordinating results for p-valent functions associated with the Srivastava–Saigo–Owa fractional differintegral operator. *Afr. Mat.* **2018**. [CrossRef]
16. Mostafa, A.O.; Aouf, M.K. Some applications of differential subordination of p-valent functions associated with Cho-Kwon-Srivastava operator. *Acta Math. Sin. (Engl. Ser.)* **2009**, *25*, 1483–1496. [CrossRef]
17. Wang, Z.; Shi, L. Some properties of certain extended fractional differintegral operator. *RACSAM* **2017**, 1–11. [CrossRef]
18. Hallenbeck, D.Z.; Ruscheweyh, S. Subordination by convex functions. *Proc. Am. Math. Soc.* **1975**, *52*, 191–195. [CrossRef]
19. Pommerenke, C. *Univalent Functions*; Vandenhoeck & Ruprecht: Göttingen, Germany, 1975.
20. Stankiewicz, J.; Stankiewicz, Z. Some applications of Hadamard convolution in the theory of functions. *Ann. Univ. Mariae Curie-Sklodowska Sect. A* **1986**, *40*, 251–265.
21. Obradović, M.; Owa, S. On certain properties for some classes of starlike functions. *J. Math. Anal. Appl.* **1990**, *145*, 357–364. [CrossRef]
22. Takahashi, N.; Nunokawa, M. A certain connection between starlike and convex functions. *Appl. Math. Lett.* **2003**, *16*, 653–655. [CrossRef]
23. Choi, J.H.; Saigo, M.; Srivastava, H.M. Some inclusion properties of a certain family of integral operators. *J. Math. Anal. Appl.* **2002**, *276*, 432–445. [CrossRef]

axioms

MDPI

Article

New Definitions about $A^{\mathcal{I}}$-Statistical Convergence with Respect to a Sequence of Modulus Functions and Lacunary Sequences

Ömer Kişi [1,*], Hafize Gümüş [2] and Ekrem Savas [3]

1 Department of Mathematics, Faculty of Science, Bartin University, Bartin 74100, Turkey
2 Department of Mathematics, Faculty of Eregli Education, Necmettin Erbakan University, Eregli, Konya 42060, Turkey; hgumus@konya.edu.tr
3 Department of Mathematics, Istanbul Ticaret University, Üsküdar, Istanbul 34840, Turkey; ekremsavas@yahoo.com
* Correspondence: okisi@bartin.edu.tr; Tel.: +90-378-501-1000

Received: 19 February 2018; Accepted: 10 April 2018; Published: 13 April 2018

Abstract: In this paper, using an infinite matrix of complex numbers, a modulus function and a lacunary sequence, we generalize the concept of $\mathcal{I}$-statistical convergence, which is a recently introduced summability method. The names of our new methods are $A^{\mathcal{I}}$-lacunary statistical convergence and strongly $A^{\mathcal{I}}$-lacunary convergence with respect to a sequence of modulus functions. These spaces are denoted by $S_{\theta}^{A}(\mathcal{I},F)$ and $N_{\theta}^{A}(\mathcal{I},F)$, respectively. We give some inclusion relations between $S^{A}(\mathcal{I},F)$, $S_{\theta}^{A}(\mathcal{I},F)$ and $N_{\theta}^{A}(\mathcal{I},F)$. We also investigate Cesáro summability for $A^{\mathcal{I}}$ and we obtain some basic results between $A^{\mathcal{I}}$-Cesáro summability, strongly $A^{\mathcal{I}}$-Cesáro summability and the spaces mentioned above.

Keywords: lacunary sequence; statistical convergence; ideal convergence; modulus function; $\mathcal{I}$-statistical convergence

MSC: 40A35, 40A05

---

1. Introduction

As is known, convergence is one of the most important notions in mathematics. Statistical convergence extends the notion. After giving the definition of statistical convergence, we can easily show that any convergent sequence is statistically convergent, but not conversely. Let E be a subset of $\mathbb{N}$, and the set of all natural numbers $d(E) := \lim_{n\to\infty} \frac{1}{n} \sum_{j=1}^{n} \chi_E(j)$ is said to be a natural density of E whenever the limit exists. Here, χ_E is the characteristic function of E.

In 1935, statistical convergence was given by Zygmund in the first edition of his monograph [1]. It was formally introduced by Fast [2], Fridy [3], Salat [4], Steinhaus [5] and later was reintroduced by Schoenberg [6]. It has become an active research area in recent years. This concept has applications in different fields of mathematics such as number theory [7], measure theory [8], trigonometric series [1], summability theory [9], etc.

Following this very important definition, the concept of lacunary statistical convergence was defined by Fridy and Orhan [10]. In addition, Fridy and Orhan gave the relationships between the lacunary statistical convergence and the Cesàro summability. Freedman and Sember [9] established the connection between the strongly Cesàro summable sequences space $|\sigma_1|$ and the strongly lacunary summable sequence space N_{θ}.

$\mathcal{I}$-convergence has emerged as a generalized form of many types of convergences. This means that, if we choose different ideals, we will have different convergences. Koystro et al. [11] introduced this concept in a metric space. Also, Das et al. [12], Koystro et al. [13], Savaş and Das [14] studied ideal convergence. We will explain this situation with two examples later. Before defining $\mathcal{I}$-convergence, the definitions of ideal and filter will be needed.

An ideal is a family of sets $\mathcal{I} \subseteq 2^{\mathbb{N}}$ such that (i) $\emptyset \in \mathcal{I}$, (ii) $A, B \in \mathcal{I}$ implies $A \cup B \in \mathcal{I}$, (iii), and, for each $A \in \mathcal{I}$, each $B \subseteq A$ implies $B \in \mathcal{I}$. An ideal is called non-trivial if $\mathbb{N} \notin \mathcal{I}$ and a non-trivial ideal is called admissible if $\{n\} \in \mathcal{I}$ for each $n \in \mathbb{N}$.

A filter is a family of sets $\mathcal{F} \subseteq 2^{\mathbb{N}}$ such that (i) $\emptyset \notin \mathcal{F}$, (ii) $A, B \in \mathcal{F}$ implies $A \cap B \in \mathcal{F}$, (iii) For each $A \in \mathcal{F}$, each $A \subseteq B$ implies $B \in \mathcal{F}$.

If $\mathcal{I}$ is an ideal in $\mathbb{N}$, then the collection

$$F(\mathcal{I}) = \{A \subset \mathbb{N} : \mathbb{N} \backslash A \in \mathcal{I}\}$$

forms a filter in $\mathbb{N}$ that is called the filter associated with $\mathcal{I}$.

The notion of a modulus function was introduced by Nakano [15]. We recall that a modulus f is a function from $[0, \infty)$ to $[0, \infty)$ such that (i) $f(x) = 0$ if and only if $x = 0$; (ii) $f(x+y) = f(x) + f(y)$ for $x, y \geq 0$; (iii) f is increasing; and (iv) f is continuous from the right at 0. It follows that f must be continuous on $[0, \infty)$. Connor [16], Bilgin [17], Maddox [18], Kolk [19], Pehlivan and Fisher [20] and Ruckle [21] have used a modulus function to construct sequence spaces. Now, let S be the space of sequences of modulus functions $F = (f_k)$ such that $\lim_{x \to 0^+} \sup_k f_k(x) = 0$. Throughout this paper, the set of all modulus functions determined by F is denoted by $F = (f_k) \in S$ for every $k \in \mathbb{N}$.

In this paper, we aim to unify these approaches and use ideals to introduce the notion of $A^{\mathcal{I}}$-lacunary statistically convergence with respect to a sequence of modulus functions.

2. Definitions and Notations

First, we recall some of the basic concepts that will be used in this paper.

Let $A = (a_{ki})$ be an infinite matrix of complex numbers. We write $Ax = (A_k(x))$, if $A_k(x) = \sum_{i=1}^{\infty} a_{ki} x_k$ converges for each k.

Definition 1. *A number sequence $x = (x_k)$ is said to be statistically convergent to the number L if for every $\varepsilon > 0$,*

$$\lim_{n \to \infty} \frac{1}{n} |\{k \leq n : |x_k - L| \geq \varepsilon\}| = 0.$$

In this case, we write $st - \lim x_k = L$. As we said before, statistical convergence is a natural generalization of ordinary convergence i.e., if $\lim x_k = L$, then $st - \lim x_k = L$ (Fast, [2]).

By a lacunary sequence, we mean an increasing integer sequence $\theta = \{k_r\}$ such that $k_0 = 0$ and $h_r = k_r - k_{r-1} \to \infty$ as $r \to \infty$. Throughout this paper, the intervals determined by θ will be denoted by $I_r = (k_{r-1}, k_r]$.

Definition 2. *A sequence $x = (x_k)$ is said to be lacunary statistically convergent to the number L if, for every $\varepsilon > 0$,*

$$\lim_{r \to \infty} \frac{1}{h_r} |\{k \in I_r : |x_k - L| \geq \varepsilon\}| = 0.$$

In this case, we write $S_\theta - \lim x_k = L$ or $x_k \to L(S_\theta)$ (Fridy and Orhan, [10]).

Definition 3. *The sequence space N_θ is defined by*

$$N_\theta = \left\{ (x_k) : \lim_{r \to \infty} \frac{1}{h_r} \sum_{k \in I_r} |x_k - L| = 0 \right\}$$

(Fridy and Orhan, [10]).

Definition 4. *Let $\mathcal{I} \subset 2^{\mathbb{N}}$ be a proper admissible ideal in $\mathbb{N}$. The sequence (x_n) of elements of $\mathbb{R}$ is said to be $\mathcal{I}$-convergent to $L \in \mathbb{R}$ if, for each $\varepsilon > 0$, the set*

$$A(\varepsilon) = \{n \in \mathbb{N} : |x_n - L| \geq \varepsilon\} \in \mathcal{I}$$

(Kostyrko et al. [11]).

Example 1. *Define the set of all finite subsets of $\mathbb{N}$ by $\mathcal{I}_f$. Then, $\mathcal{I}_f$ is a non-trivial admissible ideal and $\mathcal{I}_f$-convergence coincides with the usual convergence.*

Example 2. *Define the set $\mathcal{I}_d$ by $\mathcal{I}_d = \{A \subset \mathbb{N} : d(A) = 0\}$. Then, $\mathcal{I}_d$ is an admissible ideal and $\mathcal{I}_d$-convergence gives the statistical convergence.*

Following the line of Savas et al. [22], some authors obtained more general results about statistical convergence by using A matrix and they called this new method $A^{\mathcal{I}}$-statistical convergence (see, e.g., [17,23]).

Definition 5. *Let $A = (a_{ki})$ be an infinite matrix of complex numbers and (f_k) be a sequence of modulus functions in S. A sequence $x = (x_k)$ is said to be $A^{\mathcal{I}}$-statistically convergent to $L \in X$ with respect to a sequence of modulus functions, for each $\varepsilon > 0$, for every $x \in X$ and $\delta > 0$,*

$$\left\{n \in \mathbb{N} : \frac{1}{n}|\{k \leq n : f_k(|A_k(x) - L|) \geq \varepsilon\}| \geq \delta\right\} \in \mathcal{I}.$$

In this case, we write $x_k \to L\left(S^A(\mathcal{I}, F)\right)$ (Yamancı et al. [23]).

3. Inclusions between $S^A(\mathcal{I}, F)$, $S_\theta^A(\mathcal{I}, F)$ and $N_\theta^A(\mathcal{I}, F)$ Spaces

We now consider our main results. We begin with the following definitions.

Definition 6. *Let $A = (a_{ki})$ be an infinite matrix of complex numbers, $\theta = \{k_r\}$ be a lacunary sequence and $F = (f_k)$ be a sequence of modulus functions in S. A sequence $x = (x_k)$ is said to be $A^{\mathcal{I}}$-lacunary statistically convergent to $L \in X$ with respect to a sequence of modulus functions, for each $\varepsilon > 0$, for each $x \in X$ and $\delta > 0$,*

$$\left\{r \in \mathbb{N} : \frac{1}{h_r}|\{k \in I_r : f_k(|A_k(x) - L|) \geq \varepsilon\}| \geq \delta\right\} \in \mathcal{I}.$$

Definition 7. *Let $A = (a_{ki})$ be an infinite matrix of complex numbers, $\theta = \{k_r\}$ be a lacunary sequence and $F = (f_k)$ be a sequence of modulus functions in S. A sequence $x = (x_k)$ is said to be strongly $A^{\mathcal{I}}$-lacunary convergent to $L \in X$ with respect to a sequence of modulus functions, if, for each $\varepsilon > 0$, for each $x \in X$,*

$$\left\{r \in \mathbb{N} : \frac{1}{h_r}\sum_{k \in I_r} f_k(|A_k(x) - L|) \geq \varepsilon\right\} \in \mathcal{I}.$$

We shall denote by $S_\theta^A(\mathcal{I}, F)$, $N_\theta^A(\mathcal{I}, F)$ the collections of all $A^{\mathcal{I}}$-lacunary statistically convergent and strongly $A^{\mathcal{I}}$-lacunary convergent sequences, respectively.

Theorem 1. *Let $A = (a_{ki})$ be an infinite matrix of complex numbers and (f_k) be a sequence of modulus functions in S. $\left(S_\theta^A(\mathcal{I}, F)\right) \cap m(X)$ is a closed subset of $m(X)$ if X is a Banach space where $m(X)$ is the space of all bounded sequences of X.*

Proof. Suppose that $(x^n) \subset \left(S_\theta^A(\mathcal{I}, F)\right) \cap m(X)$ is a convergent sequence and it converges to $x \in m(X)$. We need to show that $x \in \left(S_\theta^A(\mathcal{I}, F)\right) \cap m(X)$. Assume that $x^n \to L_n\left(S_\theta^A(\mathcal{I}, F)\right)$, $\forall n \in \mathbb{N}$.

Take a sequence $\{\varepsilon_r\}_{n\in\mathbb{N}}$ of strictly decreasing positive numbers converging to zero. We can find an $r \in \mathbb{N}$ such that $\|x - x^j\|_\infty < \frac{\varepsilon_r}{4}$ for all $j \geq r$. Choose $0 < \delta < \frac{1}{5}$.

Now,

$$A = \left\{ r \in \mathbb{N} : \frac{1}{h_r} \left| \left\{ k \in I_r : f_k\left(|A_k(x^n) - L_n| \right) \geq \frac{\varepsilon_r}{4} \right\} \right| < \delta \right\} \in \mathcal{F}(\mathcal{I})$$

and

$$B = \left\{ r \in \mathbb{N} : \frac{1}{h_r} \left| \left\{ k \in I_r : f_k\left(\left|A_k\left(x^{n+1}\right) - L_{n+1}\right| \right) \geq \frac{\varepsilon_r}{4} \right\} \right| < \delta \right\} \in \mathcal{F}(\mathcal{I}).$$

Since $A \cap B \in \mathcal{F}(\mathcal{I})$ and $\emptyset \notin \mathcal{F}(\mathcal{I})$, we can choose $r \in A \cap B$. Then,

$$\frac{1}{h_r} \left| \left\{ k \in I_r : f_k\left(|A_k(x^n) - L_n| \right) \geq \frac{\varepsilon_r}{4} \vee f_k\left(\left|A_k\left(x^{n+1}\right) - L_{n+1}\right| \right) \geq \frac{\varepsilon_r}{4} \right\} \right| \leq 2\delta < 1.$$

Since $h_r \to \infty$ and $A \cap B \in \mathcal{F}(\mathcal{I})$ is infinite, we can actually choose the above r so that $h_r > 5$. Hence, there must exist a $k \in I_r$ for which we have simultaneously, $\left|x_k^n - L_n\right| < \frac{\varepsilon_r}{4}$ and $\left|x_k^{n+1} - L_{n+1}\right| < \frac{\varepsilon_r}{4}$.

Then, it follows that

$$\begin{aligned} |L_n - L_{n+1}| &\leq |L_n - x_k^n| + \left|x_k^n - x_k^{n+1}\right| + \left|x_k^{n+1} - L_{n+1}\right| \\ &\leq |x_k^n - L_n| + \left|x_k^{n+1} - L_{n+1}\right| + \|x - x^n\|_\infty + \left\|x - x^{n+1}\right\|_\infty \\ &\leq \tfrac{\varepsilon_r}{4} + \tfrac{\varepsilon_r}{4} + \tfrac{\varepsilon_r}{4} + \tfrac{\varepsilon_r}{4} = \varepsilon_r. \end{aligned}$$

This implies that $\{L_n\}_{n\in\mathbb{N}}$ is a Cauchy sequence in X. Since X is a Banach space, we can write $L_n \to L \in X$ as $n \to \infty$. We shall prove that $x_k \to L\left(S_\theta^A(\mathcal{I}, F)\right)$. Choose $\varepsilon > 0$ and $r \in \mathbb{N}$ such that $\varepsilon_r < \frac{\varepsilon}{4}$, $\|x - x_n\|_\infty < \frac{\varepsilon}{4}$. Now, since

$$\begin{aligned} &\tfrac{1}{h_r} \left| \{ k \in I_r : f_k\left(|A_k(x) - L| \right) \geq \varepsilon \} \right| \\ &\leq \tfrac{1}{h_r} \left| \{ k \in I_r : f_k\left(|A_k(x - x_n)| \right) + f_k\left(|A_k(x^n) - L_n| \right) + f_k\left(|L_n - L| \right) \geq \varepsilon \} \right| \\ &\leq \tfrac{1}{h_r} \left| \left\{ k \in I_r : f_k\left(|A_k(x^n) - L_n| \right) \geq \tfrac{\varepsilon}{2} \right\} \right|. \end{aligned}$$

It follows that

$$\begin{aligned} &\left\{ r \in \mathbb{N} : \tfrac{1}{h_r} \left| \{ k \in I_r : f_k\left(|A_k(x) - L| \right) \geq \varepsilon \} \right| \geq \delta \right\} \\ &\subset \left\{ r \in \mathbb{N} : \tfrac{1}{h_r} \left| \left\{ k \in I_r : f_k\left(|A_k(x) - L| \right) \geq \tfrac{\varepsilon}{2} \right\} \right| \geq \delta \right\} \end{aligned}$$

for given $\delta > 0$. This shows that $x_k \to L\left(S_\theta^A(\mathcal{I}, F)\right)$ and this completes the proof of the theorem. □

Theorem 2. *Let $A = (a_{ki})$ be an infinite matrix of complex numbers, $\theta = \{k_r\}$ be a lacunary sequence and (f_k) be a sequence of modulus functions in S. Then, we have*

(i) If $x_k \to L\left(N_\theta^A(\mathcal{I}, F)\right)$, then $x_k \to L\left(S_\theta^A(\mathcal{I}, F)\right)$ and $N_\theta^A(\mathcal{I}, F) \subset S_\theta^A(\mathcal{I}, F)$ is proper for every ideal $\mathcal{I}$;

(ii) If $x \in m(X)$, the space of all bounded sequences of X and $x_k \to L\left(S_\theta^A(\mathcal{I}, F)\right)$, then $x_k \to L\left(N_\theta^A(\mathcal{I}, F)\right)$;

(iii) $S_\theta^A(\mathcal{I}, F) \cap m(X) = N_\theta^A(\mathcal{I}, F) \cap m(X)$.

Proof. (i) Let $\varepsilon > 0$ and $x_k \to L\left(N_\theta^A(\mathcal{I},F)\right)$. Then, we can write

$$\begin{aligned}\sum_{k\in I_r} f_k\left(|A_k(x)-L|\right) &\geq \sum_{\substack{k\in I_r\\ f_k(|A_k(x)-L|)\geq\varepsilon}} f_k\left(|A_k(x)-L|\right)\\ &\geq \varepsilon\left|\{k\in I_r : f_k\left(|A_k(x)-L|\right)\geq\varepsilon\}\right|.\end{aligned}$$

Thus, for given $\delta > 0$,

$$\frac{1}{h_r}\left|\{k\in I_r : f_k\left(|A_k(x)-L|\right)\geq\varepsilon\}\right|\geq\delta \Longrightarrow \frac{1}{h_r}\sum_{k\in I_r} f_k\left(|A_k(x)-L|\right)\geq\varepsilon\delta,$$

i.e.,

$$\left\{r\in\mathbb{N} : \frac{1}{h_r}\left|\{k\in I_r : f_k\left(|A_k(x)-L|\right)\geq\varepsilon\}\right|\geq\delta\right\}\subseteq\left\{r\in\mathbb{N} : \frac{1}{h_r}\sum_{k\in I_r} f_k\left(|A_k(x)-L|\right)\geq\varepsilon\delta\right\}.$$

Since $x_k \to L\left(N_\theta^A(\mathcal{I},F)\right)$, the set on the right-hand side belongs to $\mathcal{I}$ and so it follows that $x_k \to L\left(S_\theta^A(\mathcal{I},F)\right)$.

To show that $\left(S_\theta^A(\mathcal{I},F)\right) \subsetneq \left(N_\theta^A(\mathcal{I},F)\right)$, take a fixed $K\in\mathcal{I}$. Define $x=(x_k)$ by

$$(x_k)=\begin{cases} ku, & \text{for } k_{r-1}<k\leq k_{r-1}+\left[\sqrt{h_r}\right], r=1,2,3..., r\notin K,\\ ku, & \text{for } k_{r-1}<k\leq k_{r-1}+\left[\sqrt{h_r}\right], r=1,2,3..., r\in K,\\ \theta, & \text{otherwise},\end{cases}$$

where $u\in X$ is a fixed element with $\|u\|=1$ and θ is the null element of X. Then, $x\notin m(X)$ and for every $0<\varepsilon<1$ since

$$\frac{1}{h_r}\left|\{k\in I_r : f_k\left(|A_k(x)-0|\right)\geq\varepsilon\}\right| = \frac{\left[\sqrt{h_r}\right]}{\sqrt{h_r}}\to 0.$$

As $r\to\infty$ and $r\notin K$, for every $\delta>0$,

$$\left\{r\in\mathbb{N} : \frac{1}{h_r}\left|\{k\in I_r : f_k\left(|A_k(x)-0|\right)\geq\varepsilon\}\right|\geq\delta\right\}\subset M\cup\{1,2,...,m\}$$

for some $m\in\mathbb{N}$. Since $\mathcal{I}$ is admissible, it follows that $x_k\to\theta\left(S_\theta^A(\mathcal{I},F)\right)$. Obviously,

$$\frac{1}{h_r}\sum_{k\in I_r} f_k\left(|A_k(x)-\theta|\right)\to\infty,$$

i.e., $x_k \nrightarrow \theta\left(N_\theta^A(\mathcal{I},F)\right)$. Note that, if $K\in\mathcal{I}$ is finite, then $x_k\nrightarrow\theta\left(S_\theta^A\right)$. This example shows that $A^{\mathcal{I}}$-lacunary statistical convergence is more general than lacunary statistical convergence.

(ii) Suppose that $x\in l_\infty$ and $x_k\to L\left(S_\theta^A(\mathcal{I},F)\right)$. Then, we can assume that

$$f_k\left(|A_k(x)-L|\right)\leq M$$

for each $x\in X$ and all k.

Given $\varepsilon > 0$, we get

$$\begin{aligned}\frac{1}{h_r}\sum_{k\in I_r} f_k\left(|A_k(x)-L|\right) &= \frac{1}{h_r}\sum_{\substack{k\in I_r\\ f_k(|A_k x-L|)\geq\varepsilon}} f_k\left(|A_k(x)-L|\right)\\ &\quad+\frac{1}{h_r}\sum_{\substack{k\in I_r\\ f_k(|A_k x-L|)<\varepsilon}} f_k\left(|A_k(x)-L|\right)\\ &\leq \frac{M}{h_r}\left|\{k\in I_r : f_k\left(|A_k(x)-L|\right)\geq\varepsilon\}\right|+\varepsilon.\end{aligned}$$

Note that

$$A(\varepsilon)=\left\{r\in\mathbb{N}:\frac{1}{h_r}\left|\{k\in I_r : f_k\left(|A_k(x)-L|\right)\geq\varepsilon\}\right|\geq\frac{\varepsilon}{M}\right\}\in\mathcal{I}.$$

If $n\in(A(\varepsilon))^c$, then

$$\frac{1}{h_r}\sum_{k\in I_r} f_k|A_k(x)-L|<2\varepsilon.$$

Hence,

$$\left\{r\in\mathbb{N}:\frac{1}{h_r}\sum_{k\in I_r} f_k|A_k(x)-L|\geq 2\varepsilon\right\}\subset A(\varepsilon)$$

and thus belongs to $\mathcal{I}$. This shows that $x_k\to L\left(N_\theta^A(\mathcal{I},F)\right)$.

(iii) This is an immediate consequence of (i) and (ii). □

Theorem 3. *Let $A=(a_{ki})$ be an infinite matrix of complex numbers and (f_k) be a sequence of modulus functions in S. If $\theta=\{k_r\}$ is a lacunary sequence with $\liminf_r q_r>1$, then*

$$x_k\to L\left(S^A(\mathcal{I},F)\right)\Rightarrow x_k\to L\left(S_\theta^A(\mathcal{I},F)\right).$$

Proof. Suppose first that $\liminf_r q_r>1$, then there exists $\delta>0$ such that $q_r\geq 1+\delta$ for sufficiently large r, which implies that

$$\frac{h_r}{k_r}\geq\frac{\delta}{1+\delta}.$$

If $x_k\to L\left(S_\theta^A(\mathcal{I},F)\right)$, then for every $\varepsilon>0$, for each $x\in X$ and for sufficiently large r, we have

$$\begin{aligned}\frac{1}{k_r}\left|\{k\leq k_r : f_k\left(|A_k(x)-L|\right)\geq\varepsilon\}\right| &\geq \frac{1}{k_r}\left|\{k\in I_r : f_k\left(|A_k(x)-L|\right)\geq\varepsilon\}\right|\\ &\geq \frac{\delta}{1+\delta}\frac{1}{h_r}\left|\{k\in I_r : f_k\left(|A_k(x)-L|\right)\geq\varepsilon\}\right|.\end{aligned}$$

Then, for any $\delta>0$, we get

$$\begin{aligned}&\left\{r\in\mathbb{N}:\tfrac{1}{h_r}\left|\{k\in I_r : f_k\left(|A_k(x)-L|\right)\geq\varepsilon\}\right|\geq\delta\right\}\\ &\subseteq\left\{r\in\mathbb{N}:\tfrac{1}{k_r}\left|\{k\leq k_r : f_k\left(|A_k(x)-L|\right)\geq\varepsilon\}\right|\geq\tfrac{\delta\alpha}{(\alpha+1)}\right\}\in\mathcal{I}.\end{aligned}$$

This completes the proof. □

For the next result, we assume that the lacunary sequence θ satisfies the condition that, for any set $C\in\mathcal{F}(\mathcal{I})$, $\cup\{n : k_{r-1}<n\leq k_r,\ r\in C\}\in\mathcal{F}(\mathcal{I})$.

Theorem 4. *Let* $A = (a_{ki})$ *be an infinite matrix of complex numbers and* (f_k) *be a sequence of modulus functions in S. If* $\theta = \{k_r\}$ *is a lacunary sequence with* $\limsup_r q_r < \infty$, *then*

$$x_k \to L\left(S_\theta^A(\mathcal{I},F)\right) \text{ implies } x_k \to L\left(S^A(\mathcal{I},F)\right).$$

Proof. If $\limsup_r q_r < \infty$, then, without any loss of generality, we can assume that there exists a $0 < M < \infty$ such that $q_r < M$ for all $r \geq 1$. Suppose that $x_k \to L\left(S_\theta^A(\mathcal{I},F)\right)$, and for $\varepsilon, \delta, \delta_1 > 0$ define the sets

$$C = \left\{r \in \mathbb{N} : \frac{1}{h_r}\left|\{k \in I_r : f_k(|A_k(x) - L|) \geq \varepsilon\}\right| < \delta\right\}$$

and

$$T = \left\{n \in \mathbb{N} : \frac{1}{n}\left|\{k \leq n : f_k(|A_k(x) - L|) \geq \varepsilon\}\right| < \delta_1\right\}.$$

It is obvious from our assumption that $C \in \mathcal{F}(\mathcal{I})$, the filter associated with the ideal $\mathcal{I}$. Further observe that

$$K_j = \frac{1}{h_j}\left|\{k \in I_j : f_k(|A_k(x) - L|) \geq \varepsilon\}\right| < \delta$$

for all $j \in C$. Let $n \in \mathbb{N}$ be such that $k_{r-1} < n \leq k_r$ for some $r \in C$. Now,

$$\begin{aligned}
\frac{1}{n}\left|\{k \leq n : f_k|A_k(x) - L| \geq \varepsilon\}\right| &\leq \frac{1}{k_{r-1}}\left|\{k \leq k_r : f_k(|A_k(x) - L|) \geq \varepsilon\}\right| \\
&= \frac{1}{k_{r-1}}\left\{\left|\{k \in I_1 : f_k(|A_k(x) - L|) \geq \varepsilon\}\right|\right\} \\
&\quad + \frac{1}{k_{r-1}}\left\{\left|\{k \in I_2 : f_k(|A_k x - L|) \geq \varepsilon\}\right|\right\} \\
&\quad + \ldots + \frac{1}{k_{r-1}}\left\{\left|\{k \in I_r : f_k|A_k(x) - L| \geq \varepsilon\}\right|\right\} \\
&= \frac{k_1}{k_{r-1}}\frac{1}{h_1}\left|\{k \in I_1 : f_k|A_k(x) - L| \geq \varepsilon\}\right| \\
&\quad + \frac{k_2 - k_1}{k_{r-1}}\frac{1}{h_2}\left|\{k \in I_2 : f_k(|A_k(x) - L|) \geq \varepsilon\}\right| \\
&\quad + \ldots + \frac{k_r - k_{r-1}}{k_{r-1}}\frac{1}{h_r}\left|\{k \in I_r : f_k(|A_k(x) - L|) \geq \varepsilon\}\right| \\
&= \frac{k_1}{k_{r-1}}K_1 + \frac{k_2 - k_1}{k_{r-1}}K_2 + \ldots + \frac{k_r - k_{r-1}}{k_{r-1}}K_r \\
&\leq \left\{\sup_{j \in C} K_j\right\}\frac{k_r}{k_{r-1}} < M\delta.
\end{aligned}$$

Choosing $\delta_1 = \dfrac{\delta}{M}$ and in view of the fact that $\cup\{n : k_{r-1} < n \leq k_r,\ r \in C\} \subset T$ where $C \in \mathcal{F}(\mathcal{I})$, it follows from our assumption on θ that the set T also belongs to $\mathcal{F}(\mathcal{I})$ and this completes the proof of the theorem. □

Combining Theorems 3 and 4, we get the following theorem.

Theorem 5. *Let $A = (a_{ki})$ be an infinite matrix of complex numbers and (f_k) be a sequence of modulus functions in S. If $\theta = \{k_r\}$ is a lacunary sequence with $1 < \liminf_r q_r \leq \limsup_r q_r < \infty$, then*

$$x_k \to L\left(S_\theta^A(\mathcal{I}, F)\right) = x_k \to L\left(S_\theta^A(\mathcal{I}, F)\right).$$

4. Cesàro Summability for $A^{\mathcal{I}}$

Definition 8. *Let $A = (a_{ki})$ be an infinite matrix of complex numbers and (f_k) be a sequence of modulus functions in S. A sequence $x = (x_k)$ is said to be $A^{\mathcal{I}}$-Cesàro summable to L if, for each $\varepsilon > 0$ and for each $x \in X$,*

$$\left\{n \in \mathbb{N} : \left|\frac{1}{n}\sum_{k=1}^{n} f_k\left(A_k(x) - L\right)\right| \geq \varepsilon\right\} \in \mathcal{I}.$$

In this case, we write $x_k \to L\left((\sigma_1)_\theta^A(\mathcal{I}, F)\right)$.

Definition 9. *Let $A = (a_{ki})$ be an infinite matrix of complex numbers and (f_k) be a sequence of modulus functions in S. A sequence $x = (x_k)$ is said to be strongly $A^{\mathcal{I}}$-Cesàro summable to L if, for each $\varepsilon > 0$ and for each $x \in X$,*

$$\left\{n \in \mathbb{N} : \frac{1}{n}\sum_{k=1}^{n} f_k\left(|A_k(x) - L|\right) \geq \varepsilon\right\} \in \mathcal{I}.$$

In this case, we write $x_k \to L\left(|\sigma_1|_\theta^A(\mathcal{I}, F)\right)$.

Theorem 6. *Let θ be a lacunary sequence. If $\liminf_r q_r > 1$, then*

$$x_k \to L\left(|\sigma_1|_\theta^A(\mathcal{I}, F)\right) \Rightarrow x_k \to L\left(N_\theta^A(\mathcal{I}, F)\right).$$

Proof. If $\liminf_r q_r > 1$, then there exists $\delta > 0$ such that $q_r \geq 1 + \delta$ for all $r \geq 1$. Since $h_r = k_r - k_{r-1}$, we have $\frac{k_r}{h_r} \leq \frac{1+\delta}{\delta}$ and $\frac{k_{r-1}}{h_r} \leq \frac{1}{\delta}$. Let $\varepsilon > 0$ and define the set

$$S = \left\{k_r \in \mathbb{N} : \frac{1}{k_r}\sum_{k=1}^{k_r} f_k\left(|A_k(x) - L|\right) < \varepsilon\right\}.$$

We can easily say that $S \in \mathcal{F}(\mathcal{I})$, which is a filter of the ideal $\mathcal{I}$,

$$\begin{aligned}
\tfrac{1}{h_r}\sum_{k\in I_r} f_k\left(|A_k(x) - L|\right) &= \tfrac{1}{h_r}\sum_{k=1}^{k_r} f_k\left(|A_k(x) - L|\right) - \tfrac{1}{h_r}\sum_{k=1}^{k_{r-1}} f_k\left(|A_k(x) - L|\right) \\
&= \tfrac{k_r}{h_r}\tfrac{1}{k_r}\sum_{k=1}^{k_r} f_k\left(|A_k(x) - L|\right) - \tfrac{k_{r-1}}{h_r}\tfrac{1}{k_{r-1}}\sum_{k=1}^{k_{r-1}} f_k\left(|A_k(x) - L|\right) \\
&\leq \left(\frac{1+\delta}{\delta}\right)\varepsilon - \frac{1}{\delta}\varepsilon'
\end{aligned}$$

for each $k_r \in S$. Choose $\eta = \left(\frac{1+\delta}{\delta}\right)\varepsilon - \frac{1}{\delta}\varepsilon'$. Therefore,

$$\left\{r \in \mathbb{N} : \frac{1}{h_r}\sum_{k\in I_r} f_k\left(|A_k(x) - L|\right) < \eta\right\} \in \mathcal{F}(\mathcal{I}),$$

and it completes the proof. □

Theorem 7. *Let $A = (a_{ki})$ be an infinite matrix of complex numbers and (f_k) be a sequence of modulus functions in S. If $(x_k) \in m(X)$ and $x_k \to L(S_\theta^A(\mathcal{I},F))$, then $x_k \to L\left((\sigma_1)_\theta^A(\mathcal{I},F)\right)$.*

Proof. Suppose that $(x_k) \in m(X)$ and $x_k \to L(S_\theta^A(\mathcal{I},F))$. Then, we can assume that

$$f_k(|A_k x - L|) \leq M$$

for all $k \in \mathbb{N}$. In addition, for each $\varepsilon > 0$, we can write

$$\begin{aligned}
\left|\frac{1}{n}\sum_{k=1}^{n} f_k(A_k(x) - L)\right| &\leq \frac{1}{n}\sum_{k=1}^{n} f_k(|A_k(x) - L|) \\
&\leq \frac{1}{n}\sum_{\substack{k=1 \\ f_k(|A_k(x)-L|)\geq\frac{\varepsilon}{2}}}^{n} f_k(|A_k(x) - L|) \\
&\quad + \frac{1}{n}\sum_{\substack{k=1 \\ f_k(|A_k(x)-L|)<\frac{\varepsilon}{2}}}^{n} f_k(|A_k(x) - L|) \\
&\leq M\frac{1}{n}|\{k \leq n : f_k(|A_k(x) - L|) \geq \varepsilon\}| + \varepsilon.
\end{aligned}$$

Consequently, if $\delta > \varepsilon > 0$, δ and ε are independent, and, putting $\delta_1 = \delta - \varepsilon > 0$, we have

$$\begin{aligned}
&\left\{n \in \mathbb{N} : \left|\frac{1}{n}\sum_{k,l=1}^{n} f_k(A_k(x) - L)\right| \geq \delta\right\} \\
&\subseteq \left\{n \in \mathbb{N} : \tfrac{1}{n}|\{k \leq n : f_k(|A_k(x) - L|) \geq \varepsilon\}| \geq \tfrac{\delta_1}{M}\right\} \in \mathcal{I}.
\end{aligned}$$

This shows that $x_k \to L\left((\sigma_1)_\theta^A(\mathcal{I},F)\right)$. □

Theorem 8. *Let θ be a lacunary sequence. If $\limsup_r q_r < \infty$, then*

$$x_k \to L\left(N_\theta^A(\mathcal{I},F)\right) \Rightarrow x_k \to L\left(|\sigma_1|_\theta^A(\mathcal{I},F)\right).$$

Proof. If $\limsup_r q_r < \infty$, then there exists $M > 0$ such that $q_r < M$ for all $r \geq 1$. Let $x_k \to L(N_\theta^A(\mathcal{I},F))$ and define the sets T and R such that

$$T = \left\{r \in \mathbb{N} : \frac{1}{h_r}\sum_{k\in I_r} f_k(|A_k(x) - L|) < \varepsilon_1\right\}$$

and

$$R = \left\{n \in \mathbb{N} : \frac{1}{n}\sum_{k=1}^{n} f_k(|A_k(x) - L|) < \varepsilon_2\right\}.$$

Let

$$A_j = \frac{1}{h_j}\sum_{k\in I_j} f_k(|A_k(x) - L|) < \varepsilon_1$$

for all $j \in T$. It is obvious that $T \in \mathcal{F}(\mathcal{I})$. Choose n as being any integer with $k_{r-1} < n < k_r$, where $r \in T$,

$$\begin{aligned}
\frac{1}{n}\sum_{k=1}^{n} f_k\left(|A_k(x) - L|\right) &\leq \frac{1}{k_{r-1}}\sum_{k=1}^{k_r} f_k\left(|A_k(x) - L|\right) \\
&= \frac{1}{k_{r-1}}\left(\sum_{k\in I_1} f_k\left(|A_k(x) - L|\right) + \sum_{k\in I_2} f_k\left(|A_k(x) - L|\right)\right. \\
&\quad \left. + \ldots + \sum_{k\in I_r} f_k\left(|A_k(x) - L|\right)\right) \\
&= \frac{k_1}{k_{r-1}}\left(\frac{1}{h_1}\sum_{k\in I_1} f_k\left(|A_k(x) - L|\right)\right) + \frac{k_2-k_1}{k_{r-1}}\left(\frac{1}{h_2}\sum_{k\in I_2} f_k\left(|A_k(x) - L|\right)\right) \\
&\quad + \ldots + \frac{k_r-k_{r-1}}{k_{r-1}}\left(\frac{1}{h_r}\sum_{k\in I_r} f_k\left(|A_k(x) - L|\right)\right) \\
&= \frac{k_1}{k_{r-1}}A_1 + \frac{k_2-k_1}{k_{r-1}}A_2 + \ldots + \frac{k_r-k_{r-1}}{k_{r-1}}A_r \\
&\leq \left(\sup_{j\in T} A_j\right)\frac{k_1}{k_{r-1}} \\
&< \varepsilon_1 M.
\end{aligned}$$

Choose $\varepsilon_2 = \frac{\varepsilon_1}{M}$ and in view of the fact that $\cup\{n : k_{r-1} < n < k_r, r \in T\} \subset R$, where $T \in \mathcal{F}(\mathcal{I})$, it follows from our assumption on θ that the set R also belongs to $\mathcal{F}(\mathcal{I})$ and this completes the proof of the theorem. □

Theorem 9. *If $x_k \to L\left(|\sigma_1|_\theta^A(\mathcal{I}, F)\right)$, then $x_k \to L\left(S^A(\mathcal{I}, F)\right)$.*

Proof. Let $x_k \to L\left(|\sigma_1|_\theta^A(\mathcal{I}, F)\right)$ and $\varepsilon > 0$ is given. Then,

$$\begin{aligned}
\sum_{k=1}^{n} f_k\left(|A_k(x) - L|\right) &\geq \sum_{\substack{k=1 \\ f_k(|A_k x - L|)\geq\varepsilon}}^{n} f_k\left(|A_k(x) - L|\right) \\
&\geq \varepsilon\left|\{k \leq n : f_k\left(|A_k(x) - L|\right) \geq \varepsilon\}\right|
\end{aligned}$$

and so

$$\frac{1}{\varepsilon n}\sum_{k=1}^{n} f_k\left(|A_k(x) - L|\right) \geq \frac{1}{n}\left|\{k \leq n : f_k\left(|A_k(x) - L|\right) \geq \varepsilon\}\right|.$$

Thus, for a given $\delta > 0$,

$$\begin{aligned}
&\left\{n \in \mathbb{N} : \frac{1}{n}\left|\{k \leq n : f_k\left(|A_k(x) - L|\right) \geq \varepsilon\}\right| \geq \delta\right\} \\
&\qquad \subseteq \left\{n \in \mathbb{N} : \frac{1}{n}\sum_{k=1}^{n} f_k\left(|A_k(x) - L|\right) \geq \varepsilon\delta\right\} \in \mathcal{I}.
\end{aligned}$$

Therefore, $x_k \to L\left(S^A(\mathcal{I}, F)\right)$. □

Theorem 10. *Let* $(x_k) \in m(X)$*. If* $x_k \to L\left(S^A(\mathcal{I},F)\right)$*. Then,* $x_k \to L\left(|\sigma_1|_\theta^A(\mathcal{I},F)\right)$*.*

Proof. Suppose that (x_k) is bounded and $x_k \to L\left(S^A(\mathcal{I},F)\right)$. Then, there is an M such that $f_k(|A_k(x) - L|) \leq M$ for all k. Given $\varepsilon > 0$, we have

$$\begin{aligned}
\frac{1}{n}\sum_{k=1}^{n} f_k(|A_k(x) - L|) &= \frac{1}{n}\sum_{\substack{k=1 \\ f_k(|A_k(x)-L|)\geq\varepsilon}}^{n} f_k(|A_k(x) - L|) \\
&+ \frac{1}{n}\sum_{\substack{k=1 \\ f_k(|A_k(x)-L|)<\varepsilon}}^{n} f_k(|A_k(x) - L|) \\
&\leq \frac{1}{n}M\,|\{k \leq n : f_k(|A_k(x) - L|) \geq \varepsilon\}| \\
&\quad + \frac{1}{n}\varepsilon\,|\{k \leq n : f_k(|A_k(x) - L|) < \varepsilon\}| \\
&\leq \frac{M}{n}\,|\{k \leq n : f_k(|A_k(x) - L|) \geq \varepsilon\}| + \varepsilon.
\end{aligned}$$

Then, for any $\delta > 0$,

$$\begin{aligned}
&\left\{n \in \mathbb{N} : \frac{1}{n}\sum_{k=1}^{n} f_k(|A_k(x) - L|) \geq \delta\right\} \\
&\qquad \subseteq \left\{n \in \mathbb{N} : \frac{1}{n}\,|\{k \leq n : f_k(|A_k(x) - L|) \geq \varepsilon\}| \geq \frac{\delta}{M}\right\} \in \mathcal{I}.
\end{aligned}$$

Therefore, $x_k \to L\left(|\sigma_1|_\theta^A(\mathcal{I},F)\right)$. □

5. Conclusions

$\mathcal{I}$-statistical convergence gained a different perspective after identification of the $A^{\mathcal{I}}$-statistical convergence with an infinite matrix of complex numbers. Some authors have studied this new method with different sequences. Our results in this paper were developed with lacunary sequences. By also using a modulus function, we obtain more interesting and general results. These definitions can be adapted to many different concepts such as random variables in order to have different results.

Acknowledgments: We would like to thanks the referees for their valuable comments.

Author Contributions: Both authors completed the paper together. Both authors read and approved the final manuscript.

Conflicts of Interest: The authors declare no conflict of interest.

References

1. Zygmund, A. *Trigonometric Series;* Cambridge University Press: Cambridge, UK, 1979.
2. Fast, H. Sur la convergence statistique. *Colloq. Math.* **1951**, *2*, 241–244.
3. Fridy, J.A. On statistical convergence. *Analysis* **1985**, *5*, 301–313.
4. Salat, T. On statistically convergent sequences of real numbers. *Math. Slovaca* **1980**, *30*, 139–150.
5. Steinhaus, H. Sur la convergence ordinaire et la convergence asymptotique. *Colloq. Math.* **1951**, *2*, 73–74.
6. Schoenberg, I.J. The integrability of certain functions and related summability methods. *Am. Math. Mon.* **1959**, *66*, 361–375.
7. Erdös, P.; Tenenbaum, G. Sur les densities de certaines suites d'entiers. *Proc. Lond. Math. Soc.* **1989**, *3*, 417–438.

8. Miller, H.I. A measure theoretical subsequence characterization of statistical convergence. *Trans. Am. Math. Soc.* **1995**, *347*, 1811–1819.
9. Freedman, A.R.; Sember, J.J. Densities and summability. *Pac. J. Math.* **1981**, *95*, 10–11.
10. Fridy, J.A.; Orhan, C. lacunary statistical convergence. *Pac. J. Math.* **1993**, *160*, 43–51.
11. Kostyrko, P.; Macaj, M.; Salat, T. $\mathcal{I}$-convergence. *Real Anal. Exchang.* **2000**, *26*, 669–686.
12. Das, P.; Ghosal, S. Some further results on $\mathcal{I}$-Cauchy sequences and condition (AP). *Comput. Math. Appl.* **2010**, *29*, 2597–2600.
13. Kostyrko, P.; Macaj, M.; Salat, T.; Sleziak, M. $\mathcal{I}$-convergence and extremal $\mathcal{I}$-limit points. *Math. Slovaca* **2005**, *55*, 443–464.
14. Savas, E.; Das, P. A generalized statistical convergence via ideals. *Appl. Math. Lett.* **2011**, *24*, 826–830.
15. Nakano, H. Concave modulars. *J. Math. Soc. Jpn.* **1953**, *5*, 29–49.
16. Connor, J.S. On strong matrix summability with respect to a modulus and statistical convergence. *Can. Math. Bull.* **1989**, *32*, 194–198.
17. Bilgin, T. Lacunary strong A-convergence with respect to a modulus. *Mathematica* **2001**, *4*, 39–46.
18. Maddox, I.J. Sequence spaces defined by a modulus. *Math. Proc. Camb. Philos. Soc.* **1986**, *100*, 161–166.
19. Kolk, E. On strong boundedness and summability with respect to a sequence moduli. *Tartu Ülikooli Toimetised* **1993**, *960*, 41–50.
20. Pehlivan, S.; Fisher, B. On some sequence spaces. *Indian J. Pure Appl. Math.* **1994**, *25*, 1067–1071.
21. Ruckle, W.H. FK spaces in which the sequence of coordinate vectors is bounded. *Can. J. Math.* **1973**, *25*, 973–978.
22. Das, P.; Savas, E.; Ghosal, S. On generalizations of certain summability methods using ideals. *Appl. Math. Lett.* **2011**, *24*, 1509–1514.
23. Yamancı, U.; Gürdal, M.; Saltan, S. $A^{\mathcal{I}}$-statistical convergence with respect to a sequence of modulus functions. *Contemp. Anal. Appl. Math.* **2014**, *2*, 136–145.

axioms

MDPI

Article

Special Numbers and Polynomials Including Their Generating Functions in Umbral Analysis Methods

Yilmaz Simsek

Department of Mathematics, Faculty of Science University of Akdeniz, TR-07070 Antalya, Turkey; ysimsek@akdeniz.edu.tr

Received: 20 January 2018; Accepted: 30 March 2018; Published: 1 April 2018

Abstract: In this paper, by applying umbral calculus methods to generating functions for the combinatorial numbers and the Apostol type polynomials and numbers of order k, we derive some identities and relations including the combinatorial numbers, the Apostol-Bernoulli polynomials and numbers of order k and the Apostol-Euler polynomials and numbers of order k. Moreover, by using p-adic integral technique, we also derive some combinatorial sums including the Bernoulli numbers, the Euler numbers, the Apostol-Euler numbers and the numbers $y_1(n,k;\lambda)$. Finally, we make some remarks and observations regarding these identities and relations.

Keywords: Apostol-Bernoulli polynomials and numbers; Apostol-Euler polynomials and numbers; Sheffer sequences; Appell sequences; Fibonacci numbers; umbral algebra, p-adic integral

MSC: 05A40; 11B68; 11B73; 11B83; 11S80; 26C05; 30B10

---

1. Introduction

In order to give the results presented in this paper, we use two techniques which are p-adic integral technique and the umbral calculus technique. In [1–5], we constructed generating functions for families of combinatorial numbers which are used in counting techniques and problems and also computing negative order of the first and the second kind Euler numbers and other combinatorial sums. In this paper, by applying umbral algebra and umbral analysis methods and their operators to generating functions of the combinatorial numbers and the Apostol type polynomials and numbers, we give many identities and relations including the Fibonacci numbers, the combinatorial numbers, the Apostol-Bernoulli polynomials and numbers of higher order and the Apostol-Euler polynomials and numbers of higher order.

Throughout this paper, we use the following notations, definitions and relations.

Here and in the following, let $\mathbb{C}$, $\mathbb{R}$, $\mathbb{Z}$, and $\mathbb{N}$ be the sets of complex numbers, real numbers, integers, and positive integers, respectively, and let $\mathbb{N}_0 := \mathbb{N} \cup \{0\}$. We assume $0^0 = 1$.

Moreover, throughout this paper, $\log z$ is tacitly assumed to denote the principal branch of the many-valued function $\log z$ with the imaginary part $(\log z)$ constrained by

$$-\pi < \mathrm{Im}(\log z) \leq \pi$$

(cf. [6–9]).

The Apostol-Bernoulli polynomials $\mathcal{B}_n^{(k)}(x,\lambda)$ of order k are defined by

$$F_{\mathcal{B}}(t,x;\lambda,k) = \left(\frac{t}{\lambda e^t - 1}\right)^k e^{tx} = \sum_{n=0}^{\infty} \mathcal{B}_n^{(k)}(x,\lambda)\frac{t^n}{n!}, \tag{1}$$

where λ is an arbitrary (real or complex) parameter and $x \in \mathbb{R}$, and $|t| < 2\pi$ when $\lambda = 1$ and $|t| < \frac{2\pi}{|\log \lambda|}$ when $\lambda \neq 1$. Moreover, $\mathcal{B}_n^{(k)}(\lambda) := \mathcal{B}_n^{(k)}(0,\lambda)$ denote the Apostol-Bernoulli numbers of order. $B_n^{(k)} := B_n^{(k)}(1)$ denote the Bernoulli numbers of order k and also $B_n := B_n^{(1)}$ denote the Bernoulli numbers (cf. see,for details, [6,8–14], and the references cited therein).

The Apostol-Euler polynomials $\mathcal{E}_n^{(k)}(x,\lambda)$ of order k are defined by

$$F_{\mathcal{E}}(t,x;\lambda,k) = \left(\frac{2}{\lambda e^t + 1}\right)^k e^{tx} = \sum_{n=0}^{\infty} \mathcal{E}_n^{(k)}(x,\lambda)\frac{t^n}{n!}, \tag{2}$$

where λ is an arbitrary (real or complex) parameter and $x \in \mathbb{R}$, and $|t| < \pi$ when $\lambda = 1$ and $|t| < \frac{\pi}{|\log \lambda|}$ when $\lambda \neq 1$. Moreover, $\mathcal{E}_n^{(k)}(\lambda) := \mathcal{E}_n^{(k)}(0,\lambda)$ denote the Apostol-Euler numbers of order k. $E_n^{(k)} := E_n^{(k)}(1)$ denote the Euler numbers of order k and also $E_n := E_n^{(1)}$ denote the Euler numbers (cf. see, for details, [6,8–15], and the references cited therein).

The λ-array polynomials $S_v^n(x;\lambda)$ are defined by

$$F_S(t,x,v;\lambda) = \frac{(\lambda e^t - 1)^v}{v!} e^{xt} = \sum_{n=0}^{\infty} S_v^n(x;\lambda)\frac{t^n}{n!} \tag{3}$$

where $v \in \mathbb{N}_0$ and $\lambda \in \mathbb{C}$ (cf. [16]). Furthermore,

$$S_2(n,v;\lambda) := S_v^n(0;\lambda)$$

where, as usual, $S_2(n,v;\lambda)$ denote the λ-Stirling numbers (cf. [8,12]). Substituting $\lambda = 1$ into (3), we have the array polynomials:

$$S_v^n(x) := S_v^n(x;1)$$

(cf. [16–18] and (Theorem 2 [19])).

In (cf. Equation (8) [1]), we defined the combinatorial numbers $y_1(n,k;\lambda)$ by means of the following generating function:

$$F_{y_1}(t,k;\lambda) = \frac{1}{k!}(\lambda e^t + 1)^k = \sum_{n=0}^{\infty} y_1(n,k;\lambda)\frac{t^n}{n!} \tag{4}$$

where $k \in \mathbb{N}_0$ and $\lambda \in \mathbb{C}$.

By using (4), we have

$$y_1(n,k;\lambda) = \frac{1}{k!}\sum_{j=0}^{k}\binom{k}{j} j^n \lambda^j \tag{5}$$

where $n \in \mathbb{N}_0$ (cf. Equation (9) [1]).

Relationships between the λ-array polynomials $S_v^n(x;\lambda)$ and the numbers $y_1(n,k;\lambda)$ and the Stirling numbers of the second kind $S_2(n,k)$ are given below, respectively:

$$S_k^n(0;\lambda) = S_2(n,v;\lambda) = (-1)^k y_1(n,k;-\lambda)$$

and

$$S_2(n,k) = (-1)^k y_1(n,k;-1) \tag{6}$$

(cf. [1,17,20–25]).

The Fibonacci numbers F_j are defined by the following generating function

$$\frac{t}{1-t-t^2} = \sum_{n=0}^{\infty} F_n t^n$$

(cf. (p. 229. [26])). We need the following well-known formulas for the Fibonacci numbers. Let $\lambda = \frac{1+\sqrt{5}}{2}$ and $\beta = \frac{1-\sqrt{5}}{2}$. Let $j \in \mathbb{N}$, we have

$$\lambda^j = \lambda F_j + F_{j-1} \tag{7}$$

and

$$\beta^j = \beta F_j + F_{j-1}$$

(cf. (p. 78, Lemma 5.1. [26])). Using the above identities, one easily derives the following Binet's formula:

$$F_j = \frac{\lambda^j - \beta^j}{\lambda - \beta}.$$

Substituting $-n$ with $n \in \mathbb{N}$, into the above equation, we easily have

$$F_{-n} = (-1)^{n+1} F_n$$

(cf. (p. 84 [26])).

1.1. p-Adic Integrals

In the last section, we will give some combinatorial sums with p-adic integrals technique. Hence, let us give definitions of these integrals and a few properties of them.

Let $f(x) \in C^1(\mathbb{Z}_p \to \mathbb{K})$, a set of continuous derivative functions, and $\mathbb{K}$ is a field with a complete valuation.

The Volkenborn integral (the bosonic p-adic integral) is defined by

$$\int_{\mathbb{Z}_p} f(x) d\mu_1(x) = \lim_{N\to\infty} \frac{1}{p^N} \sum_{x=0}^{p^N-1} f(x), \tag{8}$$

where $\mu_1(x) = \mu_1\left(x + p^N\mathbb{Z}_p\right)$ is the Haar distribution on $\mathbb{Z}_p$:

$$\mu_1\left(x + p^N\mathbb{Z}_p\right) = \frac{1}{p^N},$$

(cf. [27,28]). On the other hand, the p-adic fermionic integral is defined by

$$\int_{\mathbb{Z}_p} f(x)\, d\mu_{-1}(x) = \lim_{N\to\infty} \sum_{x=0}^{p^N-1} (-1)^x f(x) \tag{9}$$

where

$$\mu_{-1}(x) = \mu_{-1}\left(x + p^N\mathbb{Z}_p\right) = \frac{(-1)^x}{p^N}$$

(cf. [29]).

The Bernoulli numbers and the Euler numbers are related to the following p-adic integrals representations, respectively,

$$B_n = \int_{\mathbb{Z}_p} x^n d\mu_1(x), \tag{10}$$

(cf. [27,28]) and

$$E_n = \int_{\mathbb{Z}_p} x^n d\mu_{-1}(x), \tag{11}$$

(cf. [27]).

1.2. Umbral Algebra and Calculus

Throughout this section, we use the notations and definitions of the Roman's book (cf. [13]). Let $\mathbb{P} = \mathbb{C}[x]$ be the algebra of polynomials in the single variable x over the field of complex numbers. Let $\mathbb{P}^*$ be the vector space of all linear functionals on $\mathbb{P}$. Let $\langle L \mid p(x) \rangle$ be the action of a linear functional L on a polynomial $p(x)$. Let $\mathcal{F}$ denote the algebra of formal power series

$$f(t) = \sum_{k=0}^{\infty} a_k \frac{t^k}{k!},$$

(cf. [13]). Furthermore, for all $n \in \mathbb{N}_0$, one has

$$\langle f(t) \mid x^n \rangle = a_n \tag{12}$$

and also

$$\langle f(t)g(t) \mid p(x) \rangle = \langle f(t) \mid g(t)p(x) \rangle, \tag{13}$$

where $f(t)$, $g(t)$ are in $\mathcal{F}$ (cf. [13]).

For $p(x) \in \mathbb{P}$, as a *linear functional*, we have

$$\langle e^{yt} \mid p(x) \rangle = p(y). \tag{14}$$

and as a *linear operator*, we have

$$e^{yt} p(x) = p(x+y) \tag{15}$$

(cf. [13]). The Sheffer polynomials for pair $(g(t), f(t))$, where $g(t)$ must be invertible and $f(t)$ must be delta series. The Sheffer polynomials for pair $(g(t), t)$ is the Appell polynomials or Appell sequences for $g(t)$. The Appell polynomials are defined by means of the following generating function

$$\sum_{k=0}^{\infty} \frac{s_k(x)}{k!} t^k = \frac{1}{g(t)} e^{xt}, \tag{16}$$

(cf. [13]). Some properties of the Appell polynomials are given as follows.

$$s_n(x) = g(t)^{-1} x^n, \tag{17}$$

(p. 86, Theorem 2.5.5 [13]), derivative formula

$$t s_n(x) = n s_{n-1}(x) \tag{18}$$

(cf. p. 86, Theorem 2.5.6 [13]); and see also [6,30,31]).

We summarize the results presented in this paper as follows:

In Section 2, by applying the umbral algebra and umbral calculus methods to generating functions of the special numbers and polynomials, we derive some identities and relations including the numbers $y_1(n,k;\lambda)$, combinatorial sums, the Fibonacci numbers, Apostol-Bernoulli type numbers and polynomials, and the Apostol-Euler type numbers and polynomials. Finally, we give some remarks and observations.

In Section 3, by using the p-adic integrals, we give many combinatorial sums related to the Bernoulli numbers, the Euler numbers, the Apostol-Euler numbers and the numbers $y_1(n,k;\lambda)$.

2. Identities Including the Numbers $y_1(n,k;\lambda)$, Combinatorial Sums, and Apostol-Euler Type Numbers and Polynomials

In this section, by using the umbral algebra and umbral calculus methods, we derive many identities and relations containing the numbers $y_1(n,k;\lambda)$, combinatorial sums, the Fibonacci numbers, Apostol-Bernoulli type numbers and polynomials, and the Apostol-Euler type numbers and polynomials.

Theorem 1.
$$\frac{1}{k!}\sum_{j=0}^{k}\binom{k}{j}\lambda^{j}\sum_{v=0}^{m}\binom{m}{v}x^{m-v}j^{v}=\frac{1}{k!}\sum_{j=0}^{k}\binom{k}{j}\lambda^{j}\left(x+j\right)^{m}$$

or

$$\sum_{v=0}^{m}\frac{(m)_{v}}{v!}x^{m-v}y_{1}\left(v,k;\lambda\right)=\frac{1}{k!}\sum_{j=0}^{k}\binom{k}{j}\lambda^{j}\left(x+j\right)^{m} \tag{19}$$

Proof. By applying linear operators in (15) and (18) to (4), respectively, we obtain

$$\frac{1}{k!}\left(\lambda e^{t}+1\right)^{k}x^{m}=\frac{1}{k!}\sum_{j=0}^{k}\binom{k}{j}\lambda^{j}\left(x+j\right)^{m} \tag{20}$$

and

$$\begin{aligned}\frac{1}{k!}\left(\lambda e^{t}+1\right)^{k}x^{m} &= \sum_{n=0}^{\infty}y_{1}\left(n,k;\lambda\right)\frac{1}{n!}t^{n}x^{m} \qquad (21)\\ &= \begin{cases}0, & n>m\\ y_{1}\left(n,k;\lambda\right), & n=m\\ \sum_{v=0}^{m}\frac{(m)_{v}}{v!}x^{m-v}y_{1}\left(v,k;\lambda\right) & n<m.\end{cases}\end{aligned}$$

Combining (20) with (21), we get the desired results. □

Remark 1. *Substituting $x=0$ into (19), we arrive at (5).*

By applying the action of a linear operator $\left(\lambda e^{t}+1\right)^{k}$ to the Apostol-Euler polynomial $\mathcal{E}_{n}^{(a)}\left(x,\lambda\right)$, we obtain the following result.

Theorem 2.
$$\sum_{j=0}^{k}\binom{k}{j}\lambda^{j}\mathcal{E}_{n}^{(a)}\left(x+j,\lambda\right)=\sum_{j=0}^{k}\binom{k}{j}\sum_{v=0}^{j}(-1)^{j-v}\binom{j}{v}2^{v}\mathcal{E}_{n}^{(a-v)}\left(x,\lambda\right). \tag{22}$$

Proof. By applying the action of a linear operator $\left(\lambda e^{t}+1\right)^{k}$ to the Apostol-Euler polynomial $\mathcal{E}_{n}^{(a)}\left(x,\lambda\right)$, we obtain

$$\left(\lambda e^{t}+1\right)^{k}\mathcal{E}_{n}^{(a)}\left(x,\lambda\right)=\sum_{j=0}^{k}\binom{k}{j}\lambda^{j}e^{tj}\mathcal{E}_{n}^{(a)}\left(x,\lambda\right). \tag{23}$$

Applying linear operators in (15) to the above equation, we have

$$\left(\lambda e^{t}+1\right)^{k}\mathcal{E}_{n}^{(a)}\left(x,\lambda\right)=\sum_{j=0}^{k}\binom{k}{j}\lambda^{j}\mathcal{E}_{n}^{(a)}\left(x+j,\lambda\right). \tag{24}$$

Combining the following relation with (23)

$$\mathcal{E}_{n}^{(a)}\left(x,\lambda\right)=\left(\frac{2}{\lambda e^{t}+1}\right)^{a}x^{n}$$

(cf. p. 101 [13]), we have

$$\left(\lambda e^{t}+1\right)^{k}\mathcal{E}_{n}^{(a)}(x,\lambda)=\sum_{j=0}^{k}\binom{k}{j}\sum_{v=0}^{j}(-1)^{j-v}\binom{j}{v}2^{v}\left(\frac{2}{\lambda e^{t}+1}\right)^{a-v}x^{n}.$$

After some elementary calculation in the above equation, we have

$$\left(\lambda e^{t}+1\right)^{k}\mathcal{E}_{n}^{(a)}(x,\lambda)=\sum_{j=0}^{k}\binom{k}{j}\sum_{v=0}^{j}(-1)^{j-v}\binom{j}{v}2^{v}\mathcal{E}_{n}^{(a-v)}(x,\lambda). \tag{25}$$

Combining (24) and (25), we arrive at the desired result. □

Substituting $\lambda=k=1$ into (22), we arrive at the following corollary, which was proved by Roman (p. 103, Equation (4.2.11) [13]), see also (cf. [32]).

Corollary 1.

$$2E_{n}^{(a-1)}(x)=E_{n}^{(a)}(x+1)+E_{n}^{(a)}(x)$$

We assume that, $\lambda\neq 1$ and $a\in\mathbb{N}$, we have the following well-known relationships between the polynomials $\mathcal{B}_{n}^{(a)}(x,\lambda)$ and $\mathcal{E}_{n}^{(a)}(x,-\lambda)$:

$$\begin{aligned}\sum_{n=0}^{\infty}\mathcal{E}_{n}^{(a)}(x,-\lambda)\frac{t^{n}}{n!} &= \left(\frac{2}{-\lambda e^{t}+1}\right)^{a}e^{tx}\\ &= \left(-\frac{2}{t}\right)^{a}\sum_{n=0}^{\infty}\mathcal{B}_{n}^{(a)}(x,\lambda)\frac{t^{n}}{n!}.\end{aligned}$$

Therefore

$$(n)_{a}\mathcal{E}_{n-a}^{(a)}(x,-\lambda)=(-2)^{a}\mathcal{B}_{n}^{(a)}(x,\lambda)$$

or

$$\mathcal{E}_{n}^{(a)}(x,\lambda)=\frac{(-2)^{a}}{(n+a)_{a}}\mathcal{B}_{n+a}^{(a)}(x,-\lambda).$$

Substituting the above relation into (22), we get the following result.

Theorem 3.

$$\begin{aligned}&\sum_{j=0}^{k}\binom{k}{j}\lambda^{j}\frac{(-2)^{a}}{(n+a)_{a}}\mathcal{B}_{n+a}^{(a)}(x+j,-\lambda)\\ =\ &\sum_{j=0}^{k}\binom{k}{j}\sum_{v=0}^{j}(-1)^{j+a}\binom{j}{v}\frac{2^{a}}{(n+a-v)_{a-v}}\mathcal{B}_{n+a-v}^{(a-v)}(x,-\lambda).\end{aligned} \tag{26}$$

Setting $k=1$ in (26), we get the following corollary.

Corollary 2.

$$\mathcal{B}_{n+a}^{(a)}(x,-\lambda)+\lambda\mathcal{B}_{n+a}^{(a)}(x+1,-\lambda)=-(n+a)\mathcal{B}_{n+a-1}^{(a-1)}(x,-\lambda). \tag{27}$$

Remark 2. *Another proof of the Equation (27) is given by Dere et al. [6] and see also (cf. [32]).*

Remark 3. *Substituting $n+a=m$ and $\lambda=-1$ into (27), we get*

$$B_{m}^{(a)}(x+1)=B_{m}^{(a)}(x)+mB_{m-1}^{(a-1)}(x)$$

(cf. p. 95, Equation (4.2.6) [13]).

The following theorem was proved in (cf. [1]). Here, we give a proof different from that in (cf. [1]).

Theorem 4. *Let n and k be nonnegative integers. Then we have*

$$y_1(n,k;\lambda) = \frac{2^k}{k!}\mathcal{E}_n^{(-k)}(0,\lambda).$$

Proof. Using (12), we obtain

$$y_1(n,k;\lambda) = \frac{1}{k!}\left\langle \left(\lambda e^t + 1\right)^k \mid x^n \right\rangle. \tag{28}$$

From the above equation, we have

$$\begin{aligned} y_1(n,k;\lambda) &= \frac{2^k}{k!}\left\langle \left(\frac{\lambda e^t+1}{2}\right)^k \mid x^n \right\rangle \\ &= \frac{2^k}{k!}\left\langle t^0 \mid \left(\frac{\lambda e^t+1}{2}\right)^{-k} x^n \right\rangle \\ &= \frac{2^k}{k!}\mathcal{E}_n^{(-k)}(0,\lambda). \end{aligned}$$

Therefore, we arrive at the desired result. □

Theorem 5.

$$\frac{2^k}{k!}\sum_{j=0}^{n}\binom{n}{j}\mathcal{E}_j^{(a)}(\lambda)\,\mathcal{E}_{n-j}^{(-k)}(\lambda) = \sum_{j=0}^{n}\binom{n}{j}\mathcal{E}_j^{(a)}(\lambda)\,y_1(n-j,k;\lambda)$$

Proof. We set

$$\begin{aligned} \frac{1}{k!}\left\langle \left(\lambda e^t+1\right)^k \mid \mathcal{E}_n^{(a)}(\lambda) \right\rangle &= \frac{1}{k!}\left\langle \left(\lambda e^t+1\right)^k \mid \left(\sum_{j=0}^{n}\binom{n}{j}\mathcal{E}_j^{(a)}(\lambda)\right)x^{n-j} \right\rangle \\ &= \frac{1}{k!}\sum_{j=0}^{n}\binom{n}{j}\mathcal{E}_j^{(a)}(\lambda)\left\langle \left(\lambda e^t+1\right)^k \mid x^{n-j} \right\rangle \end{aligned}$$

Combining the above equation with (28), we get

$$\frac{1}{k!}\left\langle \left(\lambda e^t+1\right)^k \mid \mathcal{E}_n^{(a)}(\lambda) \right\rangle = \frac{1}{k!}\sum_{j=0}^{n}\binom{n}{j}\mathcal{E}_j^{(a)}(\lambda)\,y_1(n-j,k;\lambda). \tag{29}$$

On the other hand

$$\begin{aligned} &\frac{1}{k!}\sum_{j=0}^{n}\binom{n}{j}\mathcal{E}_j^{(a)}(\lambda)\left\langle \left(\lambda e^t+1\right)^k \mid x^{n-j} \right\rangle \\ = &\frac{2^k}{k!}\sum_{j=0}^{n}\binom{n}{j}\mathcal{E}_j^{(a)}(\lambda)\left\langle t^0 \mid \left(\frac{\lambda e^t+1}{2}\right)^{-k} x^{n-j} \right\rangle. \end{aligned} \tag{30}$$

Therefore, combining (29) with (30), we arrive at the desired result. □

Theorem 6.

$$y_1(m,k;\lambda) = \frac{1}{2k!}\sum_{j=0}^{k}\binom{k}{j}\frac{\lambda^j}{2}\left(E_m(j+1)+E_m(j)\right).$$

Proof. We set the following functional equation

$$F_{y1}(t,k;\lambda) = \frac{1}{2k!}\sum_{j=0}^{k}\binom{k}{j}\lambda^j\left(F_{\mathcal{E}}(t,j+1;1,k)+F_{\mathcal{E}}(t,j;1,1)\right).$$

By combining the above equation with (4) and (2), we get

$$\sum_{n=0}^{\infty} y_1(n,k;\lambda)\frac{t^n}{n!} = \sum_{n=0}^{\infty}\left(\frac{1}{2k!}\sum_{j=0}^{k}\binom{k}{j}\lambda^j\left(E_n(j+1)+E_n(j)\right)\right)\frac{t^n}{n!}.$$

Comparing the coefficients of $\frac{t^n}{n!}$ on both sides of the above equation yields the desired result. □

Theorem 7. *Let $m \in \mathbb{N}$. Then we have*

$$S_2(m-1,k;\lambda) = \frac{1}{mk!}\sum_{j=0}^{k}(-1)^{k-j}\binom{k}{j}\lambda^j\left(\mathcal{B}_m(j+1,\lambda)-\mathcal{B}_m(j,\lambda)\right).$$

Proof. We also set the following functional equation

$$F_S(t,0,k;\lambda) = \frac{1}{tk!}\sum_{j=0}^{k}(-1)^{k-j}\binom{k}{j}\lambda^j\left(F_{\mathcal{B}}(t,j+1;\lambda,1)-F_{\mathcal{B}}(t,j;\lambda,1)\right).$$

By combining the above equation with (1) and (3), we get

$$\begin{aligned}&\sum_{m=0}^{\infty} S_2(m,k;\lambda)\frac{t^m}{m!}\\ =\;&\sum_{m=0}^{\infty}\left(\frac{1}{tk!}\sum_{j=0}^{k}(-1)^{k-j}\binom{k}{j}\lambda^j\left(\mathcal{B}_m(j+1,\lambda)-\mathcal{B}_m(j,\lambda)\right)\right)\frac{t^m}{m!}.\end{aligned}$$

Comparing the coefficients of $\frac{t^m}{m!}$ on both sides of the above equation yields the desired result. □

Theorem 8. *Let $2\lambda = 1+\sqrt{5}$. Then we have*

$$\sum_{j=0}^{n}\binom{n}{j}\mathcal{E}_j^{(a)}(\lambda)\,y_1(n-j,k;\lambda) = \frac{1}{k!}\sum_{j=0}^{k}\binom{k}{j}\left(\lambda F_j+F_{j-1}\right)\mathcal{E}_n^{(a)}(j,\lambda).$$

Proof. We define the following functional equation:

$$F_{y1}(t,k;\lambda)\,F_{\mathcal{E}}(t,0;\lambda,a) = \frac{1}{k!}\sum_{j=0}^{k}\binom{k}{j}\lambda^j\left(F_{\mathcal{E}}(t,j;\lambda,a)\right).$$

By combining the above equation with (4), (2), and (7), we get

$$\begin{aligned}&\sum_{n=0}^{\infty}\sum_{j=0}^{n}\binom{n}{j}\mathcal{E}_j^{(a)}(\lambda)\,y_1(n-j,k;\lambda)\frac{t^n}{n!}\\ =\;&\sum_{n=0}^{\infty}\frac{1}{k!}\sum_{j=0}^{k}\binom{k}{j}\left(\lambda F_j+F_{j-1}\right)\mathcal{E}_n^{(a)}(j,\lambda)\frac{t^n}{n!},\end{aligned}$$

where $\lambda = \frac{1+\sqrt{5}}{2}$. Comparing the coefficients of $\frac{t^n}{n!}$ on both sides of the above equation yields the desired result. □

3. Combinatorial Sums via p-Adic Integral

In this section, by using the p-adic integrals, we derive some combinatorial sums containing the Bernoulli numbers, the Euler numbers, the Apostol-Euler numbers and the numbers $y_1(n,k;\lambda)$.

Theorem 9.

$$B_n = \frac{k!}{2^k}\sum_{j=0}^{n}\binom{n}{j} y_1(n-j,k;\lambda)\sum_{v=0}^{j}\binom{j}{v}\mathcal{E}_v^{(k)}(\lambda)\,B_{j-v}.$$

Proof. Combining (2) with (4), we set the following functional equation:

$$F_{y1}(t,k;\lambda)\,F_{\mathcal{E}}(t,x;\lambda,k) = \frac{2^k}{k!}e^{tx}.$$

By using the above equation, we get

$$\sum_{n=0}^{\infty}\mathcal{E}_n^{(k)}(x,\lambda)\frac{t^n}{n!}\sum_{n=0}^{\infty} y_1(n,k;\lambda)\frac{t^n}{n!} = \frac{2^k}{k!}\sum_{n=0}^{\infty}x^n\frac{t^n}{n!}.$$

Therefore

$$\sum_{n=0}^{\infty}\sum_{j=0}^{n}\binom{n}{j} y_1(n-j,k;\lambda)\sum_{v=0}^{j}\binom{j}{v}x^{j-v}\mathcal{E}_v^{(k)}(\lambda)\frac{t^n}{n!} = \frac{2^k}{k!}\sum_{n=0}^{\infty}x^n\frac{t^n}{n!}.$$

Comparing the coefficients of $\frac{t^n}{n!}$ on both sides of the above equation yields the following relation:

$$\sum_{j=0}^{n}\binom{n}{j} y_1(n-j,k;\lambda)\sum_{v=0}^{j}\binom{j}{v}x^{j-v}\mathcal{E}_v^{(k)}(\lambda) = \frac{2^k}{k!}x^n. \tag{31}$$

By applying the Volkenborn integral to (31), we get

$$\sum_{j=0}^{n}\binom{n}{j} y_1(n-j,k;\lambda)\sum_{v=0}^{j}\binom{j}{v}\mathcal{E}_v^{(k)}(\lambda)\int_{\mathbb{Z}_p}x^{j-v}d\mu_1(x) = \frac{2^k}{k!}\int_{\mathbb{Z}_p}x^n d\mu_1(x).$$

Combining the above equation with (10), we arrive at the desired result. □

Remark 4. *Replacing x by k and λ by λ^2, the Equation (31) is reduced to the following relation:*

$$\sum_{j=0}^{n}\binom{n}{j} y_1\left(n-j,k;\lambda^2\right)\mathcal{E}_j^{(k)}\left(k,\lambda^2\right) = \frac{2^k}{k!}k^n. \tag{32}$$

Since

$$\lambda^k 2^n\mathcal{E}_n^{(k)}(k,\lambda^2) = \sum_{m=0}^{n}\binom{n}{m}k^m E_{n-m}^{*(k)}(\lambda)$$

where $E_n^{*(k)}(\lambda)$ denote the Apostol-type Euler numbers of the second kind of order k (cf. [25,33]), the Equation (32) yields

$$\sum_{j=0}^{n}\binom{n}{j}\frac{y_1(n-j,k;\lambda^2)}{2^j}\sum_{m=0}^{j}\binom{j}{m}k^m E_{j-m}^{*(k)}(\lambda) = \frac{(2\lambda)^k}{k!}k^n.$$

By combining the above equation with the following identity

$$\sum_{m=0}^{n}\binom{n}{m}2^{m}y_{1}\left(m,k;\lambda^{2}\right)E_{n-m}^{*(k)}(\lambda)=\frac{(2\lambda)^{k}}{k!}k^{n}$$

(cf. [33]), we get the following combinatorial sums

$$\sum_{j=0}^{n}\binom{n}{j}\frac{y_{1}\left(n-j,k;\lambda^{2}\right)}{2^{j}}\sum_{m=0}^{j}\binom{j}{m}k^{m}E_{j-m}^{*(k)}(\lambda)=\sum_{m=0}^{n}\binom{n}{m}2^{m}y_{1}\left(m,k;\lambda^{2}\right)E_{n-m}^{*(k)}(\lambda).$$

Theorem 10.

$$E_{n}=\frac{k!}{2^{k}}\sum_{j=0}^{n}\binom{n}{j}y_{1}(n-j,k;\lambda)\sum_{v=0}^{j}\binom{j}{v}\mathcal{E}_{v}^{(k)}(\lambda)E_{j-v}.$$

Proof. By applying the fermionic p-adic integral to (31), we have

$$\sum_{j=0}^{n}\binom{n}{j}y_{1}(n-j,k;\lambda)\sum_{v=0}^{j}\binom{j}{v}\mathcal{E}_{v}^{(k)}(\lambda)\int_{\mathbb{Z}_{p}}x^{j-v}d\mu_{-1}(x)=\frac{2^{k}}{k!}\int_{\mathbb{Z}_{p}}x^{n}d\mu_{-1}(x).$$

Combining the above equation with (11), we arrive at the desired result. □

Theorem 11.

$$\sum_{j=0}^{n}\binom{n}{j}y_{1}(n-j,k;\lambda)\sum_{v=0}^{j}\binom{j}{v}\frac{\mathcal{E}_{v}^{(k)}(\lambda)}{j+1-v}=\frac{2^{k}}{(n+1)k!}.$$

Proof. Integrate Equation (31) with respect to x from 0 to 1, we obtain

$$\sum_{j=0}^{n}\binom{n}{j}y_{1}(n-j,k;\lambda)\sum_{v=0}^{j}\binom{j}{v}\mathcal{E}_{v}^{(k)}(\lambda)\int_{0}^{1}x^{j-v}dx=\frac{2^{k}}{k!}\int_{0}^{1}x^{n}dx.$$

After some calculations, we get the desired result. □

Theorem 12.

$$\sum_{j=0}^{n}\binom{n}{j}y_{1}(j,k;\lambda)\left(B_{n-j}-E_{n-j}\right)=\frac{1}{k!}\sum_{j=0}^{k}\binom{k}{j}\lambda^{j}\left(B_{n}(j)-E_{n}(j)\right).$$

Proof. Setting

$$e^{tx}F_{y1}(t,k;\lambda)=\frac{1}{k!}\sum_{j=0}^{k}\binom{k}{j}\lambda^{j}e^{(x+j)t}.$$

Combining (4), we have

$$\sum_{n=0}^{\infty}\sum_{j=0}^{n}\binom{n}{j}x^{n-j}y_{1}(j,k;\lambda)\frac{t^{n}}{n!}=\sum_{n=0}^{\infty}\frac{1}{k!}\sum_{j=0}^{k}\binom{k}{j}\lambda^{j}(x+j)^{n}\frac{t^{n}}{n!}.$$

Comparing the coefficients of $\frac{t^{n}}{n!}$ on both sides of the above equation yields the following relation:

$$\sum_{j=0}^{n}\binom{n}{j}x^{n-j}y_{1}(j,k;\lambda)=\frac{1}{k!}\sum_{j=0}^{k}\binom{k}{j}\lambda^{j}(x+j)^{n}. \tag{33}$$

By applying the bosonic p-adic integral to (33) and combining with (10), we have

$$\sum_{j=0}^{n}\binom{n}{j}B_{n-j}y_1(j,k;\lambda)=\frac{1}{k!}\sum_{j=0}^{k}\binom{k}{j}\lambda^j B_n(j). \tag{34}$$

By applying the fermionic p-adic integral to (33) and combining with (11), we obtain

$$\sum_{j=0}^{n}\binom{n}{j}E_{n-j}y_1(j,k;\lambda)=\frac{1}{k!}\sum_{j=0}^{k}\binom{k}{j}\lambda^j E_n(j). \tag{35}$$

Subtracting both sides of Equations (34) and (35), after some elementary calculations, we arrive at the desired result. □

Theorem 13.

$$\sum_{j=0}^{n}\binom{n}{j}\frac{y_1(j,k;\lambda)}{n+1-j}=\frac{1}{k!}\sum_{j=0}^{k}\binom{k}{j}\lambda^j\frac{(j+1)^{n+1}-j^{n+1}}{n+1}.$$

Proof. Integrate Equation (33) with respect to x from 0 to 1, we obtain

$$\sum_{j=0}^{n}\binom{n}{j}y_1(j,k;\lambda)\int_0^1 x^{n-j}dx=\frac{1}{k!}\sum_{j=0}^{k}\binom{k}{j}\lambda^j\int_0^1(x+j)^n dx.$$

After some calculations, we get the desired result. □

Acknowledgments: The present investigation was supported by the Scientific Research Project Administration of Akdeniz University. The author would like to thank to all referees for their valuable comments.

Conflicts of Interest: The authors declare no conflict of interest.

References

1. Simsek, Y. New families of special numbers for computing negative order Euler numbers and related numbers and polynomials. *arXiv* **2018**, *12*, arxiv:1604.05601.
2. Simsek, Y. Analysis of the Bernstein basis functions: an approach to combinatorial sums involving binomial coefficients and Catalan numbers. *Math. Method Appl. Sci.* **2015**, *38*, 3007–3021.
3. Simsek, Y. Identities and relations related to combinatorial numbers and polynomials. *Proc. Jangjeon Math. Soc.* **2017**, *20*, 127–135.
4. Simsek, Y. Apostol type Daehee numbers and polynomials. *Adv. Stud. Contemp. Math.* **2016**, *26*, 555–566.
5. Simsek, Y. Construction of some new families of Apostol-type numbers and polynomials via Dirichlet character and p-adic q-integrals. *Turk. J. Math.* **2018**, *42*, 557–577, doi:10.3906/mat-1703-114.
6. Dere, R.; Simsek, Y.; Srivastava, H.M. A unified presentation of three families of generalized Apostol type polynomials based upon the theory of the umbral calculus and the umbral algebra. *J. Number Theory* **2013**, *133*, 3245–3263.
7. Srivastava, H.M.; Kurt, B.; Simsek, Y. Some families of Genocchi type polynomials and their interpolation functions. *Integral Transforms Spec. Funct.* **2012**, *24*, 919–938.
8. Srivastava, H.M. Some generalizations and basic (or q-) extensions of the Bernoulli, Euler and Genocchi polynomials. *Appl. Math. Inform. Sci.* **2011**, *5*, 390–444.
9. Srivastava, H.M.; Choi, J. *Zeta and q-Zeta Functions and Associated Series and Integrals*; Elsevier Science Publishers: Amsterdam, The Netherlands; London, UK; New York, NY, USA, 2012.
10. Luo, Q.-M.; Srivastava, H.M. Some generalizations of the Apostol-Bernoulli and Apostol-Euler polynomials. *J. Math. Anal. Appl.* **2005**, *308*, 290–302.
11. Luo, Q.-M.; Srivastava, H.M. Some relationships between the Apostol-Bernoulli and Apostol-Euler polynomials. *Comput. Math. Appl.* **2006**, *51*, 631–642.

12. Luo, Q.-M.; Srivastava, H.M. Some generalizations of the Apostol-Genocchi polynomials and the Stirling numbers of the second kind. *Appl. Math. Comput.* **2011**, *217*, 5702–5728.
13. Roman, S. *The Umbral Calculus*; Dover Publication Inc.: New York, NY, USA, 2005.
14. Srivastava, H.M.; Kim, T.; Simsek, Y. q-Bernoulli numbers and polynomials associated with multiple q-zeta functions and basic L-series. *Russ. J. Math. Phys.* **2005**, *12*, 241–268.
15. Khan, N.U.; Usman, T.; Choi, J. A new generalization of Apostol type Laguerre-Genocchi polynomials. *C. R. Acad. Sci. Paris Ser. I* **2017**, doi:10.1016/j.crma.2017.04.010
16. Simsek, Y. Generating functions for generalized Stirling type numbers, array type polynomials, Eulerian type polynomials and their applications. *Fixed Point Theory Appl.* **2013**, *2013*, 1–28.
17. Bayad, A.; Simsek, Y.; Srivastava, H.M. Some array type polynomials associated with special numbers and polynomials. *Appl. Math. Comput.* **2014**, 244, 149–157.
18. Cakic, N.P.; Milovanovic, G.V. On generalized Stirling numbers and polynomials. *Math. Balk.* **2004**, *18*, 241–248.
19. Simsek, Y. *Interpolation Function of Generalized q-Bernstein Type Polynomials and Their Application*; Curves and Surfaces 2011, LNCS 6920; Boissonnat, J.-D.; Chenin, P.; Cohen, A.; Gout, C.; Lyche, T.; Mazure, M.-L.; Schumaker, L.L., Eds.; Springer: Berlin/Heidelberg, Germany, 2012; pp. 647–662.
20. Bona, M. *Introduction to Enumerative Combinatorics*; The McGraw-Hill Companies Inc.: New York, NY, USA, 2007.
21. Riordan, J. *Introduction to Combinatorial Analysis*; Princeton University Press: Princeton, NJ, USA, 1958.
22. Simsek, Y. On parametrization of the q-Bernstein Basis functions and Their Applications. *J. Inequal. Spec. Funct.* **2017**, *8*, 158–169.
23. Spivey, M.Z. Combinatorial Sums and Finite Differences. *Discrete Math.* **2007**, *307*, 3130–3146.
24. Yuluklu, E.; Simsek, Y.; Komatsu, T. Identities Related to Special Polynomials and Combinatorial Numbers. *Filomat* **2017**, *31*, 4833–4844.
25. Simsek, Y. Computation methods for combinatorial sums and Euler type numbers related to new families of numbers. *Math. Method Appl. Sci.* **2017**, *40*, 2347–2361.
26. Koshy, T. *Fibonacci and Lucas Numbers with Applications; A Wiley-Interscience Publication*; John Wiley & Sons, Inc.: New York, NY, USA; Chichester, UK; Weinheim, Germany; Brisbane, Australial; Singapore; Toronto, ON, Canada, 2001.
27. Kim, T. q-Volkenborn integration. *Russ. J. Math. Phys.* **2002**, *19*, 288–299.
28. Schikhof, W.H. *Ultrametric Calculus: An Introduction to p-adic Analysis*; Cambridge Studies in Advanced Mathematics 4; Cambridge University Press: Cambridge, UK, 1984.
29. Kim, T. q-Euler numbers and polynomials associated with p-adic q-integral and basic q-zeta function. *Trend Math. Inf. Cent. Math. Sci.* **2006**, *9*, 7–12.
30. Dattoli, G.; Migliorati, M.; Srivastava, H.M. Sheffer polynomials, monomiality principle, algebraic methods and the theory of classical polynomials. *Math. Comput. Model.* **2007**, 45, 1033–1041.
31. Dere, R.; Simsek, Y. Genocchi polynomials associated with the Umbral algebra. *Appl. Math. Comput.* **2011**, *218*, 756–761.
32. Komatsu, T.; Simsek, Y. Identities related to the Stirling numbers and modified Apostol-type numbers on Umbral Calculus. *Miskolc Math. Notes* **2017**, *18*, 905–916, doi:10.18514/MMN.2017.1458.
33. Kucukoglu, I.; Simsek, Y. Identities and derivative formulas for the combinatorial and Apostol-Euler type numbers by their generating functions. preprint.

MDPI
St. Alban-Anlage 66
4052 Basel
Switzerland
Tel. +41 61 683 77 34
Fax +41 61 302 89 18
www.mdpi.com

Axioms Editorial Office
E-mail: axioms@mdpi.com
www.mdpi.com/journal/axioms

www.ingramcontent.com/pod-product-compliance
Lightning Source LLC
Chambersburg PA
CBHW051843210326
41597CB00033B/5752
* 9 7 8 3 0 3 8 9 7 4 0 0 0 *